Ulrich Foelsche
Gottfried Kirchengast
Andrea Steiner

Atmosphere and Climate

Studies by Occultation Methods

Ulrich Foelsche
Gottfried Kirchengast
Andrea Steiner
Editors

Atmosphere and Climate

Studies by Occultation Methods

With 134 Figures, 21 in color

Springer

Dr. Ulrich Foelsche
Prof. Gottfried Kirchengast
Dr. Andrea Steiner

Wegener Center for Climate
and Global Change (WegCenter)
University of Graz
Leechgasse 25
8010 Graz
and
Institute for Geophysics, Astrophysics,
and Meteorology (IGAM)
University of Graz
Universitätsplatz 5
8010 Graz
Austria

Library of Congress Control Number: 2006924676

ISBN-10 3-540-34116-1 Springer Berlin Heidelberg New York
ISBN-13 978-3-540-34116-1 Springer Berlin Heidelberg New York

Springer is a part of Springer Science+Business Media
springer.com

Printed in The Netherlands

Cover design: E. Kirchner, Heidelberg
Typesetting: Camera-ready by Editors
Production: Almas Schimmel
Printing: Krips bv, Meppel
Binding: Stürtz AG, Würzburg

Printed on acid-free paper 30/3141/as 5 4 3 2 1 0

Preface

Since the early use of the occultation measurement principle for sounding planetary atmospheres and ionospheres, its exploitation in atmospheric remote sensing has seen tremendous advances. In this book we focus on sensors on Low Earth Orbit (LEO) satellites, which exploit solar, lunar, stellar, GNSS (Global Navigation Satellite Systems), and LEO-crosslink signals for observing the Earth's atmosphere and climate.

The methods all share the key properties of self-calibration, high accuracy and vertical resolution, global coverage, and (if using radio signals) all-weather capability. The atmospheric parameters obtained extend from the fundamental variables temperature, density, pressure and water vapor via trace gases, aerosols and cloud liquid water to ionospheric electron density. Occultation data are therefore of high value in a wide range of fields including climate monitoring and research, atmospheric physics and chemistry, operational meteorology, and ionospheric physics.

The 2nd International Workshop on Occultations for Probing Atmosphere and Climate – OPAC-2 – was held September 13–17, 2004, in Graz, Austria. OPAC-2 aimed at providing a casual forum and stimulating atmosphere fertilizing scientific discourse, co-operation initiatives, and mutual learning and support amongst members of all different occultation communities. The workshop was attended by 40 participants from 12 different countries who actively contributed to a scientific programme of high quality and to an excellent workshop atmosphere, which was judged by the participants to have fully met the aims expressed.

The programme included 7 tutorial lectures and 15 invited presentations, complemented by about 30 contributed ones, including 11 posters, and an occultation software demonstration. It covered occultation science from occultation methodology in general via different occultation methods and new concepts to use and applications of occultation data in atmosphere and climate science. The detailed programme and all further workshop information will continue to be available online at the OPAC-2 website at http://www.uni-graz.at/opac2.

This book was compiled based on selected papers presented at OPAC-2 and well represents in its six chapters the broad scope of the workshop. Results from the radio occultation experiment onboard CHAMP, which is now over five years in orbit, are collected in chapter 1 while chapter 2 comprises results from the stellar occultation experiment GOMOS onboard ENVISAT. Wave optics algorithms turned out to be very useful for the processing of radio occultation data in the lower troposphere; they are covered in chapter 3. Chapter 4 deals with future occultation missions and with the novel LEO-LEO crosslink concept. Radio occultation data are now increasingly used in numerical weather prediction and atmospheric studies as well as in climate monitoring and change research. This is reflected by the significant amount of articles in chapter 5 and chapter 6, respectively.

We cordially thank all OPAC-2 colleagues, who contributed as authors and coauthors to the book, for the effort and diligent work invested into their papers and for largely observing the length target. All papers were subjected to a peer review process, involving two independent expert reviewers per paper from the community of OPAC-2 participants and beyond. We also very much thank these reviewers for their important service to coherently ensure scientific correctness and high quality of the book from first to last page.

The reviewers, in alphabetical order, were C. O. Ao, G. Beyerle, C. Boone, M. Borsche, F. Cuccoli, A. de la Torre, U. Foelsche, A. Gobiet, M. E. Gorbunov, K. Hocke, A. S. Jensen, G. Kirchengast, E. R. Kursinski, E. Kyrölä, K. B. Lauritsen, R. Leitinger, S. S. Leroy, J.-P. Luntama, A. Löscher, M. S. Lohmann, R. Notarpietro, A. G. Pavelyev, M. Petitta, P. Poli, C. Retscher, S. Schweitzer, V. F. Sofieva, A. K. Steiner, M. Stendel, S. Syndergaard, J. Tamminen, A. von Engeln, and J. Wickert.

Special thanks are, furthermore, due to M. Sc. Barbara Pirscher for her tireless support in the final copy editing and formatting of the book and to Dr. Wolfgang Engel, Mrs. Helen Rachner, and Mrs. Agata Oelschläger from Springer Verlag, Heidelberg, for the kind offer to issue this book as Springer publication and the related technical support. Many thanks also to all others who provided support in one or another way, in representation of which we thank the sponsors of OPAC-2 (see the OPAC-2 website noted above for details) and the sponsors of the START Program No. Y103-N03 (Federal Ministry for Education, Science, and Culture; Austrian Science Fund) for providing the material support enabling the realization of the book.

We hope that, in the spirit of the OPAC-2 aims, the book will become a useful reference for the members of the occultation-related community but also for members of the science community at large interested in the present status and future promises of the field of occultations for probing atmosphere and climate.

Graz, January 2006

Ulrich Foelsche
Gottfried Kirchengast
Andrea K. Steiner

Table of Contents

Radio Occultation with CHAMP

Mission Status, Retrieval, Validation, and Error Analysis

GPS Radio Occultation with CHAMP and GRACE: Recent Results

J. Wickert, T. Schmidt, G. Beyerle, G. Michalak, R. König, S. Heise, and C. Reigber

GeoForschungsZentrum Potsdam (GFZ), Department 1, Geodesy & Remote Sensing, Potsdam, Germany
wickert@gfz-potsdam.de

Abstract. The German CHAMP (CHAllenging Minisatellite Payload) satellite provides continuously GPS radio occultation data since February 2001. The measurements are analyzed by an operational orbit and occultation processing system at GFZ. In total ~170 000 high quality globally distributed vertical profiles of refractivity, temperature and water vapor are provided as of October 2004. The ground infrastructure from GFZ allows for the demonstration of a rapid data analysis since February 2003. The average delay between each measurement and provision of atmospheric excess phase data was reduced to ~4 hours by mid April 2004 and is continuously reached. The complete set of the available refractivity profiles is compared with corresponding analysis data from the European Centre for Medium Range Weather Forecasts (ECMWF) between 0 km and 30 km altitude. The comparison shows nearly bias-free refractivity between ~7 km and 30 km, the standard deviation is ~1 %. The known negative refractivity bias of the CHAMP data in relation to ECMWF is significantly reduced in comparison to earlier product versions by applying the Full Spectrum Inversion (FSI) method for the data analysis in the lower troposphere. First radio occultation measurements from the GRACE-B (Gravity Recovery And Climate Experiment) satellite are available for a 25 h period on July 28/29, 2004. The stability of the satellite clock from GRACE-B is significantly improved in relation to CHAMP. This allows for precise occultation analysis using 30 s clock solutions applying a zero difference technique. Thus the disadvantageous use of a reference GPS satellite link to eliminate the clock error from GRACE-B can be avoided.

1 Introduction

Atmospheric profiling aboard the German CHAMP (Reigber et al. 2005) satellite was activated on February 11, 2001 (Wickert et al. 2001b). The experiment brought significant progress (Hajj et al. 2004; Kuo et al. 2004; Wickert et al. 2005c) for the innovative GPS (Global Positioning System) radio occultation (RO) technique (e.g., Kursinski et al. 1997) in relation to

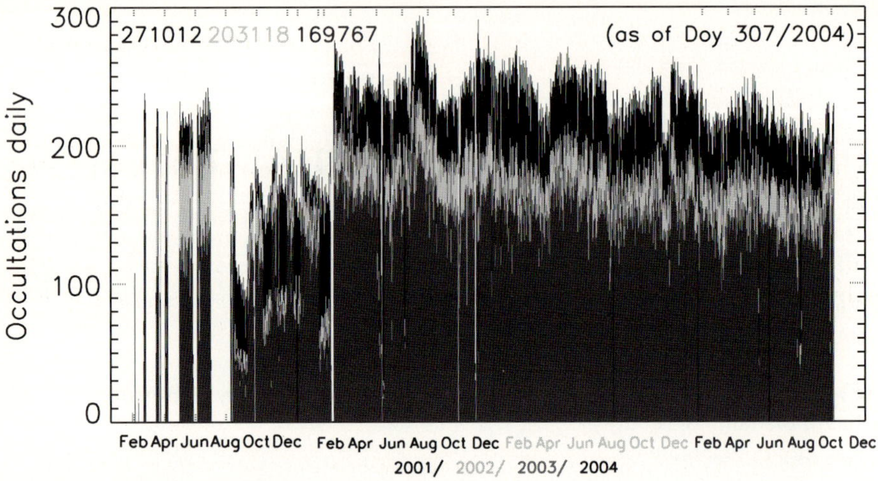

Fig. 1. Number of daily CHAMP occultations (duration >20 s) as of November 2, 2004. The total height of the columns corresponds to the number of daily measurements. The height of the light gray color indicates the number of atmospheric excess phases. The height of the dark gray columns corresponds to the number of vertical atmospheric profiles provided to the CHAMP data center (ISDC, Information System and Data Center) at GFZ.

the pioneering GPS/MET (GPS/METeorology) mission (Ware et al. 1996; Rocken et al. 1997). Main advantages of the calibration-free RO method are global coverage, high vertical resolution and all-weather capability combined with high accuracy. These properties allow for various applications in atmospheric/ionospheric research (e.g., Hajj et al. 2000; Ratnam et al. 2004; Wickert et al. 2004b; Wang et al. 2004; Wickert 2004; Kuo et al. 2005), weather forecast (e.g., Kuo et al. 2000; Healy et al. 2005; Healy and Thepaut 2005) and climate change detection (e.g., Randel et al. 2003; Schmidt et al. 2004, 2005a; Foelsche et al. 2005). Together with CHAMP, several upcoming RO missions will provide thousands of occultations daily and will extend the prospects of this promising technique (e.g., EQUARS (EQUatorial Atmosphere Research Satellite, Takahashi et al. (2004)); COSMIC (Constellation Observing System for Meteorology, Ionosphere and Climate, Rocken et al. (2000)) or Metop (e.g., Loiselet et al. 2000; Larsen et al. 2005)). We review recent results from the CHAMP RO experiment and present first results from the activation of GPS RO aboard the U.S. American/German GRACE mission (Dunn et al. 2003; Tapley and Reigber 2004).

2 Status of the CHAMP RO Experiment

Occultation measurements were performed during 1 238 days since February 2001 as of November 2, 2004; giving a total of 271 012 recorded events (~219 daily). For ∼74.9 % of the occultations (203 118) atmospheric excess phases are available (see Fig. 1). Vertical profiles of atmospheric parameters were derived for 169 767 occultations (∼62.6 %). The yield of the profiles in relation to the number of measurements is currently investigated in more detail within the framework of the Radio Occultation Sensor Evaluation activity (ROSE), jointly initiated by GFZ, Jet Propulsion Laboratory (JPL) and University Corporation for Atmospheric Research (UCAR) (Ao et al. 2003b; Wickert et al. 2005a). ROSE is aimed to evaluate and optimize the quality of CHAMP's analysis results and to improve the involved occultation processing systems.

3 Operational Data Analysis

The occultation data aboard CHAMP are recorded by the "BlackJack" GPS flight receiver provided by JPL. The ground infrastructure of GFZ is used for a fully automated data analysis. Details on the infrastructure and on the orbit and occultation processing system can be found in Wickert et al. (2004a,c); König et al. (2005); Schmidt et al. (2005b). A Near-Real-Time (NRT) provision of atmospheric excess phases is continuously demonstrated since February 2003. An average delay of ∼5 hours between each measurement and provision of corresponding analysis results was reached. Optimized GPS ground station data handling for the precise orbit determination reduced this delay to ∼4 hours since mid April 2004. For some measurements per day the delay is ∼2.5 h (see Fig. 2). Further reduction is possible due to the use of a polar satellite receiving antenna at Ny Aalesund, Spitsbergen, (access to the satellite data every ∼1.5 h) and a global low latency GPS ground network (access to the ground data every ∼15 min), but requires further optimization concerning precise satellite orbit generation and occultation processing. The demonstration of NRT data analysis is an important milestone for the future assimilation of GPS RO data in numerical weather models. A positive impact of CHAMP data on global weather forecasts was already shown by Healy et al. (2005) and Healy and Thepaut (2005).

CHAMP data are analyzed using the standard double difference method to eliminate satellite clock errors (Wickert et al. 2001a). Atmospheric bending angles are derived from the time derivative of the excess phase after appropriate filtering. The ionospheric correction is performed by linear combination of the L1 and L2 bending angle profiles (Vorob'ev and Krasil'nikova 1994). The Full Spectrum Inversion (FSI) technique (Jensen et al. 2003), a wave optics based analysis method, is applied below 15 km to correct for the effect of lower troposphere multipath.

Vertical profiles of atmospheric refractivity are derived from the ionosphere corrected bending angle profiles by Abel inversion. For dry air, the density profiles are obtained from the relationship between density and refractivity. Pressure and temperature ("dry temperature") are obtained applying the hydrostatic equation and the equation of state for an ideal gas. More details

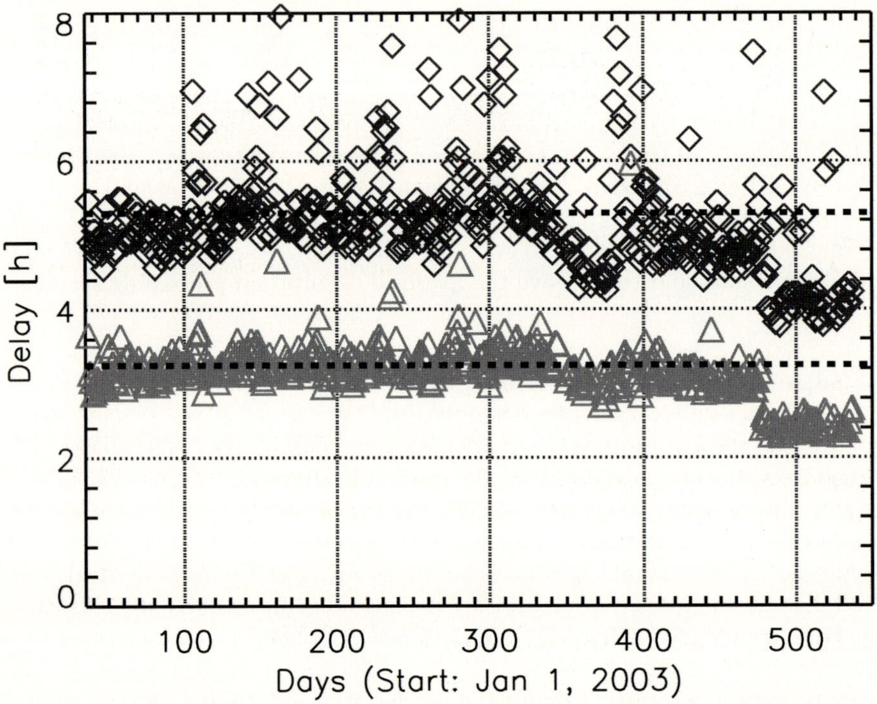

Fig. 2. Time delay between CHAMP occultation measurements and availability of analysis results at GFZ from February 2003 until mid May 2004. Black diamonds indicate the daily mean of the time delay between each measurement and the availability of the corresponding calibrated atmospheric excess phases. An average of ~5 hours for nearly the entire period is reached. The minimum time delays are marked by gray triangles. Due to improvements in the satellite orbit provision the mean delay was reduced to ~4 hours since end of April 2004.

on the retrieval are given by Wickert et al. (2004c). Basics of the GPS radio occultation technique and the derivation of atmospheric parameters are described, e.g., by Kursinski et al. (1997). The refractivity and dry temperature profiles (Product:CH-AI-3-ATM) are provided via the CHAMP data center at GFZ (http://isdc.gfz-potsdam.de/champ/).

Background information from ECMWF is used to derive vertical humidity profiles from the CHAMP refractivities. Two methods for the water vapor derivation were implemented to the operational data analysis. In addition to a standard 1Dvar retrieval (Healy and Eyre 2000) a new direct method (DWVP), introduced by Heise et al. (2005), is implemented. Here the background temperature and pressure information are used to calculate water vapor pressure p_{w} directly from the refractivity measurements using the Smith-Weintraub formula (Smith and Weintraub 1953). Both methods come to statistically comparable results and reveal a bias of less than 0.2 g/kg and a

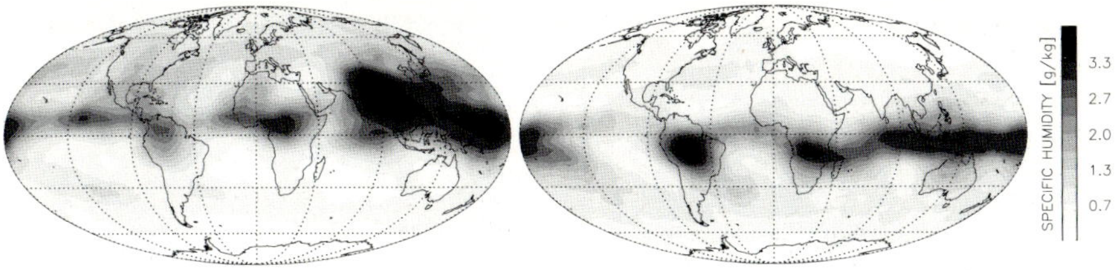

Fig. 3. Mean global distribution of specific humidity at 500 hPa derived from CHAMP occultation data. Left: Northern summer 2002, Right: Northern winter 2002/2003 (for details see Heise et al. (2005)).

standard deviation of less than 1 g/kg specific humidity in relation to radiosonde measurements in the mid troposphere. As an application example for the operational water vapor retrieval with CHAMP data, Fig. 3 shows the seasonal mean of the global water vapor distribution for northern summer (2002) and winter (2002/2003) at 500 hPa. The specific humidity data are derived using the DWVP method.

4 Recent Validation Results for CHAMP

The complete set of CHAMP measurements was reprocessed using the recent version (005) of GFZ occultation analyses software. The resulting set of refractivity profiles (~170 000) is compared with corresponding analysis data from ECMWF (Gaussian grid with 0.5° × 0.5° resolution at the Equator, 60 altitude levels) between 0 km and 30 km.

The comparison shows nearly bias-free refractivity between 10 km and 30 km (see Fig. 4). The standard deviation is ~1 %. The deviations show different characteristics in latitude (e.g., wave-like vertical structures of the bias above the south polar region with a period of ~6 km). This fact suggests weaknesses of the analyzed data, introduced by the ECMWF assimilation scheme (e.g., Gobiet et al. 2005). Our validation results are, as expected, in good agreement with earlier validation studies (ECMWF and radiosonde data) using the previous product version 004 (Schmidt et al. 2004; Wickert et al. 2004c), since FSI is applied only below 15 km.

The major advantage of the recent version (in relation to earlier ones) of GFZ analysis software is the implementation of the FSI method (Jensen et al. 2003) to eliminate the effect of atmospheric multipath to the occultation data. The resulting bias and rms of the comparison with ECMWF is depicted in Fig. 5. The negative refractivity bias of the CHAMP data depends on latitude and is most pronounced in the tropics, where it reaches a value of 5 % at 1 km. However in mid latitude and polar regions the CHAMP data are nearly bias free throughout the entire troposphere. The rms also depends on the latitude. In the tropics values of ~3 % are observed. In mid latitudes and

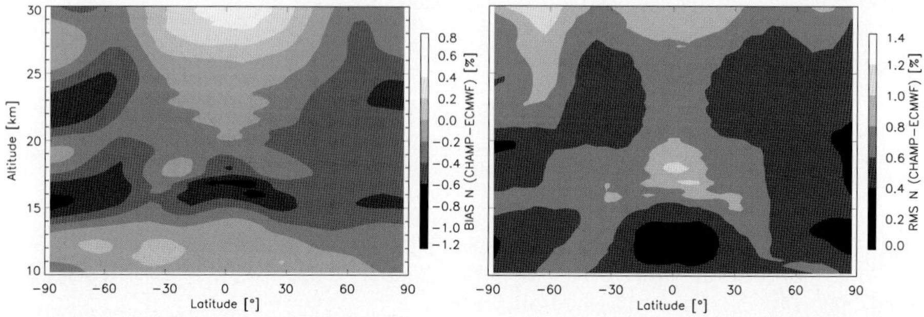

Fig. 4. Comparison of CHAMP refractivity data with corresponding ECMWF analyses (zonal means with 5° resolution) in the upper troposphere/stratosphere (left: bias; right: rms) between May 14, 2001 and November 3, 2004 (~ 170 000 profiles).

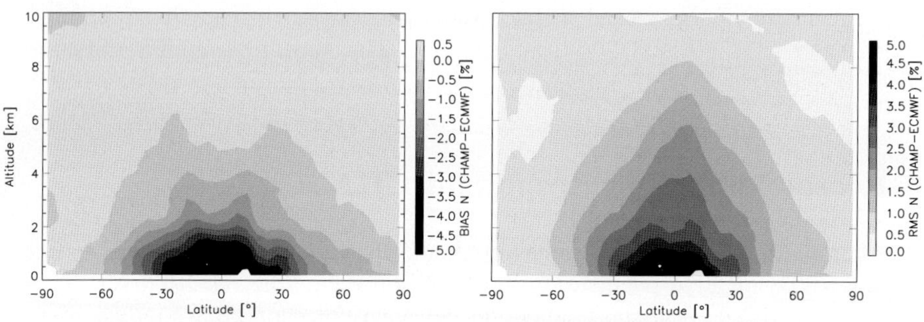

Fig. 5. Comparison of CHAMP refractivity data with corresponding ECMWF analyses (zonal means with 5° resolution) in the troposphere (left: bias; right: rms) between May 14, 2001 and November 3, 2004 (~ 170 000 profiles).

polar regions the rms is below 1 % almost down to the Earth's surface. The negative refractivity bias is a known phenomenon of the CHAMP data and is discussed in more detail by Ao et al. (2003a); Beyerle et al. (2003a,b, 2005b). Causes of the bias are, beside multipath propagation, also signal tracking errors of the GPS receiver and critical refraction, a physical limitation of the RO technique. Further progress in reducing the bias is expected by the application of advanced signal tracking methods ("open loop" technique, see, e.g., (Sokolovskiy 2001; Beyerle et al. 2005b)) and improved signal strength due to the use of more advanced occultation antenna configuration (foreseen, e.g., for COSMIC or Metop). We note, that first investigations of the global map of fractional refractivity errors between CHAMP and ECMWF also reveal complex zonal and meridional structures (Beyerle et al. 2005b).

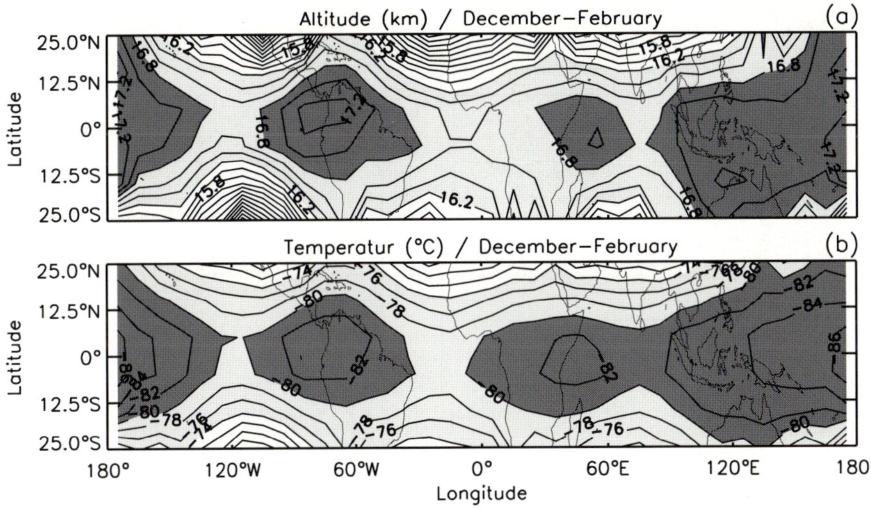

Fig. 6. Lapse rate tropopause altitude (panel a) and temperature (panel b) for northern winter. Contour interval: 0.2 km (a) and 2 K (b), respectively. The plots are derived using CHAMP data between 2001 and 2004 (3 winter seasons; update from (Schmidt et al. 2004)).

5 Monitoring of Tropopause Parameters

CHAMP temperature data in the tropopause region are not affected by background temperature fields. In conjunction with its high vertical resolution the RO technique provides a nearly perfect tool for precise monitoring of tropopause characteristics (altitude, pressure and temperature) on a global scale.

The tropopause has received increasing interest from climate change researchers during the last three decades. Changes in tropopause parameters (altitude, pressure and temperature) were used as indicators for climate change. Hereby it was shown, that these parameters have the potential to provide a more clear signal for the global warming than the surface temperature (e.g., Sausen and Santer 2003). Therefore, we use the CHAMP data to monitor tropopause parameters.

As an example for these studies Fig. 6 shows the latitudinal-longitudinal structure of the tropical tropopause derived from CHAMP RO data for the Northern Hemisphere winter months. The plots represent a mean over 3 winter seasons. The results are consistent with climatologies based on radiosonde measurements and meteorological analyses, as discussed by Schmidt et al. (2004): The highest lapse rate tropopause altitudes during December to February of >17.0 km are observed in the tropical Western Pacific and the northern part of South America (Fig. 6a). The coldest temperatures (Fig. 6b), less than −82 °C, are correlated with the maximum lapse rate tropopause altitudes. In

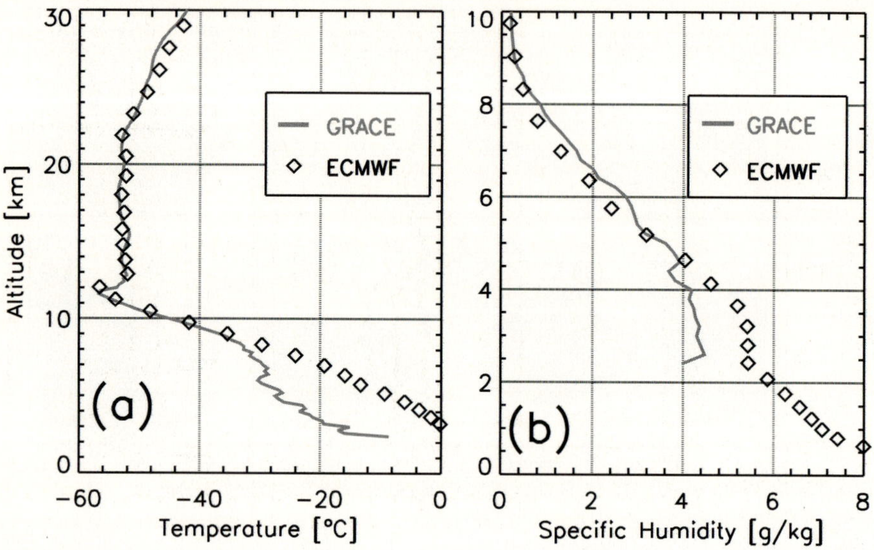

Fig. 7. Vertical profiles of (a) dry temperature and (b) specific humidity, derived from the first occultation measurement aboard GRACE, compared with corresponding ECMWF analysis (55.31°N, 22.32°E), July 28, 2004, 06:10 UTC (from (Wickert et al. 2005b)).

the Western Pacific region tropopause temperatures reach values of less than −86 °C.

6 First Results from GRACE

The "BlackJack" GPS receiver (provided by JPL) aboard the GRACE-B satellite (aft-looking antenna to observe setting occultations) was activated for the first time in atmospheric sounding mode from July 28, 06:00 UTC until July 29, 07:00 UTC, 2004. 120 occultations (parallel tracking of occultation and reference satellite ≥20 s) were recorded during this 25 h interval. The data were analyzed using the orbit and occultation processing system for CHAMP (see Sect. 3). The quality of the GRACE orbits was evaluated by comparisons with Satellite Laser Ranging (SLR) data. The rms was around 4cm to 5 cm and is slightly lower than the rms for CHAMP orbits (currently 5 cm).

The location of the first occultation measurement from GRACE is not far from the geographical center of Europe (54.85°N, 25.32°E, nearby Vilnius, Lithuania). Figure 7 shows the retrieved profiles of dry temperature and specific humidity, and the corresponding ECMWF profiles for the first occultation measurement from GRACE.

The significantly improved stability of the satellite clock from GRACE-B in relation to CHAMP (absence of periodic clock adjustments) allows for

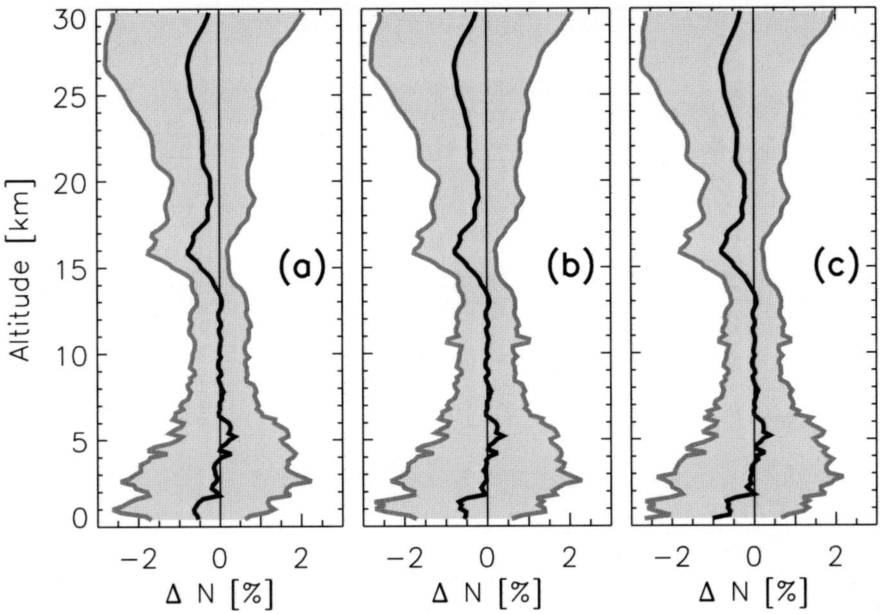

Fig. 8. Statistical comparison between 87 refractivity profiles, derived from GRACE measurements and 6-hourly ECWMF analyses during July 28 and 29, 2004. The differences (GRACE - ECMWF) are plotted for the profiles derived using (a) double differencing, (b) single differencing and (c) zero differencing (please note the difference compared to Fig. 4 of Wickert et al. (2005b), for details see text).

the application of a zero differencing technique. This avoids the disturbing influence of the additional link to a referencing GPS satellite and reduces the onboard data amount. The method was first applied and is described in more detail by Beyerle et al. (2005a). For the operational GRACE data analysis we apply an implementation of the zero difference technique using the 30 s clock solutions (Wickert et al. 2005b) provided by the precise orbit determination facility from GFZ (König et al. 2005). We processed the first GRACE occultations applying the standard double, the single (Wickert et al. 2002) and the zero differencing technique and compared the resulting vertical refractivity profiles with ECMWF (see Fig. 8). The comparison results for these three datasets are nearly identical and show no discernable differences. This finding does not confirm the early results from Wickert et al. (2005b), which indicated a better agreement of the zero difference results with ECMWF. In contrast to these preliminary results we've analyzed the first measurements from GRACE-B within the available study taking into account the reset of the GRACE-B clock by 14 ms on July 28, 16:00 UTC.

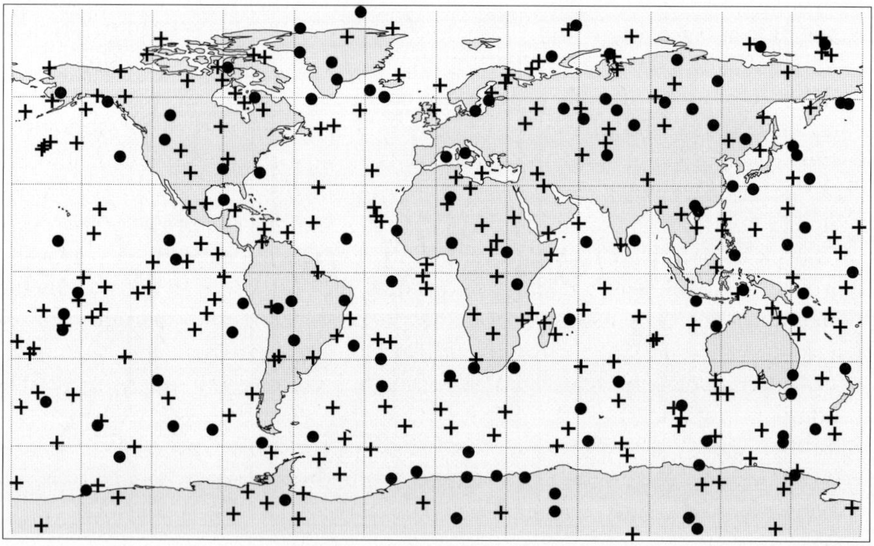

Fig. 9. Locations of the first occultations from GRACE (120 events, filled circles) and those from CHAMP (218 events, crosses) between July 28 (06:00 UTC) and July 29 (07:00), 2004 (from (Wickert et al. 2005b)).

7 Atmospheric Sounding with CHAMP and GRACE

CHAMP and GRACE will form a satellite configuration for precise atmospheric sounding after the operational activation of the GRACE occultations. The GRACE measurements will be analyzed by the operational processing system at GFZ and the analysis data will be provided to the international scientific community via the data center at GFZ (http://isdc.gfz-potsdam.de).

Considering RO data from CHAMP and GRACE, the number of continuously available occultation measurements in comparison to the current stage (only CHAMP) can be doubled. A first impression of the global occultation distribution from the CHAMP/GRACE constellation is given by Fig. 9. In total 338 occultations (218 from CHAMP, 120 from GRACE) were recorded during the 25 h activation of the GRACE occultations. The number of GRACE-B measurements was intentionally reduced during the test (reduction of the viewing angle of the occultation antenna to 40°). Working in nominal mode, the same number of measurements as from CHAMP can be expected.

8 Conclusions and Outlook

After more than three and a half years of GPS radio occultation with CHAMP about 270 000 occultation measurements are available. As the mission is expected to last at least until 2007, the first long-term dataset of GPS RO data is anticipated. The data and results of the operational data analysis are available at the data center at GFZ (http://isdc.gfz-potsdam.de/champ). A NRT occultation data transfer and analysis is continuously demonstrated since February 2003. The average delay of ∼5 hours between each measurement aboard CHAMP and data product provision was reduced to ∼4 hours since mid April 2004. The long-term dataset of CHAMP is the base for the preparation of future occultation missions and the related processing systems, impact studies to improve the global weather forecasts, studies with relevance to detect climate change and other applications for atmospheric/ionospheric research. Validation results indicate that refractivities in the upper troposphere and lower stratosphere agree well with ECMWF. Mean deviations are below 0.5 % and standard deviations below 1 %. These deviations show latitudinal dependent characteristics. A negative refractivity bias in the lower troposphere is observed. This was significantly reduced by using advanced retrieval methods, which were implemented to generate the current version of data products (005). CHAMP dry temperature data are used for a precise monitoring of tropical tropopause characteristics. First measurements from the GRACE-B satellite are analyzed. The GRACE-B clock is stable enough to be modeled by 30 s clock solutions which can be used for the precise occultation processing applying zero differencing. A significant improvement of the GRACE-B analysis results by applying zero differencing, as indicated by the early results, cannot be confirmed. CHAMP and GRACE will form a promising constellation for operational sounding of the Earth's atmosphere on a global scale. Recent information on the status of the RO experiments aboard CHAMP and GRACE, the operational data analysis at GFZ and current validation results can be obtained via WWW (http://www.gfz-potsdam.de/gasp).

Acknowledgements. We thank all members of the CHAMP and GRACE team for their work, which is the basis for our investigations. The German Ministry of Education and Research (BMBF) supports the GRACE project within the GEOTECHNOLOGIEN geoscientific Research and Development programme under grant 03F0326A. The German Weather Service provided ECMWF analysis data. We thank Dr. Sean Healy for source code and help to implement the 1Dvar water vapor retrieval.

References

Ao CO, Meehan TK, Hajj GA, Mannucci AJ, Beyerle G (2003a) Lower-troposphere refractivity bias in GPS occultation retrievals. J Geophys Res 108(D18), doi:10.1029/2002JD003216

Ao CO, Schreiner WB, Wickert J (2003b) First report on the CHAMP radio occultation intercomparison study. JPL Publication 03–016

Beyerle G, Gorbunov ME, Ao CO (2003a) Simulation studies of GPS radio occultation measurements. Radio Sci 38(5), doi:10.1029/2002RS002800

Beyerle G, Wickert J, Schmidt T, Reigber C (2003b) Atmospheric sounding by GNSS radio occultation: An analysis of the negative refractivity bias using CHAMP observations. J Geophys Res, doi:10.1029/2003JD003922

Beyerle G, Schmidt T, Michalak G, Heise S, Wickert J, Reigber C (2005a) GPS radio occultation with GRACE: Atmospheric profiling utilizing the zero difference technique. Geophys Res Lett 32, L13806, doi:10.1029/2005GL023109

Beyerle G, Schmidt T, Wickert J, Heise S, Reigber C, König-Langlo G (2005b) An analysis of refractivity biases detected in GPS radio occultation data: Results from simulation studies, aerological soundings and CHAMP satellite observations. Paper available via www.arxiv.org/physics/0502052

Dunn C et al. (2003) Instrument of GRACE – GPS Augments Gravity Measurements. GPS World

Foelsche U, Kirchengast G, Gobiet A, Steiner AK, Löscher A, Wickert J, Schmidt T (2005) The CHAMPCLIM project: An overview. In: Reigber C, Schwintzer P, Lühr H, Wickert J (eds) Earth Observation with CHAMP: Results from Three Years in Orbit. Springer Verlag, Berlin, pp 615–620

Gobiet A, Foelsche U, Steiner AK, Borsche M, Kirchengast G, Wickert J (2005) Climatological validation of stratospheric temperatures in ECMWF operational analyses with CHAMP radio occultation data. Geophys Res Lett 32(12), doi:10.1029/2005GL022617

Hajj GA, Lee LC, Pi X, Romans LJ, Schreiner WS, Straus PR, Wang C (2000) COSMIC GPS ionospheric sensing and space weather. Terrestrial, Atmospheric and Oceanic Science 11:235-272

Hajj GA et al. (2004) CHAMP and SAC-C atmospheric occultation results and intercomparisons. J Geophys Res 109(D06109), doi:10.1029/2003JD003909

Healy S, Eyre J (2000) Retrieving temperature, water vapour and surface pressure information from refractive-index profiles derived by radio occultation: A simulation study. Quart J Roy Meteorol Soc 126:1661–1683

Healy S, Jupp A, Marquardt C (2005) Forecast impact experiment with GPS radio occultation measurements. Geophys Res Lett 32(3), doi:10.1029/2004GL020806

Healy SB, Thepaut J-N (2005) Assimilation experiments with CHAMP GPS radio occultation measurements. Q J R Meteorol Soc, subm

Heise S, Wickert J, Beyerle G, Schmidt T, Reigber C (2005), Global monitoring of tropospheric water vapor with GPS radio occultation aboard CHAMP. Adv Space Res, in print

Jensen AS, Lohmann M, Benzon HH, Nielsen A (2003) Full spectrum inversion of radio occultation signals. Radio Sci 38(3), doi:10.1029/2002RS002763

König R, Michalak G, Neumayer K, Schmidt R, Zhu S, Meixner H, Reigber C (2005) Recent developments in CHAMP orbit determination at GFZ. In: Reigber C, Schwintzer P, Lühr H, Wickert J (eds) Earth Observation with CHAMP: Results from Three Years in Orbit. Springer Verlag, Berlin, pp 65–70

Kuo Y-H, Sokolovskiy SV, Anthes RA, Vandenberghe F (2000) Assimilation of GPS radio occultation data for numerical weather prediction. Terrestrial, Atmospheric and Oceanic Sciences 11(1):157–186

Kuo Y-H, Wee T-K, Sokolovskiy S, Rocken C, Schreiner W, Hunt D, Anthes RA (2004) Inversion and error estimation of GPS radio occultation data. J Meteorol Soc Jpn 82(1B):507–531

Kuo Y-H, Schreiner WS, Wang J, Rossiter DL, Zhang Y (2005) Comparison of GPS radio occultation soundings with radiosondes. Geophys Res Lett 32, L05817, doi:10.1029/2004GL021443

Kursinski ER, Hajj GA, Schofield JT, Linfield RP, Hardy KR (1997) Observing Earth's atmosphere with radio occultation measurements using Global Positioning System. J Geophys Res 102(D19):23429–23465

Larsen GB, Lauritsen KB, Rubek F, Sørensen MB (2005) Processing of CHAMP radio occultation data using GRAS SAF software. In: Reigber C, Schwintzer P, Lühr H, Wickert J (eds) Earth Observation with CHAMP: Results from Three Years in Orbit. Springer Verlag, Berlin, pp 543–548

Loiselet M, Stricker N, Menard Y, Luntama J (2000) GRAS – MetOps GPS based atmospheric sounder. ESA Bulletin, May, 102, pp 38–44

Randel WJ, Wu F, Rios WR (2003) Thermal variability of the tropical tropopause region derived from GPS/MET observations. J Geophys Res 108(D1):4024, doi: 10.1029/2002JD002595

Ratnam MV, Tsuda T, Jacobi C, Aoyama Y (2004) Enhancement of gravity wave activity observed during a major southern hemisphere stratospheric warming by CHAMP/GPS measurements. Geophys Res Lett 31(L16101), doi: 10.1029/2004GL019789

Reigber C, Schwintzer P, Lühr H, Wickert J (eds) (2005) Earth Observation with CHAMP: Results from Three Years in Orbit. Springer Verlag, Berlin

Rocken C, Kuo Y-H, Schreiner W, Hunt D, Sokolovskiy S, McCormick C (2000) COSMIC system description. Terrestrial, Atmospheric and Oceanic Sciences 11(1):21–52

Rocken C et al. (1997) Analysis and validation of GPS/MET data in the neutral atmosphere. J Geophys Res 102(D25):29849–29866

Sausen R, Santer BD (2003) Use of changes in tropopause height to detect human influences on climate. Met Zeitschrift 12(3):131–136

Schmidt T, Wickert J, Beyerle G, Reigber C (2004) Tropical tropopause parameters derived from GPS radio occultation measurements with CHAMP. J Geophys Res 109(D13105), doi:10.1029/2004JD004566

Schmidt T, Heise S, Wickert J, Beyerle G, Reigber C (2005a) GPS radio occultation with CHAMP and SAC-C: global monitoring of thermal tropopause parameters. Atmosph Chemistry and Physics 5:1473–1488

Schmidt T, Wickert J, Beyerle G, König R, Galas R, Reigber C (2005b) The CHAMP atmospheric processing system for Radio occultation measurements. In: Reigber C, Schwintzer P, Lühr H, Wickert J (eds) Earth Observation with CHAMP: Results from Three Years in Orbit. Springer Verlag, Berlin, pp 597–602

Smith EK, Weintraub S (1953) The constants in the equation for atmospheric refractive index at radio frequencies. Proc IRE 41:1035–1037

Sokolovskiy SV (2001) Tracking tropospheric radio occultation signals from low Earth orbit. Radio Sci 36(3):483–498

Takahashi HD et al. (2004) Scientific satellite EQUARS for remote sensing of the equatorial atmosphere. 7th Latin-American Conference on Space Geophysics, Atibaia

Tapley B, Reigber C (2004) GRACE (Gravity Recovery and Climate Experiment). In: McGraw-Hill Yearbook of Science & Technology

Vorob'ev VV, Krasil'nikova TG (1994) Estimation of the accuracy of the refractive index recovery from Doppler shift measurements at frequencies used in the NAVSTAR system. Phys Atmos Ocean 29:602–609

Wang DY et al. (2004) Cross-validation of MIPAS/ENVISAT and GPS-RO/CHAMP temperature profiles. J Geophys Res 109(D19311), doi: 10.1029/2004JD004963

Ware R et al. (1996) GPS sounding of the atmosphere from low Earth orbit: Preliminary results. Bull Am Meteorol Soc 77(1):19–40

Wickert J (2004) Comparison of vertical refractivity and temperature profiles from CHAMP with radiosonde measurements. Scientific Technical Report 04/19, GFZ Potsdam, ISSN 1610-0956

Wickert J, Galas R, Beyerle G, König R, Reigber C (2001a) GPS ground station data for CHAMP radio occultation measurements. Phys Chem Earth (A) 26:503–511

Wickert J, Beyerle G, Hajj GA, Schwieger V, Reigber C (2002) GPS radio occultation with CHAMP: Atmospheric profiling utilizing the space-based single difference technique. Geophys Res Lett 29(8), doi:10.1029/2001GL013982

Wickert J, Galas R, Schmidt T, Beyerle G, Reigber C, Förste C, Ramatschi M (2004a) Atmospheric sounding with CHAMP: GPS ground station data for occultation processing. Phys Chem Earth 29(2-3):267–275

Wickert J, Pavelyev AG, Liou YA, Schmidt T, Reigber C, Pavelyev AA, Igarashi K, Matyugov S (2004b) Amplitude variations in GPS signals as a possible indicator of ionospheric structures. Geophys Res Lett 31, L24801, doi:10.1029/2004GL020607

Wickert J, Schmidt T, Beyerle G, König R, Reigber C, Jakowski N (2004c) The radio occultation experiment aboard CHAMP: Operational data analysis and validation of vertical atmospheric profiles. J Meteorol Soc Jpn 82(1B):381–395

Wickert J, Ao CO, Schreiner WS (2005a) GPS based atmospheric sounding with CHAMP: Comparison of data analysis and results from GFZ, JPL and UCAR. GFZ Scientific Technical Report, in preparation

Wickert J, Beyerle G, König R, Heise S, Grunwaldt L, Michalak G, Reigber C, Schmidt T (2005b) GPS radio occultation with CHAMP and GRACE: A first look at a new and promising satellite configuration for global atmospheric sounding. Ann Geophysicae 23:653–658

Wickert J, Schmidt T, Beyerle G, Heise S, Reigber C (2005c) Global atmospheric sounding with GPS radio occultation aboard CHAMP. In: Rummel R, Reigber C, Rothacher M, Boedecker G, Schreiber U, Flury J (eds) Observation of the Earth System from Space. Springer Verlag, Berlin

Wickert J et al. (2001b) Atmosphere sounding by GPS radio occultation: First results from CHAMP. Geophys Res Lett 28(17):3263–3266

Sensitivity of Stratospheric Retrievals from Radio Occultations on Upper Boundary Conditions

C. O. Ao[1], G. A. Hajj[1], B. A. Iijima[1], A. J. Mannucci[1], T. M. Schrøder[1], M. de la Torre Juárez[1], and S. S. Leroy[2]

[1] Jet Propulsion Laboratory, California Institute of Technology, Pasadena, USA
chi.o.ao@jpl.nasa.gov
[2] Department of Chemistry and Chemical Biology, Harvard University, Cambridge, USA

Abstract. The main uncertainty in the stratospheric retrievals from GPS radio occultation (RO) measurements comes from the lack of reliable measurements in the upper stratosphere and above where the bending due to the neutral atmosphere is weak and residual ionospheric effects are strong. In this work, we quantify the bias and uncertainty of the refractivity and temperature retrievals due to different upper boundary strategies using a simulation study. We use refractivity profile derived from lidar pressure and temperature profiles as the input states in generating the synthetic occultations. Random noise levels commensurate with the CHAMP RO measurements are added to the simulated data. We examine the sensitivity of stratospheric retrievals to two different upper boundary methods, one based on exponential extrapolation and the other on MSIS climatology. The simulation results show that both methods lead to comparable levels of temperature bias (less than 0.5 K below 30 km altitude), provided that the upper boundary heights are set above 55 km.

1 Introduction

GPS radio occultation (RO) has been touted as one of the most promising remote sensing techniques in climate monitoring because RO measurements are self-calibrating and are not subject to time-dependent biases due to instrumental drifts (Goody et al. 1998). Over the years, the precision, accuracy, and resolution of the measurements, especially in the altitude range of 5 km to 25 km, has been well-established through theoretical considerations (e.g., Kursinski et al. 1997), validation studies (e.g., Rocken et al. 1997), and inter-satellite comparison (Hajj et al. 2004). However, stratospheric retrievals at altitudes higher than 25 km are more uncertain. This is mainly due to the lack of reliable measurements in the upper stratosphere and above where the bending due to the neutral atmosphere is weak compared with various error sources including thermal noise, orbital and local multipath errors, and perhaps most significantly, uncorrected ionospheric effects (Kursinski et al. 1997). A solution for this problem is to replace the noisy bending angles at high altitudes (typically above 40 km to -60 km) with "modeled" bending angles obtained from a climatology such as MSIS (Picone et al. 2002). The influence of the modeled bending angles decreases as the altitude decreases. The possible problem with this approach is that the retrievals could become biased toward the adopted climatology. An alternative, climatology-independent approach is to extrapolate the data at lower altitudes upward to altitudes where the data are not trustworthy. The problem with the extrapolation approach is that it relies on questionable assumptions regarding the characteristics of the stratosphere and mesosphere.

While there have been numerous published works addressing the upper boundary treatments and the retrieval errors associated with them (Kursinski et al. 1997; Hocke 1997; Steiner et al. 1999; Healy 2001; Rieder and Kirchengast 2001; Gorbunov 2002; Gobiet and Kirchengast 2004), the present study is unique in that the retrieval errors are examined with a simulation study which is based on atmospheric profiles from lidar measurements. The lidar profiles are more representative of the real stratospheric and mesospheric conditions than available models (albeit much more localized spatially and temporally) and are relatively independent of climatology. Thus they can be used to assess more quantitatively the errors due to the use of climatology as upper boundary conditions. The focus of this paper is to evaluate and compare the sensitivity of RO retrievals to the extrapolation and climatology approaches. We consider random bending angle noise at levels that are representative of CHAMP measurements.

The rest of the paper proceeds as follows. In Sect. 2, we give more details on the data and methodology used in this study. The numerical results are presented in Sect. 3, where we show refractivity and temperature errors under different upper boundary conditions. Finally, we summarize the main findings in Sect. 4 and discuss future work.

2 Data and Methodology

The input atmosphere used in the simulations is based on one year of lidar observations from Mauna Loa, Hawaii (19.5°N, 155.6°W) (Leblanc and McDermid 2001). The retrieved lidar profiles are freely accessible from the Network for the Detection of Stratospheric Change (NDSC) web site (http://www.ndsc.ncep.noaa.gov). In the year 2001, a total of 156 profiles covering the altitude range 20 km to 90 km are available. Since the contribution of water vapor to refractivity is negligible in the stratosphere and above, the refractivity profile can be derived simply from the lidar temperature and pressure profiles with the standard expression $N = 77.6(P/T)$, where P is the pressure in hPa and T is the temperature in Kelvin.

From the refractivity profile, and assuming local spherical symmetry, the bending angle can be obtained with the forward Abel integral (e.g., Kursinski et al. 1997)

$$\alpha(a) = -2a \int_a^\infty \frac{\mathrm{d}a'}{\sqrt{a'^2 - a^2}} \frac{\mathrm{d}\ln n}{\mathrm{d}a'} \tag{1}$$

where $a = n(r)r$ is the impact parameter and $n(r) = 1 + N(r) \times 10^{-6}$ is the index of refraction. Since input profiles only reach altitudes of ≈ 90 km, we extend the refractivity profiles beyond the maximum lidar altitude with exponential extrapolation. Because the atmosphere is so tenuous at these altitudes, the results presented here are not sensitive to the exact manner in which these profiles are extended.

The next step is to add realistic level of noise to the simulated bending angles.

$$\alpha_{\mathrm{obs}}(a) = \alpha(a) + \alpha_{\mathrm{n}}(a) \tag{2}$$

Note that we have ignored the ionosphere in the computation of $\alpha(a)$. Thus we regard $\alpha_{\mathrm{obs}}(a)$ as the ionosphere-free bending angle. Any residual calibration and ionosphere errors should be modeled in $\alpha_{\mathrm{n}}(a)$. The choice of $\alpha_{\mathrm{n}}(a)$ will be discussed more below.

In the presence of noise, $\alpha_{\mathrm{obs}}(a)$ can be trusted only below certain impact parameter a_{u} where the bending signal is large compared to the noise level. However, to obtain the refractivity at $a < a_{\mathrm{u}}$, the bending angles at all impact parameters above it are required. Thus, $\alpha_{\mathrm{obs}}(a)$ for $a > a_{\mathrm{u}}$ should be replaced with external data or *a priori* model, $\alpha_{\mathrm{mod}}(a)$. The refractivity profile is obtained from the Abel inversion integral as

$$\ln n(a) = \frac{1}{\pi} \int_a^{h+R} \mathrm{d}a' \frac{\alpha_{\mathrm{obs}}(a')}{\sqrt{a'^2 - a^2}} + \frac{1}{\pi} \int_{h+R}^\infty \mathrm{d}a' \frac{\alpha_{\mathrm{mod}}(a')}{\sqrt{a'^2 - a^2}} \tag{3}$$

where R is the local radius of curvature of the Earth and $h = a_{\mathrm{u}} - R$ is the impact parameter height corresponding to the transition from data to model.

The specification of $\alpha_{\mathrm{mod}}(a)$ above h used (including the choice of h) will henceforth be referred to as the Abel boundary condition (ABC). As discussed in Sect. 1, current approaches to ABC can be grouped into two categories: *extrapolation* (EXT) and *climatology* (CLI). The ideal ABC should be able to minimize the propagation of noise in the retrieved refractivity to lower altitudes while producing little or no bias.

In the EXT approach, the bending angle in the region below h is used to extrapolate the data to higher altitudes. No other external information is needed other than the functional form assumed in the extrapolation. We use a simple exponential function $\exp(b_0 + b_1 a)$ to characterize the bending angle above h with the parameters b_0, b_1 determined from fitting the data from $\approx (h - 10$ km) to h. The exponential functional form approximates an isothermal atmosphere. The effectiveness of this approach depends on how well such an approximation works as well as how well the fitting parameters can be determined. In the CLI approach, the observed bending angle in the region above h is replaced with bending derived from a climatology such as MSIS (computed at the time and location of each occultation profile). Thus the replacement is completely independent of the observed bending angle and is not susceptible to noise in the data. On the other hand, the MSIS profiles are likely biased relative to the true states of the atmosphere. For instance, compared with the Mauna Loa lidar profiles used in the simulation, the MSIS refractivity is larger in the mesosphere, with a peak average difference of approximately 7 % at 60 km. The MSIS temperature is colder in the mesosphere, with a peak average difference of about 2 K at 60 km. The use of MSIS in modeling the bending angles at high altitudes can lead to systematic biases in the retrievals (cf. Sect. 3).

Note from Eq. (3) that we have used a "hard" boundary where the $\alpha(a)$ switches from $\alpha_{\mathrm{obs}}(a)$ to $\alpha_{\mathrm{mod}}(a)$ at a fixed impact parameter height h. An alternative approach is to adopt the so-called statistical optimization method where the measured and modeled bending angles are linearly combined to minimize the root-mean-square (rms) error in the bending angles (see e.g., (Gobiet and Kirchengast 2004) and references therein). The effective application of optimization method requires reasonable estimates of the variance and covariance characteristics of the measurements and model, which is a nontrivial task. The hard boundary is applied here because the results are much simpler to interpret.

A key ingredient in the simulation study is to come up with a realistic representation of the bending angle noise $\alpha_{\mathrm{n}}(a)$. This, however, proves difficult because of the multitude of random and systematic error sources that might contribute to the ionosphere-free bending angle (Kursinski et al. 1997). For simplicity, we assume that the noise is characterized by a random Gaussian process with standard deviation which is independent of altitude. The level of random noise can be determined through the examination of CHAMP bending angles. Figure 1 shows the rms residual of the ionosphere-free bending angle obtained by linearly detrending $\alpha_{\mathrm{obs}}(a)$ with impact parameter heights between 50 km and 55 km. Interesting seasonal and latitudinal variations can be noted, with significantly more scatters in the polar regions. From the figure, it can be concluded that bending angle residuals for most of the CHAMP occultations fall between 1 μrad and 4 μrad. For comparison, the U.S. Standard atmosphere gives a bending angle of about 5 μrad at 60 km and 8 μrad at 45 km. To obtain a reasonable upper bound on the averaged errors, we consider in the following bending angle noise with 4 μrad standard deviation and examine the retrieved refractivity and temperature profiles resulting from ABC strategies with upper boundary heights from 45 km to 60 km.

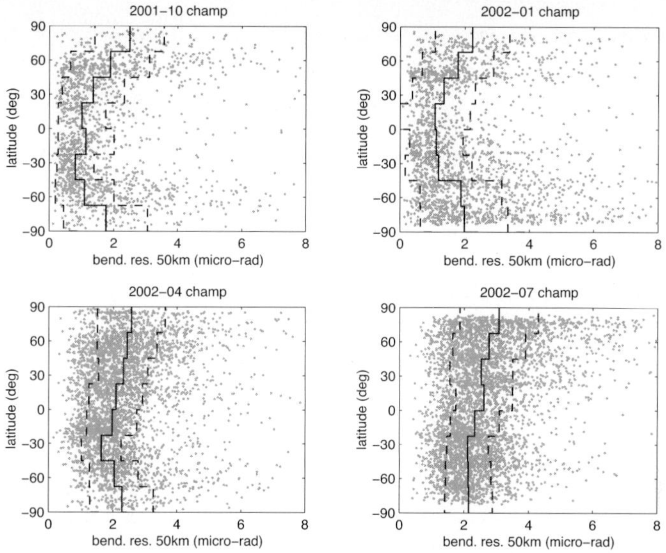

Fig. 1. Residuals of the ionosphere-corrected bending angle from four months of CHAMP data. Solid line indicates the median values within 20° latitudinal bands while dashed lines indicate the mean absolute deviation values about the median.

3 Simulation Results

3.1 Refractivity Errors

Consider first the noiseless case. Figure 2 shows the mean and rms fractional refractivity errors corresponding to different ABC strategies (EXT and CLI approaches, with upper boundary heights $h = 45, 50, 55, 60$ km). These results are obtained by averaging the errors over the 156 simulated occultations. In the noiseless case, the refractivity errors are entirely due to the inaccurate modeling of the atmosphere above h. As expected, the mean and rms errors increase as h decreases. The bias is positive for the CLI approach and mostly negative for the EXT approach.

Figure 3 shows the corresponding results for the case with 4 μrad bending angle noise. While the mean errors remain at about the same level as the noiseless case, the rms errors are now several times larger. It should be noted that lowering the upper boundary height h has relatively little impact on reducing the rms errors, indicating that the rms errors are dominated by the bending angles below 45 km impact parameter heights.

These results show that for both EXT and CLI approaches, it is far better to use an upper boundary height which is in the range of 55 km to 60 km. These strategies yield the smallest biases without introducing significantly larger rms errors. For the 4 μrad noise case, $h = 55$ km gives a refractivity bias of $-0.05\,\%$ and rms error of $0.71\,\%$ at $z = 30$ km for EXT. For CLI, the

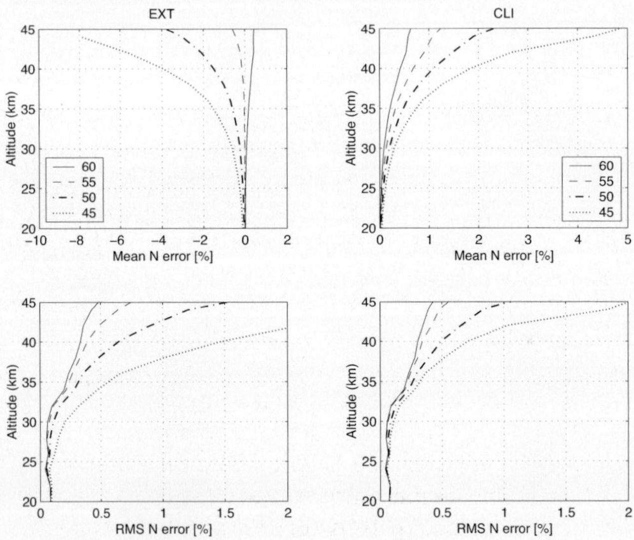

Fig. 2. Fractional refractivity errors for EXT and CLI strategies with upper boundary heights at 45, 50, 55, 60 km: noiseless case.

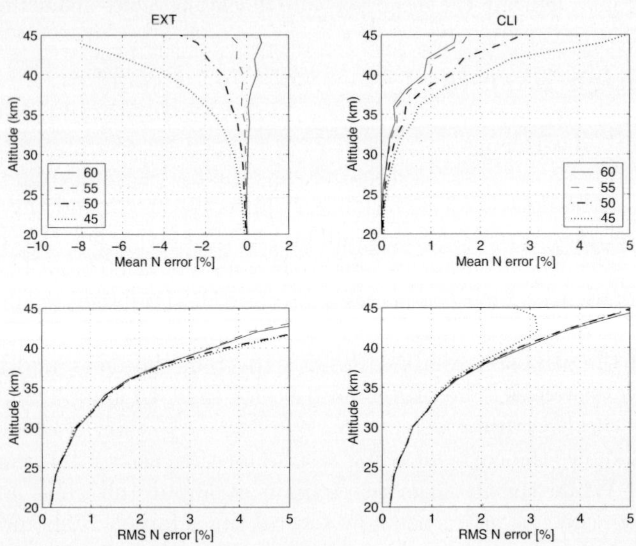

Fig. 3. Fractional refractivity errors for EXT and CLI strategies with upper boundary heights at 45, 50, 55, 60 km: 4 μrad case.

corresponding refractivity bias is 0.14 % with rms error of 0.61 %. Thus EXT and CLI results are quite comparable, with CLI yielding a smaller rms error at the cost of a larger bias.

3.2 Temperature Errors

To derive the temperature profile from refractivity profile (Kursinski et al. 1997), we initialize the hydrostatic equation with the input temperature at 40 km. Figures 4 and 5 show the temperature errors for the noiseless case and the case with 4 μrad bending angle noise respectively for different ABC approaches. Because the temperature is fixed at the initialization height of 40 km, the maximum temperature errors occur at altitudes slightly below 40 km. At 30 km, the mean temperature error for the 4 μrad noise with $h = 55$ km case is -0.50 K $\pm$ 2.71 K for EXT and 0.40 K $\pm$ 2.35 K. The rms errors become less than 1 K below 20 km for all ABC strategies.

When retrieving real data, we do not have the luxury of knowing the actual temperature. Errors in the initialization temperature introduce additional errors in the retrieved temperature profiles, although such errors decrease rapidly away from the initialization height. This is illustrated in Fig. 6, which shows the mean and rms temperature errors when the initialization temperature error is varied from 0 K to 10 K for the EXT strategy with $h = 55$ km. In the worst case with 10 K initialization error, the mean temperature error becomes 2.04 K at 30 km, but it drops to only 0.36 K at 20 km.

It should be noted that an alternative approach for deriving the temperature profile exists where temperature initialization is not required. In this method, the pressure is derived first by integrating the hydrostatic equation from the tangent height to infinity (Gobiet and Kirchengast 2004). Because the pressure is zero at infinity, this method eliminates any need to initialize

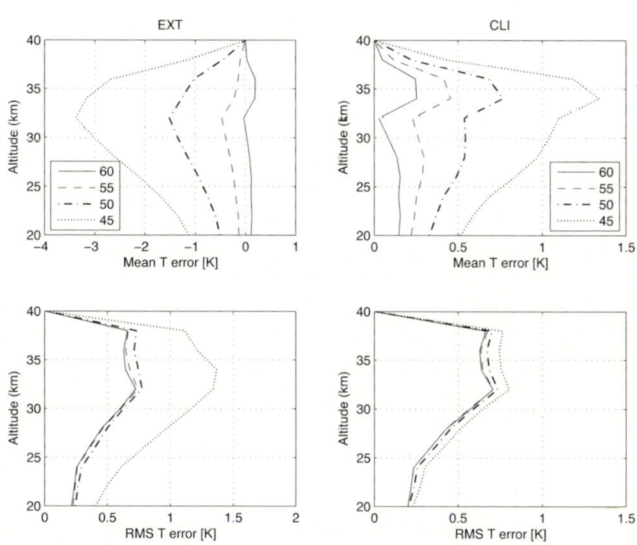

Fig. 4. Temperature errors for EXT and CLI strategies with upper boundary heights at $45, 50, 55, 60$ km: noiseless case.

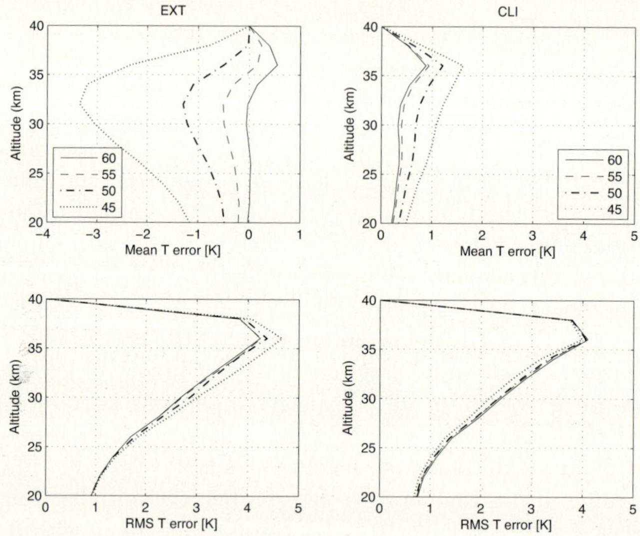

Fig. 5. Temperature errors for EXT and CLI strategies with upper boundary heights at $45, 50, 55, 60$ km: 4 μrad case.

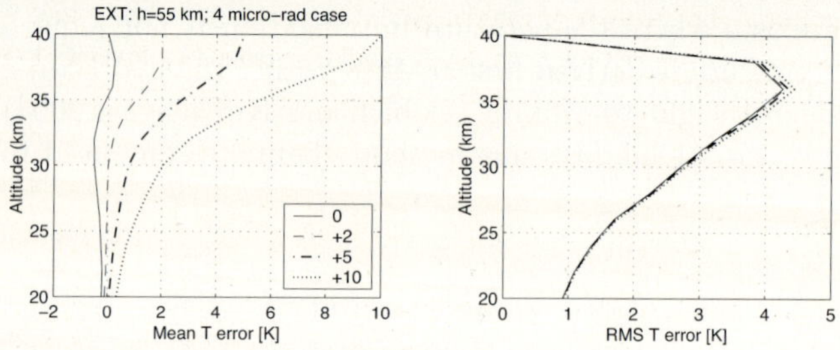

Fig. 6. Temperature errors due to temperature initialization error based on EXT approach with $h = 55$ km and 4 μrad bending angle noise.

the integration with ancillary data. However, the tradeoff is that this method requires refractivity values at very high altitudes, where they are obtained entirely from the modeled bending angles.

4 Conclusion

The simulation results presented here suggest that EXT and CLI upper boundary conditions yield comparable levels of biases and rms errors in refractivity and temperature below 40 km when upper boundary heights at ≈ 55 km to 60 km are used. Lower h (especially $h = 45$ km for EXT) leads to much larger biases without a significant benefit in reducing the rms errors. For the higher upper boundary heights, a bias on the order of 0.5 K can be expected at 30 km if the initialization temperature is exact. This bias increases to about 2 K if we assume a 10 K temperature initialization error. The simulation results also confirm the sub-Kelvin accuracy of temperature retrievals below 20 km, achieved under fairly noisy conditions and with great insensitivity to the Abel upper boundary conditions and temperature initialization.

It is important to recognize that the assumption of noise characteristics will have some impact on the EXT simulation results. Because EXT boundary condition requires the least-square fitting of noisy data, its performance will likely be worse if the noise is non-Gaussian and/or non-random.

The present study is rather limited in scope in that it only considers the atmospheric conditions at one tropical location. A more robust evaluation requires extending the study to multiple locations around the globe. In addition, only the gross statistical characteristics of the retrieval errors have been examined thus far. We have not addressed the key question of how the use of a climatology such as MSIS in the boundary condition would distort the interannual variability or long-term climate trends deduced from the occultation retrievals. However, further simulation studies of the kind performed here can be instrumental in clarifying such issues.

Acknowledgements. This research was carried out at the Jet Propulsion Laboratory, California Institute of Technology, under a contract with the National Aeronautics and Space Administration.

References

Gobiet A, Kirchengast G (2004) Advancements of Global Navigation Satellite System radio occultation retrieval in the upper stratosphere for optimal climate monitoring utility. J Geophys Res 109, D24110, doi:10.1029/2004JD005117

Goody R, Anderson J, North G (1998) Testing climate models: An approach. Bull Am Meteorol Soc 79(11):2541–2549

Gorbunov ME (2002) Ionospheric correction and statistical optimization of radio occultation data. Radio Sci 37(5), 1084, doi:10.1029/2000RS002370

Hajj GA, Ao CO, Iijima BA, Kuang D, Kursinski ER, Mannucci AJ, Meehan TK, Romans LJ, de la Torre Juárez M, Yunck TP (2004) CHAMP and SAC-C atmospheric occultation results and intercomparisons. J Geophys Res 109, D06109, doi:10.1029/2003JD003909

Healy SB (2001) Smoothing radio occultation bending angles above 40 km. Ann Geophys 19:459–468

Hocke K (1997) Inversion of GPS meteorology data. Ann Geophys 15:443–450

Kursinski ER, Hajj GA, Schofield JT, Linfield RP, Hardy KR (1997) Observing Earth's atmosphere with radio occultation measurements using the Global Positioning System. J Geophys Res 102(D19):23429–23465

Leblanc T, McDermid IS (2001) Quasi-biennial oscillation signatures in ozone and temperature observed by lidar at Mauna Loa, Hawaii (19.5°N, 155.6°W). J Geophys Res 106(D14):14869–14874

Picone JM, Hedin AE, Drob DP, Aikin AC (2002) NRLMSISE-00 empirical model of the atmosphere: Statistical comparisons and scientific issues. J Geophys Res 107(A12):1468, doi:10.1029/2002JA009430

Rieder MJ, Kirchengast G (2001) Error analysis and characterization of atmospheric profiles retrieved from GNSS occultation data. J Geophys Res 106(D23):31755–31770

Rocken C, Anthes R, Exner M, Hunt D, Sokolovskiy S, Ware R, Gorbunov M, Schreiner W, Feng D, Herman B, Kuo YH, Zou X (1997) Analysis and validation of GPS/MET data in the neutral atmosphere. J Geophys Res 102(D25):29849–29866

Steiner AK, Kirchengast G, Ladreiter HP (1999) Inversion, error analysis, and validation of GPS/MET occultation data. Ann Geophys 17:122–138

Error Characteristics of Refractivity Profiles Retrieved from CHAMP Radio Occultation Data

A. K. Steiner, A. Löscher, and G. Kirchengast

Wegener Center for Climate and Global Change (WegCenter) and Institute for Geophysics, Astrophysics, and Meteorology (IGAM), University of Graz, Austria
andi.steiner@uni-graz.at

Abstract. We present results of an empirical error analysis of refractivity profiles based on CHAMP radio occultation data. We analyzed two seasons of observations, boreal winter 2002/03 and boreal summer 2003. The processing was performed with the WegCenter/CHAMPCLIM Retrieval version 2. The error statistics is based on comparison to reference profiles calculated from ECMWF analyses fields. Bias profiles and error covariance matrices are provided, the latter separated into standard deviation profiles and error correlation matrices. Since the error characteristics contain both the observational error of the retrieved data and the model error of the ECMWF analyses we performed an estimation of the ECMWF model error and separated the observation error. The relative refractivity bias of CHAMP radio occultation data with respect to ECMWF was found to oscillate around –0.4 % at 5–25 km globally. Wavelike structures apparent at high latitudes in Southern Hemisphere winter are mainly due to the representation of the polar vortex in the ECMWF analyses. The combined relative standard deviation was found to be 0.7–1 % at 5–25 km height globally, showing larger values in winter than in summer in the upper stratosphere at mid- and high latitudes. The global observation error for CHAMP refractivity was estimated to be 0.5–0.75 % at 6–30 km. The results are compared to the findings of Kuo et al. (2004) and to those of an end-to-end simulation study being the precursor of this work (Steiner and Kirchengast 2004, 2005). Based on the simulation study we provide simple observation error covariance matrix formulations for CHAMP refractivity for convenient use in retrieval algorithms and in data assimilation systems.

1 Introduction

The assimilation of radio occultation (RO) data has the potential to significantly improve the accuracy of global and regional meteorological analysis and weather prediction, which has been confirmed by several studies (e.g., Kuo et al. 2000; Healy et al. 2005). One important issue in this respect is knowledge of radio occultation measurement errors in order to formulate adequate observation error covariance matrices for data assimilation systems.

Since refractivity seems to be the most appropriate parameter for assimilation purposes (Syndergaard et al. 2006; Healy et al. 2005) we performed an empirical error analysis of a set of refractivity profiles retrieved from CHAMP RO observations. Regarding the error analysis method, we build on the heritage of an earlier simulation study (Steiner and Kirchengast 2004, 2005) and extend it to a separate estimation of the observation error for CHAMP refractivity data.

Section 2 gives a brief description of the data set and the retrieval algorithms. In Sect. 3 the estimation of the combined error (CHAMP RO plus ECMWF) is described. The ECMWF model error is estimated in Sect. 4 and the results on the observed refractivity error are presented in Sect. 5. Summary and conclusions are drawn in Sect. 6.

2 Description of the Data Set and the Retrieval Scheme

The study is based on a CHAMP level 2 data set comprising two seasons of radio occultation observations, DJF 2002/2003 (December-January-February) and JJA 2003 (June-July-August). For each season more than 12 000 profiles of atmospheric excess phases were analyzed. The data sets were separated into three latitude bands, low (–30° to +30°), middle (±30° to ±60°), and high (±60° to ±90°) latitudes. In addition, we separately analyzed the Northern Hemispheric (NH) and the Southern Hemispheric (SH) region. The sample sizes for the various analyzed data sets are listed in Table 1. Refractivity profiles were processed from this data base of excess phase profiles with the CHAMPCLIM Retrieval version 2 (CCRv2), which includes an advanced upper stratospheric retrieval scheme (Gobiet and Kirchengast 2004a, 2004b). The retrieval is based on the standard geometric optics approach, thus we will not interpret the results below 5 km height. For the upper-boundary initialization of bending angles we used the MSISE-90 climatological model. The CCRv2 processing is described in detail by Steiner et al. (2004) and an overview is included in Borsche et al. (2006).

Table 1. The number of occultation events for the different data sets analyzed.

	DJF 2002/2003			JJA 2003		
	Total	*NH*	*SH*	*Total*	*NH*	*SH*
Global	**12329**	5995	6334	**12710**	5989	6721
Low lat	3790	1812	1978	3784	1841	1943
Mid lat	4310	2120	2190	4095	1983	2112
High lat	4229	2063	2166	4831	2165	2666

3 Estimation of the Combined Error

The error statistics is based on the comparison of the retrieved and smoothed (comparable to ECMWF vertical grid resolution) refractivity profiles with co-located refractivity profiles derived from 6-hourly operational meteorological analyses fields from ECMWF. The co-located vertical ECMWF profiles were calculated at a fixed mean tangent point location. When regarding the ECMWF profiles as the truth this implies that the error estimates represent an upper bound error estimate including the observation error, the model (ECMWF) error and the representativeness error. The representativeness error stems from the limited spatial and temporal measurement resolution and from the comparison of the retrieved profiles with vertical reference profiles. The latter fact is important in the troposphere below ~7 km, where higher horizontal variability is present (Foelsche and Kirchengast 2004; Syndergaard et al. 2004). Since we will not interpret results below 5 km, the representativeness errors to this end are largely negligible.

The statistical method for calculating the deviation of CHAMP from ECMWF ($x_{CHAMP} - x_{ECMWF}$), denoted as combined error (CHAMP observed error plus ECMWF model error), is described in detail in Steiner and Kirchengast (2004). Bias profiles and error covariance matrices are provided, the latter separated into standard deviation profiles and error correlation matrices.

The resulting error statistics for the combined refractivity error is shown in Fig. 1 for the global ensemble and the latitudinal data sets (horizontal panel rows), globally (left), for the Northern Hemisphere (middle), and the Southern Hemisphere (right) up to 35 km height. The relative bias (gray) and the relative standard deviation (Rel.StdDev) (black) of CHAMP RO with respect to ECMWF are shown for the JJA 2003 season (solid) and for the DJF 2002/2003 season (dashed).

The refractivity bias of CHAMP RO with respect to ECMWF oscillates around –0.4 % at 5–25 km globally as well as at mid- and high latitudes, increasing to 0.5–1 % at 35 km. Bias oscillations are seen at low latitudes, ranging from –0.4 % to 0.5 % at 5–35 km, with salient structures appearing at tropopause heights. This effect may partly be due to the higher resolved tropopause in CHAMP RO data than in ECMWF data (RO resolution ~1 km at that altitude, ECMWF analyses >1.3 km) but may also stem from a weak representation of tropopause height variability in ECMWF (Gobiet et al. 2005). The most prominent features can be seen in SH winter at high latitudes (lower left panel), which is an indication that the ECMWF field does not accurately represent the polar vortex in this region (Gobiet et al. 2005). The smallest bias occurs at NH high latitudes (lower middle panel) being about –0.3 % at 5–25 km in DJF almost vanishing in JJA.

The combined Rel.StdDev is of the order of 0.7–1 % at 5–28 km height globally and at low latitudes (NH and SH). At mid- and high latitudes the Rel.StdDev shows different behavior in the winter hemisphere and in the summer hemisphere at upper stratospheric heights, being 0.75–1 % at 20–34 km in summer becoming twice as large (~2 % at 35 km) in winter. This may partly be due to the larger atmospheric variability in winter and is subject to further investigation.

4 Estimation of the ECMWF Error

In order to separate the observed error of the CHAMP RO refractivity retrievals from the combined error, we calculated a global estimate of the ECMWF refractivity model error. M. Fisher (ECMWF Reading, UK, pers. communications, 2004) provided global error estimates of ECMWF analyses in form of standard deviations for temperature, specific humidity, and surface pressure, and of vertical error correlations for temperature and specific humidity. Temperature T (K), water

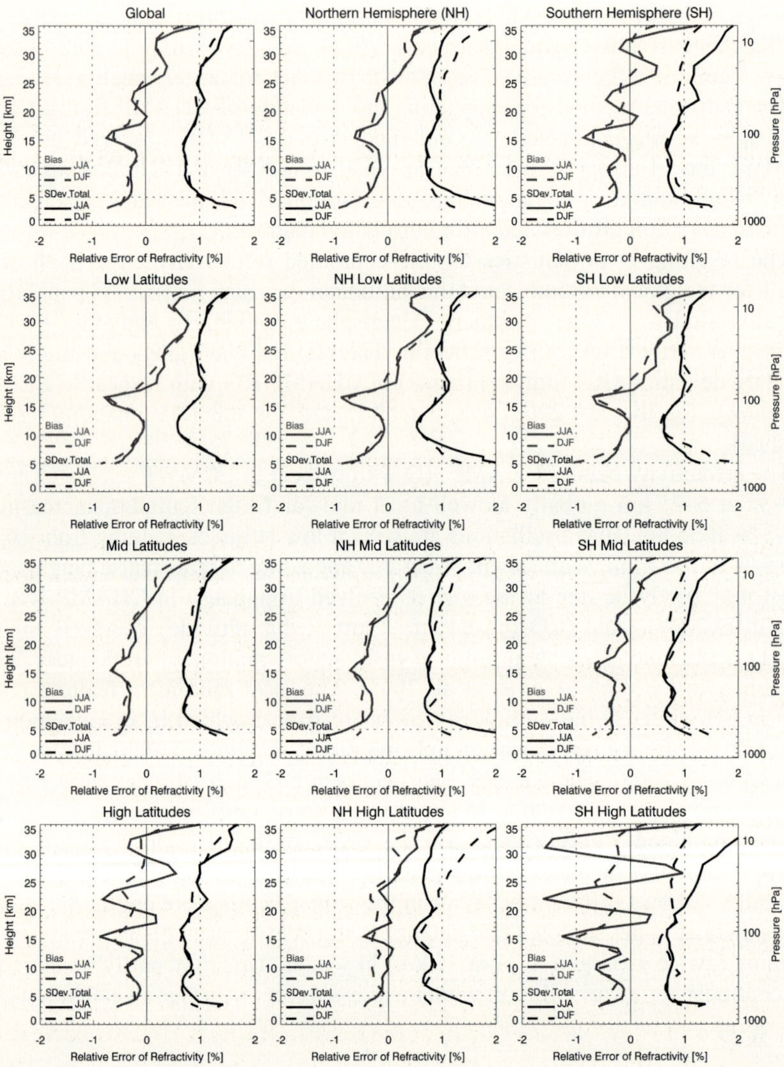

Fig. 1. Combined refractivity error as a function of height for the global and the latitudinal ensembles (horizontal panel rows), globally (left), Northern Hemisphere (middle), Southern Hemisphere (right). Relative bias (gray) and relative standard deviation (black) are shown for the JJA 2003 season (solid) and for the DJF 2002/03 season (dashed), respectively.

vapor pressure e (hPa), and total pressure p (hPa) are related to refractivity N (N units) via the Smith-Weintraub formula (Smith and Weintraub 1953),

$$N = c_1 \frac{p}{T} + c_2 \frac{e}{T^2}, \tag{1}$$

with the constants $c_1 = 77.6$ K/hPa and $c_2 = 3.73*10^5$ K^2/hPa. Water vapor pressure in Eq. 1 was substituted for specific humidity q (kg/kg) using the following relation,

$$e = \frac{p \cdot q}{a + bq}, \tag{2}$$

with a = 0.622 and b = 0.378. A simple error propagation based on Eq. 1 was applied via

$$\Delta N = \sqrt{\left(\frac{\partial N}{\partial T}\Delta T\right)^2 + \left(\frac{\partial N}{\partial q}\Delta q\right)^2 + \left(\frac{\partial N}{\partial p}\Delta p\right)^2}, \tag{3}$$

in order to calculate the standard deviation of refractivity. The pressure error at a given height was calculated by error propagation using the given standard deviation of surface pressure of 2.5 hPa. The vertical refractivity error correlations were derived by a weighted combination of temperature error correlations ($w_T = \Delta N(\Delta p,\Delta T)/\Delta N(\Delta p,\Delta q,\Delta T)$) and specific humidity error correlations ($w_q = \Delta N(\Delta q)/\Delta N(\Delta p,\Delta q,\Delta T)$).

Figure 2 displays global ECMWF error specifications for temperature (left panels), specific humidity (middle panels), and estimated refractivity (right panels). Standard deviations are shown in the upper panel row and error correlation functions for three different height levels (~10 km, ~20 km, ~30 km for T and N; ~3 km, ~6 km, ~10 km for q) are presented in the lower panel row.

We tested the sensitivity of the estimated refractivity error with respect to the temperature error input for the four cases displayed in Fig. 2, where we multiplied the temperature standard deviation (case 1xT in light gray) by 1.5 (middle gray), 2 (dark gray), and 2.5 (black). The results judged most reasonable were found for the case of doubling the temperature standard deviation (2xT case), giving a Rel.StdDev of ECMWF refractivity of the order of 0.5 % at 8–15 km increasing to 0.75 % at 30 km and to 1 % at 35 km. These results are consistent with the findings of Kuo et al. (2004) who performed an estimation of short-range forecast errors using the Hollingsworth-Lönnberg method (Hollingsworth and Lönnberg 1986). For comparison we included their estimates for low latitudes (dotted) and mid-latitudes (dashed) in Fig. 2 (upper right panel).

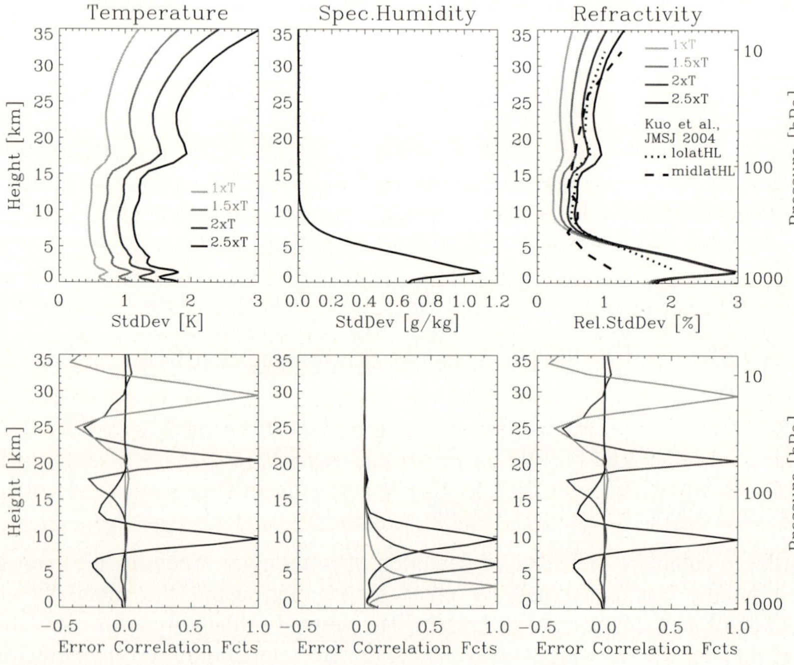

Fig. 2. ECMWF error for temperature (left) for 4 test cases (1x*T*, 1.5x*T*, 2x*T*, 2.5x*T*), specific humidity (middle), and corresponding estimated refractivity (right) in terms of standard deviation (upper panels) and error correlation functions (lower panels), the latter shown for three heights (~10 km (black), ~20 km/humidity: 6 km (gray), ~30 km/humidity: 3 km (light gray)). Estimates of short-range forecast errors for refractivity for low latitudes (dotted) and mid-latitudes (dashed) made by Kuo et al. (2004) are also shown (upper right panel).

5 Estimation of the Observation Error

The observed refractivity error was then derived by subtracting the ECMWF error from the combined error in terms of variances s^2,

$$(s_{\text{obs}})^2 = (s_{\text{combined}})^2 - (s_{\text{ECMWF}})^2 . \tag{4}$$

The results are displayed in Fig. 3 for the JJA season showing the combined error (black with diamond symbols), the ECMWF error for the 2x*T* case (gray), and the corresponding observation error (black) in terms of Rel.StdDev, respectively.

The corresponding global estimate of the observed Rel.StdDev for CHAMP refractivity (2x*T* case) is of the order of 0.5 % at 6–18 km, increasing to 0.7 % at

28 km and to 1.2 % at 35 km (left panel). At upper stratospheric heights the observed Rel.StdDev is around 0.5 % at 10–32 km in summer (middle panel) whilst it reaches 1–1.5 % in winter. Our observation error appears to be a more conservative estimate compared to the results of Kuo et al. (2004), who found the observation error of refractivity to be of the order of 0.3–0.5 % at 5–25 km.

As a further result, refractivity error correlation functions are displayed in Fig. 4 for three different heights, ~10 km, ~20 km, ~30 km, representative for troposphere, lower and upper stratosphere. Basically, these functions express the correlation of errors at these heights with the errors in the remainder of the profile. The ECMWF refractivity error correlation functions (dotted) show negative correlation features in the vicinity of the peaks whilst the correlation functions for the combined error (solid with diamond shaped symbols) show a flattening. These features suggest that the correlation wings are dominated by the observed data.

For the construction of refractivity observation error covariance matrices for data assimilation systems we therefore suggest a combination of the observed Rel.StdDev with the total error correlation matrix. We provide simple analytical formulations of refractivity error covariance matrices, which were deduced in a simulation study for a Metop/GRAS receiving system (Steiner and Kirchengast 2005). The functional formulations for Rel.StdDev and for correlation functions depend on a few parameters, which can be fitted for any given data set. Table 2 summarizes the functions. Using them for Rel.StdDev and approximating an exponential drop-off for the error correlations, a simple covariance matrix model $\boldsymbol{S}$ for the observed refractivity error can then be constructed via

$$\boldsymbol{S} = s_i s_j \exp\left[-\left|z_i - z_j\right| / L(z)\right]. \tag{5}$$

Table 2. Rel.StdDev $s(z)$ model for CHAMP refractivity with respective fitting parameters: z_{troptop} denoting the top level of the "troposphere domain", z_{stratbot} the bottom level of the "stratosphere domain", s_{utls} the Rel.StdDev between z_{troptop} and z_{stratbot}, s_0 the Rel.StdDev at ~1 km, H_{strat} the scale height of error, and $L(z)$ the correlation length, respectively.

<table>
<tr><th></th><th colspan="2">Relative Standard Deviation $s(z)$</th><th>Correlation Length $L(z)$</th></tr>
<tr><td>2 km < $z \le z_{\mathrm{troptop}}$</td><td>$s_{\mathrm{utls}}+s_0\,[1\mathrm{km}/z-1\mathrm{km}/z_{\mathrm{troptop}}]$</td><td rowspan="3">$s_{\mathrm{utls}}=0.5$ %
$s_0=4.5$ %
$z_{\mathrm{troptop}}=14$ km
global/NH: $z_{\mathrm{stratbot}}=20$ km
SH: $z_{\mathrm{stratbot}}=18$ km
global/SH: $H_{\mathrm{strat}}=15$ km
NH: $H_{\mathrm{strat}}=30$ km</td><td>$L=2$ km</td></tr>
<tr><td>$z_{\mathrm{troptop}} < z < z_{\mathrm{stratbot}}$</td><td>$s_{\mathrm{utls}}$</td><td rowspan="2">linear decrease to $L=1$ km at z=50 km</td></tr>
<tr><td>$z_{\mathrm{stratbot}} \le z < 35$ km</td><td>$s_{\mathrm{utls}}\exp[(z-z_{\mathrm{stratbot}})/H_{\mathrm{strat}}]$</td></tr>
</table>

Figure 3 visualizes the analytical functions (thin solid) for the observed Rel.StdDev (thick solid) for the JJA 2003 season. In order to fit the seasonal behavior we adjusted the scale height of error H_{strat} to 30 km for summer (NH) and the bottom level of the stratosphere domain z_{stratbot} to 18 km for winter (SH). The validity of the fit regarding the upper height limit depends on the receiving system, e.g., 35 km for CHAMP data and near 50 km for Metop/GRAS (Steiner and Kirchengast 2005).

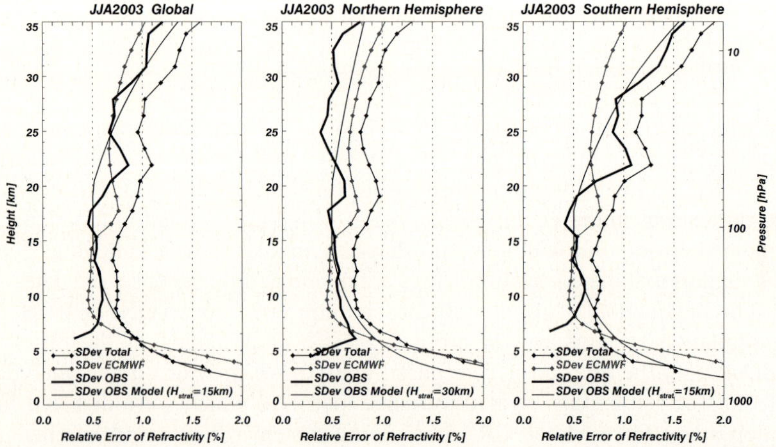

Fig. 3. Rel.StDev of refractivity for JJA 2003: combined (black diamond shaped symbols), ECMWF (gray diamond shaped symbols), observed (black thick), model (black thin); global (left), NH (middle), SH (right).

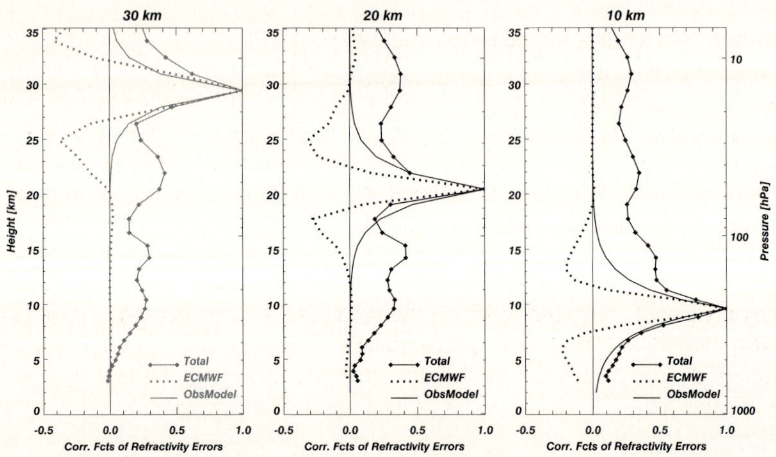

Fig. 4. Error correlation functions for the refractivity error: combined (solid line with diamond shaped symbols), ECMWF (dotted), and model (solid) shown for three heights ~30 km (light gray), ~20 km (gray), ~10 km (black).

6 Summary and Conclusions

As a follow-on study to the ensemble-based error analysis of simulated RO data (Steiner and Kirchengast 2004, 2005) we performed an error analysis for CHAMP RO refractivity profiles, processed with the IGAM CCRv2 scheme, for two seasons, DJF 2002/03 and JJA 2003. The error statistics was based on a comparison to reference profiles from ECMWF analyses fields, implying that the statistics includes both, the observation error and the ECMWF model error. In order to separate the errors, we performed an error estimation of the ECMWF error based on error propagation of temperature, humidity, and pressure error into refractivity error. Finally, the subtraction of the ECMWF error from the combined error allowed an estimation of the global observation error.

The relative refractivity bias of CHAMP RO with respect to ECMWF was found to, in general, oscillate around –0.4 % at 5–25 km globally. Wavelike structures apparent in SH winter at high latitudes are an indication that the ECMWF fields do not accurately represent the polar vortex (Gobiet et al. 2005). The smallest bias occurs at NH high latitudes in JJA. The combined Rel.StdDev of refractivity was found to be 0.7–1 % at 5–25 km height showing larger values in winter than in summer in the upper stratosphere at mid- and high latitudes. The estimated ECMWF refractivity error (2x*T* case) was found to be 0.5–0.75 % at 8–30 km. The global observation error of CHAMP refractivity was found to be 0.5 % at 6–18 km increasing to 1 % at 30 km globally, for mid- and high latitudes being ~0.5 % throughout this height range. These estimates are slightly more conservative than the 0.3–0.5 % results of Kuo et al. (2004).

In addition we analyzed refractivity error correlations. The ECMWF refractivity error correlation functions show negative correlation features in the vicinity of the peaks whilst the correlation functions for the combined error show a flattening. These features suggest that the correlation wings are dominated by the observed data. We therefore suggest a combination of the observed Rel.StdDev with the total error correlation matrix for the construction of observation error covariance matrices for data assimilation systems. These observation error covariance matrices can be approximated with simple analytical functions presented in Steiner and Kirchengast (2005); we presented parameters adjusted to the CHAMP CCRv2 performance. The refractivity error covariance formulation provided may be useful for implementation in optimal estimation parts of retrieval algorithms as well as in data assimilation systems.

Acknowledgments. The authors gratefully acknowledge the GFZ Potsdam, Germany, for the provision of CHAMP data; T. Schmidt for providing technical support and J. Wickert for scientific discussion. M. Borsche (WegCenter, Univ of Graz, Austria) is thanked for the processing of the CHAMP data sets and M. Fisher (ECMWF, Reading, UK) for providing error specifications for ECMWF analyses. The authors are thankful for valuable discussions on the topic with A. Gobiet (WegCenter, Univ of Graz, Austria), X.-Y. Huang (UCAR, Boulder, CO, USA), and C. Marquardt (EUMETSAT, Darmstadt, Germany). A.K.S. was funded from the START research award of G.K. financed by the Austrian Ministry for Education, Science, and Culture under Program Y103-N03 of the Austrian Science Fund.

References

Borsche M, Gobiet A, Steiner AK, Foelsche U, Kirchengast G, Schmidt T, Wickert J (2006) Pre-operational retrieval of radio occultation based climatologies. This issue

Foelsche U, Kirchengast G (2004) Sensitivity of GNSS occultation profiles to horizontal variability in the troposphere: A simulation study. In: Kirchengast G, Foelsche U, Steiner AK (eds) Occultations for Probing Atmosphere and Climate. Springer-Verlag, Berlin Heidelberg, pp 127–136

Gobiet A, Kirchengast G (2004a) Advancement of GNSS radio occultation retrieval in the upper stratosphere. In Kirchengast G, Foelsche U, Steiner AK (eds) Occultations for Probing Atmosphere and Climate. Springer-Verlag, Berlin Heidelberg, pp 137–148

Gobiet A, Kirchengast G (2004b) Advancements of GNSS radio occultation retrieval in the upper stratosphere for optimal climate monitoring utility. J Geophys Res 109(D24110), doi:10.1029/2004JD005117

Gobiet A, Foelsche U, Steiner AK, Borsche M, Kirchengast G, Wickert J (2005) Climatological validation of stratospheric temperatures in ECMWF operational analyses with CHAMP radio occultation data. Geophys Res Lett 32(L12806), doi:10.1029/2005GL022617

Healy SB, Jupp AM, Marquardt C (2005) Forecast impact experiment with GPS radio occultation measurements, Geophys Res Lett 32:L03804, doi:10.1029/2004GL020806

Hollingsworth A, Lönnberg P (1986) The statistical structure of short range forecast errors as determined from radiosonde data. Part I: The wind field. Tellus 38A:111–136

Kuo Y-H, Sokolovskiy SV, Anthes RA, Vandenberghe F (2000) Assimilation of GPS radio occultation data for numerical weather prediction. Terr Atmos Oceanic Sci 11:157–186

Kuo Y-H, Wee T-K, Sokolovskiy S, Rocken C, Schreiner W, Hunt D, Anthes RA (2004) Inversion and error estimation of GPS radio occultation data. J Meteorol Soc Japan 82: 507–531

Smith EK, Weintraub S (1953) The constants in the equation for atmospheric refractive index at radio frequencies. Proc of the IRE 41:1035–1037

Steiner AK, Kirchengast G (2004) Ensemble-based analysis of errors in atmospheric profiles retrieved from GNSS occultation data. In: Kirchengast G, Foelsche U, Steiner AK (eds) Occultations for Probing Atmosphere and Climate. Springer-Verlag, Berlin Heidelberg, pp 149–160

Steiner AK, Gobiet A, Foelsche U, Kirchengast G (2004) Radio occultation data processing advancements for optimizing climate utility. IGAM/Uni Graz Tech Rep for ASA No 3/2004

Steiner AK, Kirchengast G (2005) Error analysis for GNSS radio occultation data based on ensembles of profiles from end-to-end simulations. J Geophys Res 110(D15307), doi:10.1029/2004JD005251

Syndergaard S, Flittner DE, Kursinski ER, Feng DD, Herman BM, Ward DM (2004) Simulating the influence of horizontal gradients on retrieved profiles from ATOMS occultation measurements – a promising approach for data assimilation. In: Kirchengast G, Foelsche U, Steiner AK (eds) Occultations for Probing Atmosphere and Climate. Springer-Verlag, Berlin Heidelberg, pp 221–232

Syndergaard S, Kuo Y-H, Lohmann M (2006) Observation operators for the assimilation of occultation data into atmospheric models: A review. This issue

Refractivity Biases in GNSS Occultation Data

G. Beyerle[1], S. Heise[1], J. Kaschenz[1], G. König-Langlo[2], C. Reigber[1], T. Schmidt[1], and J. Wickert[1]

[1] GeoForschungsZentrum Potsdam (GFZ), Deptartment 1, Geodesy & Remote Sensing, Potsdam, Germany
gbeyerle@gfz-potsdam.de

[2] Alfred Wegener Institute for Polar and Marine Research, Bremerhaven, Germany

Abstract. An analysis of atmospheric refractivity profiles observed by the geo-research satellite CHAMP between May 2001 and October 2004 reveals a negative bias compared to ECMWF meteorological fields at altitudes below 5 km. In order to separate bias contributions caused by critical refraction from contributions induced by the receiver tracking process a comprehensive end-to-end simulation study was performed. The simulations are based on radiosonde profiles obtained aboard research vessel "POLARSTERN". Within a subset of 3039 profiles recorded on the Atlantic Ocean between 60°N and 60°S, 1202 profiles (39.6 %) are found with vertical refractivity gradients below the threshold value of $-157\ \mathrm{km}^{-1}$. Critical refraction layers occur mainly between 1 km and 2.5 km altitude, above 3 km the occurrence of critical refraction can be disregarded. End-to-end simulations using these 3039 refractivity profiles confirm that four quadrant carrier phase extraction outperforms the two quadrant method currently implemented on CHAMP. Within regions of low signal-to-noise ratios "open-loop" tracking methods yield improvements with respect to the current "fly-wheeling" method.

1 Introduction

Atmospheric sounding by means of Global Positioning System (GPS) radio occultation (RO) may contribute to improvements in numerical weather prediction and climate change studies. Validation studies of past and current RO satellite missions, however, indicate systematic deviations between retrieved refractivities and corresponding data from meteorological analyses in the lower troposphere at tropical latitudes. These biases have been identified and discussed already in the GPS/MET validation study (Rocken et al. 1997). Similar biases are observed in the CHAMP dataset (Wickert et al. 2004; Marquardt et al. 2003; Ao et al. 2003a; Hajj et al. 2004).

The CHAMP geo-research satellite (Reigber et al. 2004) was launched on July 15, 2000 and collects atmospheric occultation profiles since February 11, 2001. As of October 31, 2004 the "BlackJack" occultation receiver

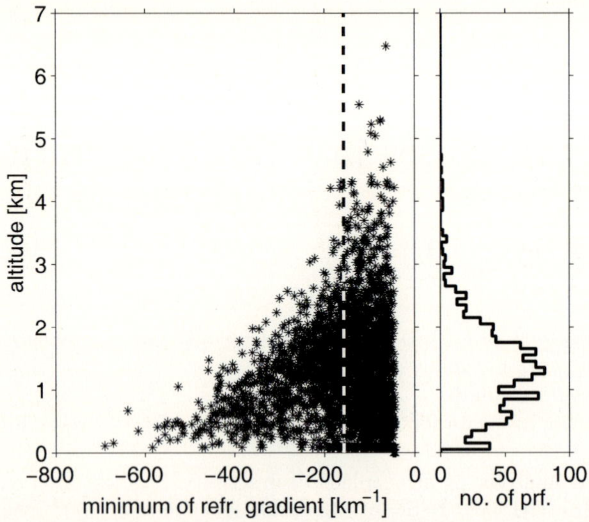

Fig. 1. Left: relationship between minimum vertical refractivity gradients and the corresponding altitudes derived from 3039 radiosonde observations. The dashed line marks the threshold value for critical refraction at -157 km^{-1}. Right: histogram distribution of the highest altitude at which critical refraction is observed.

aboard CHAMP has recorded 270 573 occultation events since activation of the operational occultation mode on May 14, 2001; 63.5 % of these events pass the quality control criteria and are processed to profiles of atmospheric refractivity (Wickert et al. 2001, 2004; Hajj et al. 2004). For more detailed accounts of the RO methodology see e.g., Melbourne et al. (1994); Kursinski et al. (1997); Hajj et al. (2002).

2 Data Analysis and Discussion

2.1 Radiosonde Observations

For more than 20 years Alfred Wegener Institute for Polar and Marine Research (AWI) performs aerological soundings of atmospheric temperature and humidity aboard research vessel "POLARSTERN" using Vaisala RS80 sondes (Vaisala 1989). As of August 28, 2004 the data archive operated by AWI contains 23 676 measurements starting on December 29, 1982. Critical refraction in the lower troposphere induced by vertical gradients of the water vapor distribution occurs predominately at low and mid latitudes. Thus, we focus on a subset of 3039 profiles collected between 60°S and 60°N.

In Fig. 1 (left panel) the correlation between minimum vertical refractivity gradient and its altitude is plotted; 39.6 % of the observed profiles exhibit critical refraction with vertical gradients below -157 km^{-1} (dashed line).

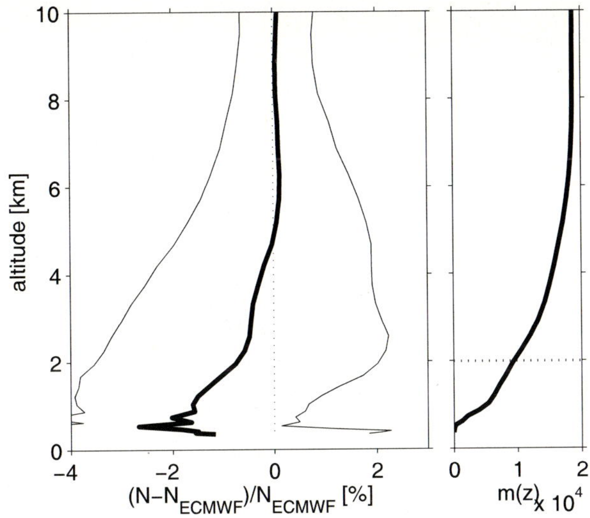

Fig. 2. Left: mean fractional refractivity deviation between CHAMP and ECMWF analyses restricted to the subset of 18 755 observations between 60°N–60°S and 45°W–15°E (thick line) and its one-sigma standard deviations (thin lines). Right: number of observed refractivity data as a function of altitude.

The right panel shows the histogram distribution of altitudes with critical refraction. Thus, critical refraction is a frequent phenomenon, at least in the marine environment, but it is restricted to altitudes below 2.5 km to 3 km.

2.2 CHAMP Occultation Data

As of October 31, 2004 (day of year 305) 270 573 radio occultation events are collected by the occultation experiment aboard CHAMP. A fraction of 63.5 % (171 942 observations) pass the quality control criteria of the operational processing software (Wickert et al. 2004) and are converted to vertical profiles of atmospheric refractivity. Version 5 of GFZ's occultation processing software (POCS) utilizes the Full Spectrum Inversion (FSI) method (Jensen et al. 2003) to obtain bending angle profiles in the troposphere. For details see Wickert et al. (2001, 2004).

The retrieved CHAMP refractivity profiles are intercompared with meteorological analysis results provided by the European Centre for Medium-Range Weather Forecasts (ECMWF). ECMWF pressure and temperature values are calculated by linear interpolation between grid points ($0.5° \times 0.5°$ resolution). Linear interpolation in time is performed between 6 h ECMWF analyses fields. The comparison between RO observation and ECMWF is performed on the 60 pressure levels provided by the ECMWF atmospheric model ranging from the ground surface up to 0.1 hPa (about 60 km altitude). Within the altitude

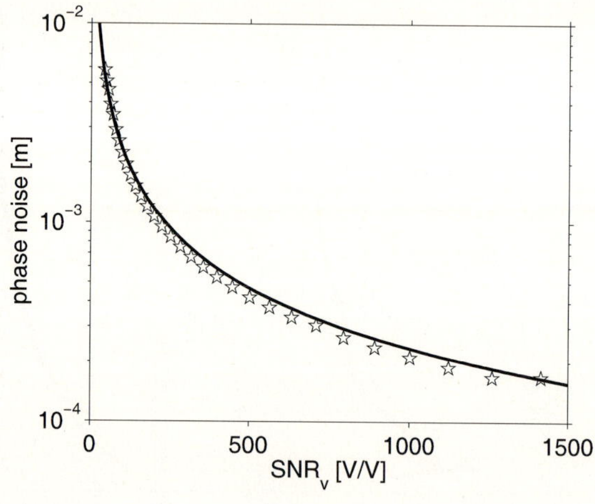

Fig. 3. Carrier phase noise $\sigma_{\rm PLL}$ extracted from 31 simulated profiles with voltage $\rm SNR_v$ between 44 V/V and 1414 V/V (stars). The full line marks the phase noise calculated from Eqn. 1.

range relevant for this study vertical spacing of the model grid points increases from about 200 m at 1 km altitude to about 700 m at 10 km altitude.

The mean fractional refractivity error and its standard deviation is shown in Fig. 2 (left panel) for a subset of 18 755 profiles observed within the region of 60°N–60°S and 45°W–15°E. In the right panel the number of retrieved data points is plotted; only 50 % of the profiles reach below an altitude of $z_{50\,\%} = 1.98$ km.

2.3 Simulation Results

The objective of the simulations is to separate receiver-induced bias contributions from those caused by critical refraction. For efficiency the simulation methodology described in Beyerle et al. (2003) has been modified following Ao et al. (2003b). From the sonde refractivity profile GPS carrier phase and amplitude data are modeled using the inverse FSI (Gorbunov and Lauritsen 2004). The simulated data are tracked with software receivers producing amplitude and phase data. Finally, bending angles are calculated with the FSI method (Jensen et al. 2003) and refractivity profiles are obtained. We consider three receiver models: first, the ideal receiver exactly reproduces the input data; second, a "fly-wheeling" receiver (Hajj et al. 2004) with two quadrant carrier phase extraction (Beyerle et al. 2003); and, third, an "open-loop" receiver using four quadrant extraction (Sokolovskiy 2001). In open-loop tracking the model is taken to be the ensemble average over 3039 Doppler profiles calculated from the dataset of sonde observations.

To validate the implementation of the signal tracking process we compare the carrier phase-locked loop thermal noise $\sigma_{\rm PLL}$ with the theoretically

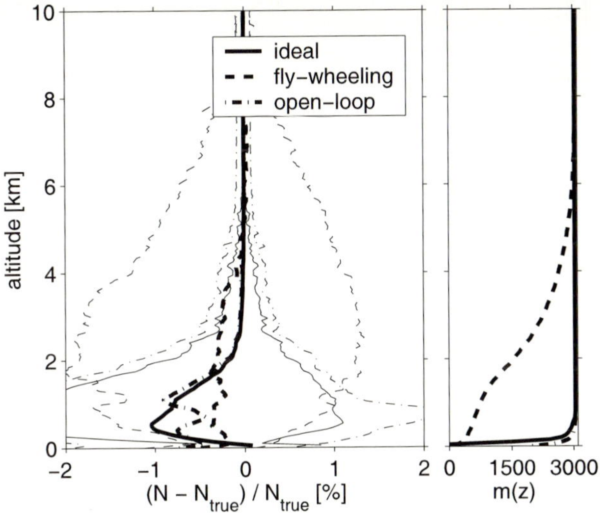

Fig. 4. Simulation results for three receiver models. The mean fractional refractivity error (left, thick lines), the corresponding standard deviations (thin lines) and the number of retrieved data points (right) are shown.

predicted value for a given voltage signal-to-nose ratio $\mathrm{SNR_v}$ (Ward 1996)

$$\sigma_{\mathrm{PLL}} = \frac{\lambda}{2\pi}\sqrt{\frac{2\,B_{\mathrm{w}}\,1\,\mathrm{s}}{(\mathrm{SNR_v})^2}\left(1+\frac{1\,\mathrm{s}}{(\mathrm{SNR_v})^2\,T}\right)} \tag{1}$$

with carrier wavelength λ, sampling time T and carrier loop bandwidth B_{w}. Setting $T = 20$ ms and $B_{\mathrm{w}} = 30$ Hz the calculated phase noise is in approximate agreement with the corresponding values extracted from simulation results for $\mathrm{SNR_v}$ values between 44 V/V and 1414 V/V (corresponding to carrier-to-noise density ratios between 30 dB Hz and 60 dB Hz) as shown in Fig. 3.

Figure 4 presents the retrieval error in terms of the fractional refractivity deviation between retrieved and true refractivity for the three receiver models. Critical refraction causes a negative bias of up to $-1\,\%$ within the planetary boundary layer (full line). About $-0.3\,\%$ bias above 2.5 km in the "fly-wheeling" profiles is attributed to receiver-induced signal tracking errors. "Open-loop" tracking (dashed-dotted lines) performs better than "fly-wheeling" in this altitude region and almost reproduces the ideal receiver profile (dashed-dotted vs. full lines).

3 Conclusions

A statistical analysis of 3039 radiosonde observations above the Atlantic Ocean suggests that critical refraction is a frequent phenomenon in the lower troposphere; it is, however, restricted to altitudes below 2.5 km to 3 km. CHAMP occultation profiles exhibit a negative bias at altitudes up to 5 km. On average, only 50 % of the profiles reach below 2 km altitude at mid and low latitudes. Simulation studies reproduce on a qualitative level the tracking behavior of the CHAMP occultation receiver and suggest a significant improvement by implementing advanced “open-loop” tracking schemes.

References

Ao CO, Hajj GA, Meehan TK, Leroy SS, Kursinski ER, de la Torre Juárez M, Iijima BA, Mannucci AJ (2003a) Backpropagation processing of GPS radio occultation data. In: Reigber C, Lühr H, Schwintzer P (eds) First CHAMP mission results for gravity, magnetic and atmospheric studies. Springer, Berlin, pp 415–422

Ao CO, Meehan TK, Hajj GA, Mannucci AJ, Beyerle G (2003b) Lower-troposphere refractivity bias in GPS occultation retrievals. J Geophys Res 108(D18):4577, doi:10.1029/2002JD003216

Beyerle G, Gorbunov ME, Ao CO (2003) Simulation studies of GPS radio occultation measurements. Radio Sci 38(5):1084, doi:10.1029/2002RS002800

Gorbunov ME, Lauritsen KB (2004) Analysis of wave fields by Fourier integral operators and their application for radio occultations. Radio Sci 39:RS4010, doi:10.1029/2003RS002971

Hajj GA, Kursinski ER, Romans LJ, Bertiger WI, Leroy SS (2002) A technical description of atmospheric sounding by GPS occultation. J Atmos Solar-Terr Phys 64(4):451–469

Hajj GA, Ao CO, Iijima BA, Kuang D, Kursinski ER, Mannucci AJ, Meehan TK, Romans LJ, de la Torre Juarez M, Yunck TP (2004) CHAMP and SAC-C atmospheric occultation results and intercomparisons. J Geophys Res 109:D06109, doi:10.1029/2003JD003909

Jensen AS, Lohmann M, Benzon HH, Nielsen A (2003) Full spectrum inversion of radio occultation signals. Radio Sci 38(3):1040, doi:10.1029/2002RS002763

Kursinski ER, Hajj GA, Schofield JT, Linfield RP, Hardy KR (1997) Observing Earth's atmosphere with radio occultation measurements using Global Positioning System. J Geophys Res 19(D19):23429–23465

Marquardt C, Schöllhammer K, Beyerle G, Schmidt T, Wickert J, Reigber C (2003) Validation and data quality of CHAMP radio occultation data. In: Reigber C, Lühr H, Schwintzer P (eds) First CHAMP mission results for gravity, magnetic and atmospheric studies, Springer-Verlag, Berlin, pp 384–396

Melbourne WG, Davis ES, Duncan CB, Hajj GA, Hardy KR, Kursinski ER, Meehan TK, Young LE, Yunck TP (1994) The application of spaceborne GPS to atmospheric limb sounding and global change monitoring. JPL Publication 94-18, Jet Propulsion Laboratory, Pasadena, CA, USA

Reigber C, Lühr H, Schwintzer P, Wickert J (eds) (2004) Earth Observation with CHAMP: Results from Three Years in Orbit. Springer-Verlag, Berlin Heidelberg New York

Rocken C, Anthes R, Exner M, Hunt D, Sokolovskiy S, Ware R, Gorbunov M, Schreiner W, Feng D, Herman B, Kuo Y-H, Zou X (1997) Analysis and validation of GPS/MET data in the neutral atmosphere. J Geophys Res 102(D25):29849–29866

Sokolovskiy SV (2001) Tracking tropospheric radio occultation signals from low Earth orbit. Radio Sci 36(3):483–498

Vaisala (1989) Upper air systems: RS 80 radiosondes. Tech rep, Vaisala GmbH, Hamburg

Ward P (1996) Satellite Signal Acquisition and Tracking. In: Kaplan ED (ed) Understanding GPS: Principles and applications. Artech House, Boston, London

Wickert J, Reigber C, Beyerle G, König R, Marquardt C, Schmidt T, Grunwaldt L, Galas R, Meehan TK, Melbourne WG, Hocke K (2001) Atmosphere sounding by GPS radio occultation: First results from CHAMP. Geophys Res Lett 28(17):3263–3266

Wickert J, Schmidt T, Beyerle G, König R, Reigber C, Jakowski N (2004) The radio occultation experiment aboard CHAMP: Operational data analysis and validation of vertical atmospheric profiles. J Meteorol Soc Jpn 82(1B):381–395

Stellar Occultation with GOMOS

Retrieval, Validation, and Error Analysis

GOMOS Ozone Profiles at High Latitudes: Comparison with Marambio and Sodankylä Sonde Measurements

J. Tamminen[1], J. A. Karhu[1], E. Kyrölä[1], S. Hassinen[1], E. Kyrö[2], A. Y. Karpechko[2], and E. Piacentini[3]

[1] Finnish Meteorological Institute, Earth Observation, Helsinki, Finland johanna.tamminen@fmi.fi
[2] Finnish Meteorological Institute, Arctic Research Center, Sodankylä, Finland
[3] Servicio Meteorológico Nacional, Buenos Aires, Argentina

Abstract. The ozone profiles measured by GOMOS are compared with the ozone soundings at two stations: Marambio (56.7°W, 64.3°S) in Antarctica and Sodankylä (23.6°E, 67.4°N) in northern Europe. The agreement between the GOMOS night measurements and the ozone soundings from Sodankylä and Marambio are found to be good. Comparisons of measurements during 2003 show that the difference between the averages at 15 km and 30 km altitude range are within ±5 % for Marambio and somewhat worse for Sodankylä. The individual comparisons show that the good vertical resolution of 2 km to 3 km together with the dense altitude grid (0.5 km to 1.7 km) of the GOMOS profiles make it possible to detect also small scale structures in the ozone profiles.

1 Introduction

Since the launch of the Envisat satellite on the 1st of March 2002 the GOMOS (Global Ozone Monitoring by Occultation of Stars) instrument has measured ozone profiles with a good vertical resolution from pole to pole (Bertaux et al. 2004; Kyrölä et al. 2004). The altitude range covered by GOMOS is typically from 15 km to 100 km, but extends occasionally down to 5 km. GOMOS is capable of making measurements both in day and night conditions, but naturally the scattered solar light causes decreased signal-to-noise ratio and the accuracy of the day measurements is worse than the night measurements. The vertical sampling resolution and the grid in which the profiles are given is typically around 1.4 km at 40 km and decreases due to refraction at lower altitudes so that around 15 km it is typically 0.6 km (e.g., Sofieva et al. 2004). The resolution of the retrieved ozone profiles is 2 km to 3 km due to the inversion algorithm which is based on Tikhonov regularization with 2 km to 3 km target resolution requirement (Tamminen et al. 2004; Sofieva et al.

2004). In this study we have considered only night measurements as they are the main GOMOS products. During year 2003 more than 100 000 profiles were measured from which about half were done in night conditions. The latitude range that can be covered by the GOMOS measurements is 80°S to 80°N, but it varies slightly, caused by the different star positions over season.

2 Co-Locating Measurements

In this study we have compared the GOMOS night measurements and ozone sonde measurements in 2003. Two high-latitude ground stations are considered: Sodankylä/Finland (23.6°E, 67.4°N) in the northern hemisphere and Marambio/Antarctica (56.7°W, 64.3°S) in the southern hemisphere. Finnish Meteorological Institute operates both stations and regular ozone soundings are performed on a weekly basis.

GOMOS measurements and ozone soundings with geographical distance less than 5 deg in latitude and 10 deg in longitude are considered. The time variation criteria that is used for co-location is ±12 hours which is quite large and may also allow natural variability in the profiles. In addition, the GOMOS data has been selected so that only measurements with duration less than 100 s have been included in this study. This step was introduced to remove measurements with large tangent point movements which differ significantly from a quasi vertical sounding. Typical tangent point movement is about 200 km to 300 km for an occultation with tangent altitude from 130 km down to 15 km and if only the measurements below 40 km are considered typical tangent point movement is about 60 km. Therefore, there is a natural geometrical difference between the limb geometry measurements of GOMOS and the actual vertical profiles measured by ozone sondes.

The number of co-located measurements which fulfill these criteria is 59 at Marambio and 23 at Sodankylä. See Fig. 1 for the geographical coverage of the measurements compared.

The quality of the GOMOS data varies depending on the type of star used as the light source. Naturally the brightest stars with the best signal-to-noise ratio lead to most accurate ozone profiles. The stellar temperature has also an impact on the accuracy: hot stars with intensity peak in UV wavelengths show best accuracy at high altitudes whereas cool stars with intensity peak in visible wavelengths are better at low altitudes; see Fig. 2 for stars used for comparison. The monthly coverage is also shown in Fig. 2. Note, that poles cannot be covered using only night measurements during summer time.

3 Comparing GOMOS with Ozone Soundings

The individual profiles measured by GOMOS agree quite well with the soundings as can be seen in Figs. 3 (Marambio) and 4 (Sodankylä). The examples

Fig. 1. Geographical locations of the GOMOS night occultations (black stars) compared here, on left: near the Marambio station and on right: near Sodankylä. The stations are denoted with gray dots.

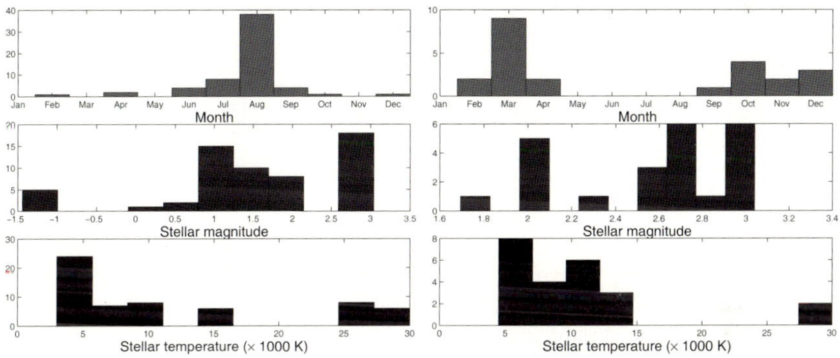

Fig. 2. Histograms of the GOMOS occultations near Marambio (on left) and near Sodankylä (on right) used for comparison. Top panel: month, middle: stellar magnitude, bottom: stellar temperature.

are chosen so that they represent both good and typical cases. In these figures the profiles are plotted as they are, i.e., no additional averaging has been performed. The error estimates (1σ) of the GOMOS profiles are shown in the figures as shaded area. The error estimate of the ozone soundings is about 5 % (not shown in the figure). Differences in both time and location are indicated in the individual figures (see figure text). The figures show that the small scale structures (< 2 km) are not captured by GOMOS measurements but many profiles repeat similar few kilometer structures.

Both Sodankylä and Marambio are often located close to the edge region of the polar vortex. The variability of ozone is especially large around these stations and this makes the comparison with sonde and satellite measurements challenging. In the dataset used for comparison also profiles with clear

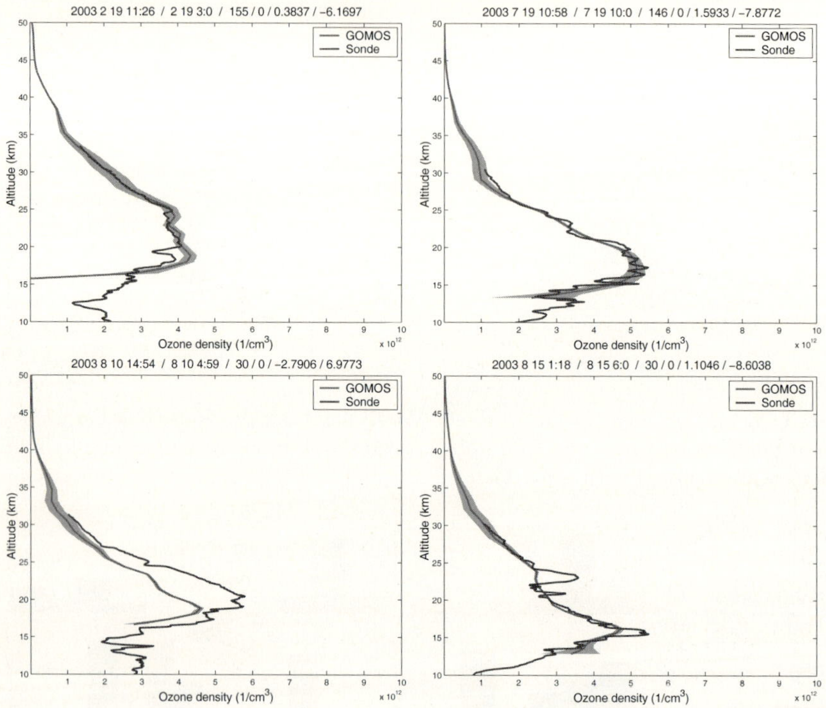

Fig. 3. Comparisons of individual profiles at Marambio. Gray line – GOMOS, black – sonde. The gray area represents GOMOS error estimates (1σ). The heading of individual figures include following information: time (yyyy mm dd hh:mm) of the sonde measurement / date and time (mm dd hh:mm) of the GOMOS measurement / star number / flag indicating night measurement / difference in latitude / difference in longitude.

disagreement have been found (see Fig. 5) and, indeed, these most obvious cases have been identified to represent conditions where either the station is inside the polar vortex and the GOMOS measurements outside the vortex or vice versa. In Fig. 5 the total ozone measured by NASAs TOMS instrument (NASA-GSFC 2005) at Marambio is around 150 DU and at GOMOS measurement around 225 DU. This huge difference is clearly seen in the profiles as well.

The averages over the profiles compared are shown in Fig. 6. The variability in the accuracy of the GOMOS measurements is not taken into account here and simple mean is used for comparison. As a more robust estimator we have also used the median. However, both of these indicators give similar results. The number of ozone values included in the comparison varies at different altitudes (see the right panel in Fig. 6) due to the data availability which is limited by the capabilities of GOMOS at low altitudes and soundings

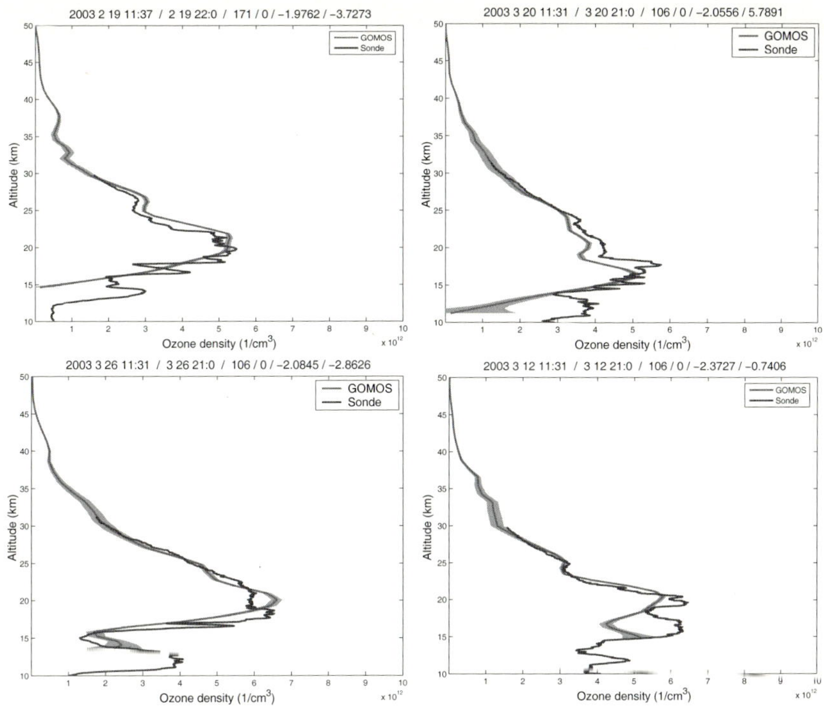

Fig. 4. Comparisons of individual profiles at Sodankylä. Gray line – GOMOS, black – sonde. The gray area represents GOMOS error estimates (1σ).

at high altitudes. The lowest altitude measured by GOMOS varies between 6 km and 27 km and is on average around 14 km. The maximum altitude measured by the balloon sondes varies between 20 km and 35 km. Therefore, the maximum number of measurements compared at Marambio is 49 which is less than the total number of profiles included in the study (59). For the comparison GOMOS profiles have been interpolated to a fixed 2 km grid which corresponds to the actual 2 km resolution of the GOMOS ozone profiles below 35 km. The sounding measurements have been averaged to this grid using a simple boxcar function. Re-processed GOMOS data is used (prosessing was done under an ESA contract by ACRI using the GOMOS prototype, GOPR, software version 6.0a for level 1b and level 2).

Within the altitude range 16 km to 32 km the bias (difference between GOMOS and sondes) at Marambio is slightly negative, around $-5\,\%$. At low altitudes, around 15 km the bias is positive, around $5\,\%$. The agreement with GOMOS and Sodankylä sonde measurements is slightly worse with negative bias up to $-10\,\%$. Above 30 km the differences seem to be larger but this is probably due to small number of sonde measurements at high altitudes. These results agree well with Meijer et al. (2004) where a negative bias

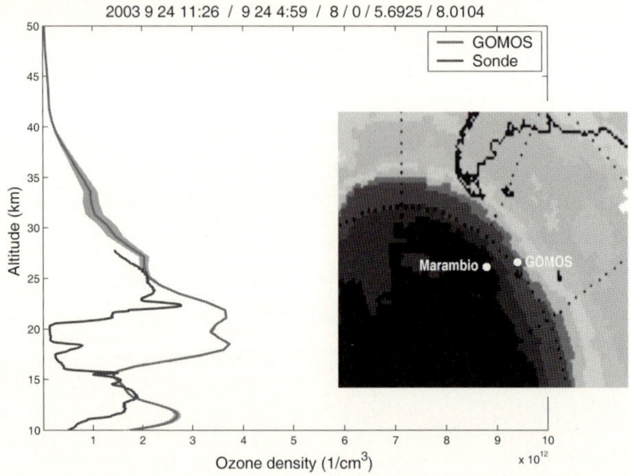

Fig. 5. Comparisons of Marambio sonde and GOMOS measurements. On right: total ozone maps by TOMS/NASA on the same day with the geographical positions (roughly) of the GOMOS measurement and the Marambio station.

around −2.5 % is found for GOMOS ozone measurements within the altitude range 14 km to 45 km by analyzing 1376 profiles around the globe in night conditions.

The reason for the better agreement at Marambio is partly due to GOMOS measurements: the stars compared at Marambio are brighter and the GOMOS measurements are closer to the station than at Sodankylä. The latter is also seen in the equivalent latitude frame. Relative differences between potential vorticity values at the station and at the co-located occultation are of 14 % at Marambio and of 17 % at Sodankylä at the 475 K isentropic level. There might also be natural reasons for the larger discrepancy at Sodankylä related to the structure of polar vortex. The equivalent latitude analysis shows that Marambio and Sodankylä are approximately equally often located within the vortex edge region. However, the Arctic vortex is more disturbed than the Antarctic one, with more frequent and stronger laminae in ozone profile generated by planetary wave breaking in the vortex edge region. Since their spatial scale is comparable to the distance between occultation and sonde station, these laminae could appear in one profile but not in the other (see, e.g., Fig. 4), providing so a larger source of error in the northern hemisphere in the dynamically active region.

4 Summary

GOMOS ozone profiles measured at night conditions in 2003 are compared with the ozone soundings of two high-latitude stations. The validation shows that the overall agreement with GOMOS ozone profiles and ozone soundings

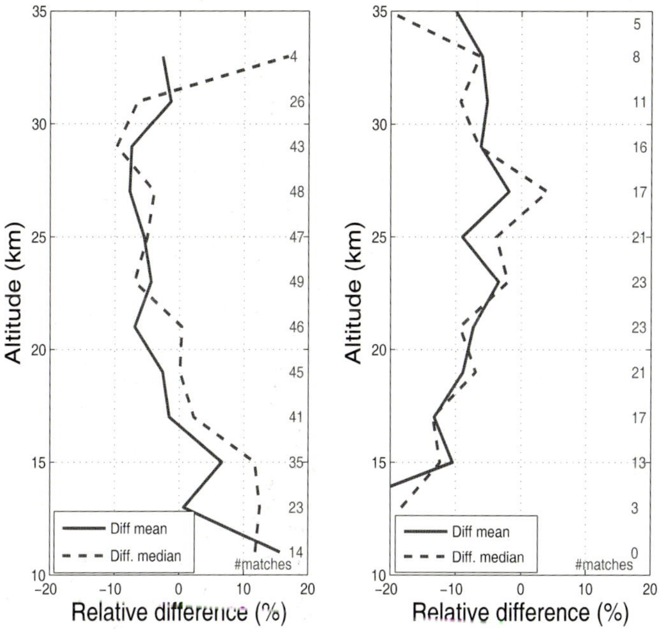

Fig. 6. Difference between GOMOS and soundings, on left Marambio and on right Sodankylä. The solid line is the relative difference between mean values and the dashed line the relative difference between median values. The numbers on right denote the number of pairs compared at each altitude.

is good: between the altitude range 15 km to 30 km the bias is around ±5 % at Marambio and up to −10 % at Sodankylä. The good vertical resolution of the GOMOS instrument shows that small-scale structures of a few kilometers can be seen in the data.

References

Bertaux JL, Hauchecorne A, Dalaudier F, Cot C, Kyrölä E, Fussen D, Tamminen J, Leppelmeier GW, Sofieva V, Hassinen S, d'Andon OF, Barrot G, Mangin A, Théodore B, Guirlet M, Korablev O, Snoeij P, Koopman R, Fraisse R (2004) First results on GOMOS/Envisat. Adv Space Res 33(7):1029–1035

Kyrölä E, Tamminen J, Leppelmeier GW, Sofieva V, Hassinen S, Bertaux JL, Hauchecorne A, Dalaudier F, Cot C, Korablev O, d'Andon OF, Barrot G, Mangin A, Théodore B, Guirlet M, Etanchaud F, Snoeij P, Koopman R, Saavedra L, Fraisse R, Fussen D, Vanhellemont F (2004) GOMOS on Envisat: An overview. Adv Space Res 33:1020–1028

Meijer YJ, Swart DPJ, Allaart M, Andersen SB, Bodeker G, Boyd I, Braathena G, Calisesia Y, Claude H, Dorokhov V, von der Gathen P, Gil M, Godin-Beekmann

S, Goutail F, Hansen G, Karpetchko A, Keckhut P, Kelder HM, Koelemeijer R, Kois B, Koopman RM, Lambert JC, Leblanc T, McDermid IS, Pal S, Kopp G, Schets H, Stubi R, Suortti T, Visconti G, Yela M (2004) Pole-to-pole validation of Envisat GOMOS ozone profiles using data from ground-based and balloon sonde measurements. J Geophys Res 109(D23305), doi:10.1029/2004JD004834

NASA-GSFC (2005) TOMS, Total Ozone Mapping Spectrometer. URL http://toms.gsfc.nasa.gov

Sofieva VF, Tamminen J, Haario H, Kyrölä E, Lehtinen M (2004) Ozone profile smoothness as a priori information in the inversion of limb measurements. Annales Geophysicae 22(10):3411–3420

Tamminen J, Kyrölä E, Sofieva VF (2004) Does a prior information improve the retrievals of stellar occultation measurements? In: Kirchengast G, Foelsche U, Steiner A (eds) Occultations for Probing Atmosphere and Climate. Springer Verlag, Berlin, pp 87–98

Ozone and Temperature Retrieval Results from GOMOS Validated with CHAMP and ECMWF

C. Retscher, G. Kirchengast, and A. Gobiet

Wegener Center for Climate and Global Change (WegCenter) and Institute for Geophysics, Astrophysics, and Meteorology (IGAM), University of Graz, Austria
christian.retscher@uni-graz.at

Abstract. Data from the Global Ozone Monitoring by Occultation of Stars (GOMOS) instrument on-board the European environmental satellite Envisat have been used for retrieval of ozone and temperature profiles by new algorithms, followed by profile validation. We discuss ozone profiles, determined from GOMOS transmission data, validated with operational GOMOS ozone profiles and ECMWF (European Centre for Medium-Range Weather Forecasts) analysis data. For ozone we developed an optimal estimation retrieval scheme, using sensibly selected channels from the Spectrometer A transmission spectra within 260 nm to 340 nm and 602 nm to 634 nm. Furthermore, profiles of a new GOMOS temperature profile retrieval are compared to CHAMP (CHAllenging Minisatellite Payload) and ECMWF analysis data. GOMOS temperatures are gained by exploiting pointing data of the Steering Front Assembly (SFA) and the Star Acquisition and Tracking Unit (SATU), which provide information on the refraction of the star light in the atmosphere and thus allow to derive refractive bending angle profiles. The bending angle profiles are then converted via refractivity and pressure profiles to temperature profiles. Bending angles were assumed to have errors of 3 μrad. Statistical optimization of observed bending angles with model bending angles was used to provide adequate data quality for the Abel transform from the stratopause region upwards, which led to a significant gain in temperature retrieval accuracy up to 40 km height due to suppressed downward propagation of errors induced by Abel transform and hydrostatic integral. Based on data from year 2002, a set of GOMOS ozone and bending angle derived temperature profiles is validated and discussed.

1 Introduction

The Global Ozone Monitoring by Occultation of Stars (GOMOS) (Bertaux et al. 1991; Kyrölä et al. 1993; Kyrölä et al. 2004) sensor on-board Envisat is a self-calibrating instrument intended to provide data on the trace gas ozone, and chemicals favoring its depletion such as NO_2, NO_3, BrO, and OClO. Atmospheric reference profiles are measured under dark and bright limb conditions obtaining very good global coverage with about 300 high-quality profiles per day and a height resolution of about 1.5 km. GOMOS records the transmission of radiation passing the atmosphere along a path from the star to the instrument. Its Spectrometer A measures ozone and ozone depleting chemicals within a wavelength range from 250 nm to 675 nm and provides a spectral resolution of 1.2 nm.

GOMOS star tracking and acquisition (SFA/SATU) data is exploited for obtaining bending angles of rays passing the atmosphere. These bending angles, highlighting the effect of atmospheric refraction, allow to derive atmospheric refractivity profiles and subsequently temperature profiles.

We present and discuss ozone and retrieved temperature profiles based on an analysis of year 2002 data. The algorithm is based on temperature dependent cross-section data of the corresponding GOMOS database, standard climatological atmospheric profiles for the trace gases, and CIRA model profiles for background temperature and pressure.

2 GOMOS Retrieval Schemes

2.1 Ozone Retrieval

The spectrometer A on the GOMOS instrument provides transmission data from 250 nm to 375 nm and 405 nm to 675 nm at a spectral sampling of ~ 0.3 nm. In the method presented here only a few available channels were selected by empirical arguments and tests on the algorithm. The selected channels were chosen to give a good signal response in the Hartley and Huggins band and as well in the Chappuis band, which is especially important for lower stratosphere ozone retrieval. The set of wavelengths selected for presentation in this work contains 14 channels with $\lambda \in [260, 280, 288, 295, 302, 309, 317, 328, 334, 337, 340, 343, 600, 605]$, where numbers are given in [nm]. A more rigorous analysis including information content theory is postponed for future enhancements of the retrieval scheme (cf., Sofieva and Kyrölä 2004; Sofieva and Kyrölä 2003).

The atmospheric transmission is defined by Beer-Bouguer-Lambert's law at each wavelength λ of interest, by incorporating ozone, NO_2, NO_3, and bulk air background information. This forward model is a best estimate of the current atmospheric state. Due to a currently low stratospheric aerosol concentration a term representing the corresponding signal extinction – the Mie scattering – was neglected, but we plan to integrate it in next steps of algorithm enhancements.

Having a forward model established, we have to find an inverse connection between measurements (transmission data) and targeted state (ozone profile) of the atmosphere. In the case presented here, ozone and NO_2 are retrieved simultaneously, forming a joint retrieval. The enhancement of the inversion scheme by incorporating other trace gases (e.g., NO_3, BrO, and OClO) targeted in the operational GOMOS retrieval is possible, but not investigated here. The measured transmission contains information of the density of the absorbing species and of the atmospheric refraction. To proceed from the transmission function one can find a solution for, e.g., the ozone density, by first separating the absorbing species by a spectral inversion and then perform a vertical inversion via an Abel transform. This is known as the two-step retrieval and is realized in the GOMOS operational processing (Paulsen 2000). Performing such a retrieval includes several restrictions one has to be aware of. Due to a line-by-line vertical retrieval, a correlation of adjacent vertical lines is not given any more after a following spectral inversion. This is not observed in nature, where depending on altitude and selected wavelength the bending of rays and thus the absorption of light by target species has a vertical correlation. This feature can be handled by a one-step inversion first performing a vertical retrieval (Vanhellemont et al. 2004). In the case of GOMOS this would include an inversion of large matrices (containing all wavelengths), which is then time consuming but can be processed in a proper way.

Here we use a different approach for the inversion of atmospheric species densities. Discrete inverse theory as discussed in detail by Rodgers (1976, 1990, 2000) provides a framework, where the forward model can be seen as an algebraic mapping of the state space into the measurement space. The physics of the measurement is approximated by the forward model and we introduce an operator $\mathbf{K}$, which here will be the Jacobian matrix (weighting function matrix) with the dimension $m \times n$ for m measurements and n elements of the state vector. The forward model reads $\mathbf{y} = \mathbf{K}\mathbf{x} + \boldsymbol{\varepsilon}$. Because of this generally non-linear equation, it is obvious that a straightforward solution for $\mathbf{x}$ by direct inversion is not feasible. The direct inverse mapping would be $\mathbf{x}_\mathrm{r} = \mathbf{K}^{-\mathrm{g}}\mathbf{y}$, where $\mathbf{K}^{-\mathrm{g}}$ denotes a general inverse matrix and $\mathbf{x}_\mathrm{r}$ is the retrieved state. As the problem of interest here is ill-posed at high altitudes due to low signal-to-noise ratio (it may also be over-determined if we use more measurements than unknown states; $m > n$), we cannot directly employ the latter but rather constrain the solution by incorporating sensible *a priori* information. The Bayesian approach is the method of choice to solve such inverse problems perturbed by noise, where we have rough but reliable prior knowledge of the behavior of a state of interest. If the problem is only moderately non-linear we can use a Gauss-Newton method for an iterative approach to find an optimal solution. Assuming Gaussian probability distributions and a linearized forward model, the primary task of a retrieval method is to find a state by satisfying optimal criteria from an ensemble of states, which best agree with the *a priori* state and with the measurement within experimental errors.

Here we make use of a fast converging iterative optimal estimation algorithm,

$$\mathbf{x}_{i+1} = \mathbf{x}_\mathrm{ap} + \left(\mathbf{K}_i^\mathrm{T}\mathbf{S}_\varepsilon^{-1}\mathbf{K}_i + \mathbf{S}_\mathrm{ap}^{-1}\right)^{-1}\mathbf{K}_i^\mathrm{T}\mathbf{S}_\varepsilon^{-1}\Big[(\mathbf{y} - \mathbf{y}_i) + \mathbf{K}_i\left(\mathbf{x}_i - \mathbf{x}_\mathrm{ap}\right)\Big], \quad (1)$$

where $\mathbf{x}_{i+1}$ is the retrieved, $\mathbf{x}_{\mathrm{ap}}$ the *a priori* profile, $\mathbf{y}$ the measurement vector, and $\mathbf{y}_i = \mathbf{K}_i\mathbf{x}_i$ the forward-modeled measurement vector. Key ingredients of Eq. 1 are the *a priori* covariance matrix $\mathbf{S}_{\mathrm{ap}}$ and the measurement error covariance matrix $\mathbf{S}_\varepsilon$. Diagonal elements of $\mathbf{S}_{\mathrm{ap}}$ for ozone are assumed to have 30 % standard deviation, while off-diagonal elements follow an exponential drop-off with a correlation length of 6 km. NO_2 variances have values of 40 % and covariances are set equally to ozone. The Jacobian (weighting) matrix $\mathbf{K}_i$ represents the mapping involved. Index i is the iteration index, which is started by using $\mathbf{x}_0 = \mathbf{x}_{\mathrm{ap}}$.

The number of needed iteration steps for Eq. 1 is found by a calculation of the cost function

$$\chi_i^2 = (\mathbf{y} - \mathbf{y}_i)^{\mathrm{T}} \mathbf{S}_\varepsilon^{-1} (\mathbf{y} - \mathbf{y}_i) + (\mathbf{x}_i - \mathbf{x}_{\mathrm{ap}})^{\mathrm{T}} \mathbf{S}_{\mathrm{ap}}^{-1} (\mathbf{x}_i - \mathbf{x}_{\mathrm{ap}}) \tag{2}$$

at each iteration step i. The cost function χ_{i+1}^2 at iteration step $i+1$ has always to be smaller than χ_i^2. The minimum criterion is met if χ_i^2 is smaller than the number of selected channels. If this criterion is not reached the retrieval stops per definition after 10 iterations.

2.2 Temperature Retrieval

The GOMOS temperature retrieval follows a different method compared to the ozone retrieval procedure. Instead of introducing a third matrix, besides ozone and NO_2, for a joint retrieval, temperature profiles are inverted by an Abel transform. Light coming from, e.g., stellar sources undergo a refractive bending before the signal gets measured by the instrument. The degree of refraction is a measure for the atmospheric density, pressure, and temperature, respectively.

GOMOS derived bending angles have their origin in an exploitation of the elevation movement of the star tracking unit (SFA/SATU) (cf., Sofieva et al. 2003). The star image is kept inside the CCD with a sampling of 100 Hz and errors of ± 10 μrad from the central position. The measured elevation angles are compared to an unrefracted ray, measured outside the atmosphere and thus giving a bending angle profile.

For the derivation of refractivity profiles from bending angles and impact parameters using the Abel transform, we assume local spherical symmetric conditions. A correction for the ellipsoidal shape of the Earth is applied.

In theory one can now start with the calculation of the refractive index, which is the next step in the temperature retrieval procedure. If the bending angle $\alpha(a)$ is given, we find for the Abel transform

$$n(r_0) = \exp\left[\frac{1}{\pi}\int_{\alpha=\alpha(a_0)}^{\alpha=0} \ln\left(\frac{a(\alpha)}{a_0} + \sqrt{\left(\frac{a(\alpha)}{a_0}\right)^2 - 1}\right) \mathrm{d}\alpha\right], \tag{3}$$

written here in a favorable form for numerical use by avoiding poles (cf., Steiner 1998; Foelsche 1999). The refractive index is denoted by n and $\alpha(a_0)$

is the bending angle dependent on the impact parameter a_0, associated with radius r_0, the bottom height of the Abelian integration, which extends over the height domain above r_0. From n follow density, pressure, and temperature profiles by applying the hydrostatic equation and the ideal gas law.

Bending angles at high altitudes, due to small refraction, are typically very small numbers with large error bars. Measurement errors and errors of discretization further decrease the quality of bending angle profiles. The signal-to-noise ratio in the measured bending angle profile becomes less than unity at altitudes above ~40 km. Due to the application of the Abelian integral and the hydrostatic equation, errors in the bending angle profile at high altitudes propagate down through all height steps and thus also low altitude temperature profiles suffer from an inappropriate initialization value as explained, e.g., by Syndergaard (1999). A sensible use of good-quality bending angle data for high-altitude initialization is therefore needed and realized by the introduction of the statistical optimization technique.

The statistical optimization (Sokolovsky and Hunt 1996) optimally combines measured and background (*a priori*) bending angle profiles leading to the most probable bending angle profile. An optimal solution can be found via

$$\alpha_{\mathrm{opt}} = \alpha_{\mathrm{b}} + \mathbf{B}\left(\mathbf{B} + \mathbf{O}\right)^{-1}\left(\alpha_{\mathrm{o}} - \alpha_{\mathrm{b}}\right), \tag{4}$$

where α_{b} is the background and α_{o} the observed bending angle profile, respectively. The matrices $\mathbf{B}$ and $\mathbf{O}$ express the background and the observation error covariances, respectively, which are found similar to $\mathbf{S}_{\mathrm{ap}}$ and $\mathbf{S}_{\varepsilon}$. The correlation length L was set to 6 km for $\mathbf{B}$, while we found $L = 1$ km appropriate for $\mathbf{O}$. As a background profile we chose a CIRA-86 climatology (e.g., Rees 1988). Background errors were assumed to be 20 % in line with radio occultation literature (e.g., Healy 2001; Gobiet and Kirchengast 2002). The observation errors were estimated from the root-mean-square deviation of the observed data from the background at high altitudes (70 km to 80 km), where noise dominates the measured signal. More details on the statistical optimization scheme used here are found in, e.g., Gobiet and Kirchengast (2002, 2004); Retscher (2004); Gobiet et al. (2005).

In radio occultation, due to the selected wavelengths in the order of centimeter to meter, an ionospheric correction (e.g., Syndergaard 2000) is necessary and thus is an important issue. Rays in the optical, UV, and NIR wavelength range are not affected by the ionosphere and thus such a correction is not an issue.

3 Retrieval Results

For the retrieval of ozone profiles we used level 1b and level 2 data from the official GOMOS data product in the validation periods in September 2002,

20 – 27, October 2002, 11 – 13, and December 2002, 02. Both data products were processed and provided by ACRI-ST (Sophia Antipolis, France). The level 2 data were used as validation reference. Here we reduced the set of available data by eliminating those with large errors in level 1b transmissions and reference level 2 ozone data (cf., Retscher 2004). Measured transmissions were smoothed by running means and then corrected for scintillation and refractive dilution effects (cf., Paulsen 2000), which has a significant positive impact on the lower stratospheric ozone retrieval quality. ECMWF T511L60 analysis data (60 height levels, spherical harmonics truncation 511) were taken for the same period, where GOMOS data were available.

The quality of transmission data is further strongly dependent on the star magnitude, the star temperature, and the obliquity of the occultation. Light coming from bright stars is able to penetrate the Earth's atmosphere down to lower heights than light stemming from weak stars. In general the penetration depth of light into the atmosphere is known to be wavelength dependent. In our analysis we found an ozone retrieval dependence mostly on the star temperature. Stars with temperatures between 8 000 K and 11 000 K clearly favor the retrieval as there the corresponding maximum of the Planck function lies within the UV wavelength range. This is reflected by a large number (passing the quality check) of contributing profiles in the error statistics of stars like Sirius (-1.44^{m}, 11 000 K), Fomalhaut (1.16^{m}, 9 700 K) and δ Velorum (1.95^{m}, 10 600 K).

Here we present the statistics of globally distributed occultation events separated by date and latitude regimes. The latitude regimes are divided into low ($0° - 30°$), mid ($30° - 60°$), and high ($60° - 90°$) latitudes. Such a selection is especially useful for looking at ozone trends in the high latitude profiles, where the ozone density is significantly diminished. Large scale stratospheric ozone depletion is mostly reported from high latitudes, where meteorological conditions favor depletion processes.

Figure 1 presents ozone validation results. The graphs include data of occultations of stars between -1.44^{m} and 3.03^{m}. Stars with magnitudes greater than 3.03^{m} show significantly larger errors and are not considered. We applied a quality check, by not considering profiles with errors larger than 50 % (compared to official GOMOS level 2 data) between altitudes of 30 km and 60 km.

We found good retrieval performance in all latitude regions, where biases against level 2 data are found $\sim$5 % from 25 km to 65 km. Below 25 km – the lower altitude boundary of the measurements – we encounter a tendency to large biases. Above 65 km most profiles tend to negative biases with values of 20 % at 70 km. This reflects the fact that data measured at such altitudes have small transmission values and large fluctuations, which affects the retrieval quality. In general, differences between the operational retrieval and the retrieval discussed here may have reasons in the selected wavelength region, as well as in the retrieval method itself, by retrieving only two species at once (NO_2 is retrieved simultaneously to support the ozone retrieval quality, but it is not considered as a retrieval product in the current version of the application) instead of performing a full inversion of measured data. Future enhancements of the code will deal more properly with these effects.

Errors from the ECMWF validation show a clear bias of about −30 %. This bias is known to be in the ECMWF product (cf., e.g., Dethof 2004; Bracher et al. 2004). Ozone is fully integrated into the ECMWF forecast model and analysis system as an additional three-dimensional model and analysis variable similar to humidity. The forecast model includes a prognostic equation for the ozone mass mixing ratio with a parameterization of sources and sinks of ozone.

A profile validation with ECMWF can be useful, if the observed biases behave almost equally throughout our profile validation. Validation at high latitudes has a bias, with values mostly between −20 % and −30 %, while the validation with mid latitudes has large positive deviations. The low latitude profile validation performs similar to the high latitude, but with larger positive biases below 25 km.

In the temperature retrieval simulation we have shown the influence of bending angle optimization on corresponding temperatures. We discuss the results of our real GOMOS SFA/SATU data temperature retrieval (cf., Retscher et al. 2004). Statistically optimized bending angles from GOMOS SFA/SATU data lead to temperature profiles and are then compared to ECMWF analyses and WegCenter retrieval results based on CHAMP profiles (cf., Gobiet et al. 2004). For this analysis, the GeoForschungsZentrum (GFZ) in Potsdam provided CHAMP data.

As already found for the ozone retrieval, the results of a GOMOS temperature retrieval strongly depends on the selected star and its magnitude as well as on its corresponding temperature. In general, one finds for bending angles and temperatures that a low star magnitude, favorably below 0, allows for retrievals into the troposphere down to heights of around 5 km, which is the theoretical minimum of GOMOS occultation data.

In our selected set of data, bending angles were available for the same period as given for ozone. The GOMOS occultation events cover different latitude regions, but especially in the high altitude region there is clear need for more occultation data. The GOMOS level 1b data product is separated into three latitude regions as done for ozone. A global set of retrieved data is given, where dependences on different latitude regions of the quality of an overall profile can be seen. GOMOS SFA/SATU (level 1b) data in general suffers from different influences. SFA/SATU data often show much larger errors than we estimated, by considering 10 μrad at 100 Hz (in our application 3 μrad at 10 Hz). The data set, which was provided for this study, only includes a small set of occulter stars, which makes it especially difficult to ensure good retrieval quality at high latitudes.

For this work we first validated GOMOS temperature profiles with CHAMP data. Phase delays from the CHAMP level 2 (version 2) (cf., Wickert et al. 2004) data served as input into our retrieval, which had ECMWF data as background information (cf., Gobiet et al. 2004). From 3 724 CHAMP profiles

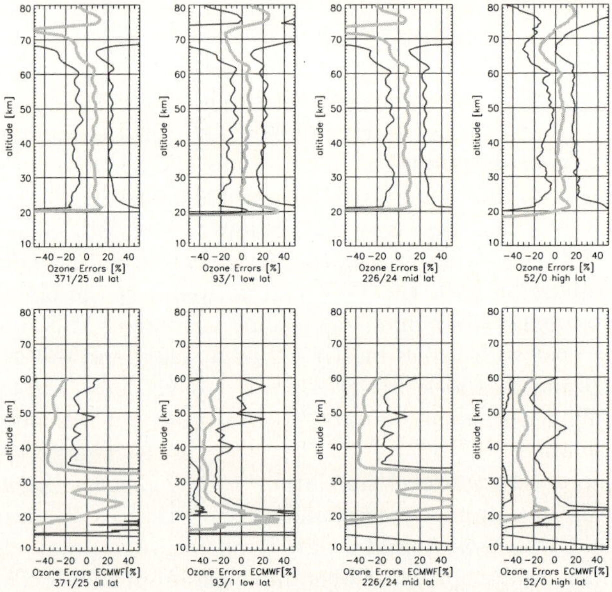

Fig. 1. Ozone error statistics for multi day ensembles in the period 2002-09-20 – 2002-12-02 (whole set 2002). In the upper panels we present results from ensembles validated with GOMOS level 2 (operational product) files and in the lower panels we validate with ECMWF analysis data. From the left, the first panel shows the overall statistics, the second, third, and fourth panel show the low, mid, and high latitude selected statistics. The heavy gray line denotes the bias profile $\hat{\mathbf{b}}$, while the enveloping fine black lines are the bias $\pm$ standard deviation profile $\hat{\mathbf{b}} \pm \hat{\mathbf{s}}$. To each error profile contributes a certain number of available retrieved profiles passing the quality check. This is given below each plot, where on the left one finds the number of accepted profiles and the right number denotes the outliers (e.g., upper left: 371 accepted, 25 rejected profiles).

we found 56 profiles, where our coincidence intervals (300 km/3 hrs) with GOMOS data were reached.

In Fig. 2 GOMOS level 1b data is validated with CHAMP data at altitudes between 20 km and 35 km. The high latitude bias profile $\hat{\mathbf{b}}$ is shifted to positive values for altitudes below 25 km, while the standard deviation is <5 K. This is due to the fact that only a small set of GOMOS data with very few selected stars for occultation was available. For altitudes higher than 25 km a negative bias of up to 2 K was found. At mid and high latitudes $\hat{\mathbf{b}}$ has a negative deviation below 25 km and performs similar to the high latitude sample. The global ensemble clearly averages over negative and positive drifts below 25 km and a positive tendency in $\hat{\mathbf{b}}$ can be seen.

For the comparison of GOMOS SFA/SATU data to ECMWF data we chose co-located vertically distributed temperatures and refractivity profiles

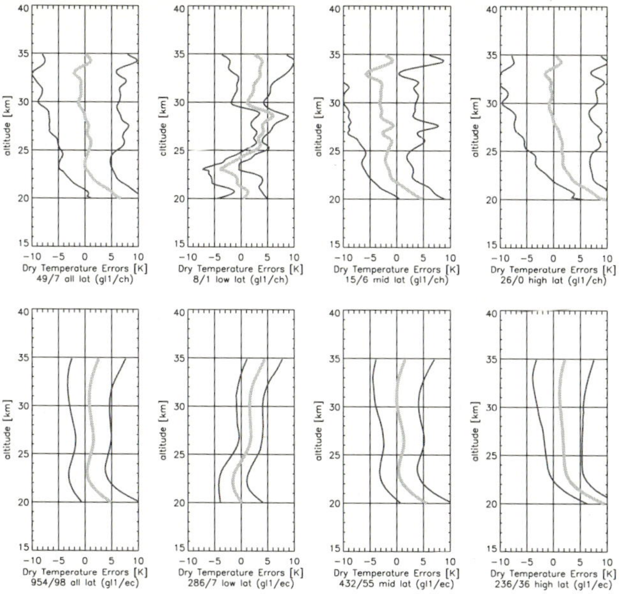

Fig. 2. Temperature error statistics for multi day ensembles in the period 2002-09-20 to 2002-12-02 (whole set 2002). In the upper panels we present results from ensembles validated with CHAMP GPS profiles and in the lower panels we validate with ECMWF analysis data. For further description of the graphs see caption of Fig. 1.

from the nearest analysis time of 6-hourly ECMWF T511L60 operational analysis data. ECMWF validation shows better results compared to the GOMOS-CHAMP validation. Typical features, as the positive bias for the low latitude profiles at altitudes >25 km are seen in the CHAMP and the ECMWF validation. At high latitudes and altitudes between 20 km and 35 km one can see the deviation to positive values of $\hat{\mathbf{b}}$. Due to interpolations, the profiles here appear more smoothed than in the GOMOS – CHAMP validation. For altitudes higher than 25 km a negative bias of >2 K was found. At mid and high latitudes $\hat{\mathbf{b}}$ has a positive deviation below 25 km and performs equally to the high latitude sample.

The validation results with ECMWF have smaller biases than the validation with CHAMP profiles. This has different reasons. First, the validation set with CHAMP profiles, especially at low and mid latitudes is too small, so that the bias profile suffers from errors at few single profile validation results (cf., Retscher 2004). A second reason is that coincidence intervals between GOMOS and CHAMP data are often too large, while for ECMWF we first select analysis data at the exact tangent point location of GOMOS.

4 Summary and Conclusions

We developed an optimal estimation algorithm for retrieval of atmospheric trace gas profiles from Envisat/GOMOS-measured transmission data. In this study ozone profiles were retrieved, which is the primary focus of the algorithm, though the whole processing chain is capable of retrieving other trace gases simultaneously.

Furthermore, we applied refractive occultation retrieval to bending angle data from GOMOS star tracker data (SFA/SATU), gaining refractivity and temperature profiles. A simultaneous exploitation of atmospheric transmission and bending angles was not yet performed in this study but will be introduced as a future enhancement. This enhancement will allow to simultaneously retrieve ozone and temperature profiles in the stratosphere as well as atmospheric refractivity and density, which can be used to improve the background fields required by the ray tracing. These improvements will further aid the ozone retrieval, given that the star tracker data are of adequate quality allowing < 5 μrad bending angle accuracy.

In conclusion, the approach adopted for an efficient retrieval of ozone and temperature profiles has yielded satisfactory results. Data from later periods than year 2002 as well as Envisat/MIPAS ozone and temperature profiles will be analyzed and validated in future.

Acknowledgements. The authors gratefully acknowledge discussions with and support by U. Foelsche, M. Schwärz, C. Rehrl, and J. Ramsauer (WegCenter, Univ. of Graz, Austria) as well as V. Sofieva and J. Tamminen (FMI, Helsinki, Finland). The European Space Agency is thanked for operating the Envisat satellite and for its related Announcement of Opportunity for free access to Envisat data. We are especially thankful to B. Theodore and G. Barrot (ACRI-ST, Sophia Antipolis, France) for providing reprocessed GOMOS data. Furthermore, we are grateful to J. Wickert and T. Schmidt (GFZ, Potsdam, Germany) for providing CHAMP occultation data. C.R. received financial support for the work from Envisat Project AO-620/Part-I funded by the Austrian Ministry for Traffic, Innovation, and Technology and carried out under contract with the Austrian Space Agency.

References

Bertaux JL, Megie G, Widemann T, Chassefiere E, Pellinen R, Kyrölä E, Korpela S, Simon P (1991) Monitoring of ozone trend by stellar occultations: the GOMOS instrument. Advances in Space Research 11:237–242

Bracher A, Eichmann K, von Savigny C, Weber M (2004) Ozone Distribution in the Arctic Winter/Spring 2002/2003 as Measured by GOMOS, MIPAS, and SCIAMACHY. In: Envisat Symposium 2004, Salzburg, Austria

Dethof A (2004) Assimilation of ozone data in the ECMWF Integrated Forecast System. In: SPARC 2004, Victoria, Canada

Foelsche U (1999) Tropospheric water vapor imaging by combination of ground-based and spaceborne GNSS sounding data. Tech rep, No 10/1999, IGAM, Univ of Graz, Austria

Gobiet A, Kirchengast G (2002) Sensitivity of atmospheric profiles retrieved from GNSS occultation data to ionospheric residual and high-altitude initialization errors. Tech rep, ESA/ESTEC No 1/2002, IGAM, Univ of Graz, Austria

Gobiet A, Kirchengast G (2004) Advancements of Global Navigation Satellite System Radio Occultation Retrieval in the Upper Stratosphere for Optimal Climate Monitoring Utility. J Geophys Res 109(D24110), doi:10.1029/20043D005117

Gobiet A, Steiner AK, Retscher C, Foelsche U, Kirchengast G (2004) Radio Occultation Data and Algorithms Validation Based on CHAMP/GPS Data. Tech rep, No 1/2004 IGAM, Univ of Graz, Austria

Gobiet A, Kirchengast G, Wickert J, Retscher C, Wang D, Hauchecorne A (2005) Evaluation of Stratospheric Radio Occultation Retrieval Using Data from CHAMP, MIPAS, GOMOS and ECMWF Analysis Fields. In: Reigber C, Lühr H, Schwintzer P, Wickert J (eds) Earth Observation with CHAMP: Results from Three Years in Orbit. Springer, Berlin, pp 531–536

Healy S (2001) Smoothing Radio Occultation Bending Angles Above 40 km. Ann Geophys 19:459

Kyrölä E, Sihvola E, Kotivuori Y, Tikka M, Tuomi T, Haario H (1993) Inverse theory for occultation measurements. I - Spectral inversion. J Geophys Res 98(17):7367–7381

Kyrölä E, Tamminen J, Leppelmeier GW, Sofieva V, Hassinen S, Bertaux JL, Hauchecorne A, Dalaudier F, Cot C, Korablev O, d Andon OF, Barrot G, Mangin A, Thodore B, Guirlet M, Etanchaud F, Snoeij P, Koopman R, Saavedra L, Fraisse R, Fussen D, Vanhellemont F (2004) GOMOS on Envisat: An overview. Advances in Space Research 33:1020–1028

Paulsen T (2000) GOMOS Level 1b/2 Algorithm Description Document for Enviview Pre-Release. Tech rep, ESA Doc, PO-TN-ESA-GM-1019, ESA/ESTEC, Noordwijk, Netherlands

Rees D (ed) (1988) COSPAR International Reference Atmosphere: 1986, Part I: Thermosphere Models, vol 8

Retscher C (2004) Stratospheric Ozone and Temperature Sounding by the Envisat/GOMOS Stellar Occultation Sensor. Tech rep, No 23/2004, IGAM, Univ of Graz, Austria

Retscher C, Kirchengast G, Gobiet A, Hauchecorne A (2004) Stratospheric Temperature and Ozone Sounding with Envisat/GOMOS Stellar Occultation. In: Kirchengast G, Foelsche U, Steiner AK (eds) Occultations for Probing Atmosphere and Climate. Springer, Berlin, pp 299–308

Rodgers C (1976) Retrieval of Atmos. Temperature and Composition from Remote Measurements of Thermal Radiation. Rev Geophys 14:609–624

Rodgers C (1990) Characterization and Error Analysis of Profiles Retrieved From Remote Sensing Measurements. J Geophys Res 95:5587–5595

Rodgers C (2000) Inverse Methods for Atmospheric Remote Sounding: Theory and Practice. World Scientific, Singapore

Sofieva VF, Kyrölä E (2003) Information approach to optimal selection of spectral channels. Journal of Geophysical Research (Atmospheres) 108(D16):15, doi: 10.1029/2002JD002980

Sofieva VF, Kyrölä E (2004) Information Approach to Channel Selection for Stellar Occultation Measurements. In: Kirchengast G, Foelsche U, Steiner AK (eds) Occultations for Probing Atmosphere and Climate, Springer, Berlin, pp 309–318

Sofieva VF, Kyrölä E, Ferraguto M, Team GC (2003) From Pointing Measurements in Stellar Occultation to Atmospheric Temperature, Pressure and Density Profiling: Simulations and First GOMOS Results. In: Geoscience and Remote Sensing Symposium, IGARSS '03, Proceedings, vol 5, pp 2990

Sokolovsky S, Hunt D (1996) Statistical Optimization Approach for GPS/Met Data Inversions. In: URSI GPS/Met Workshop, Tucson, Arizona

Steiner AK (1998) High Resolution Sounding of Key Climate Variables Using the Radio Occultation Technique. Tech rep, No 3/1998, IGAM, Univ of Graz, Austria

Syndergaard S (1999) Retrieval Analysis and Methodologies in Atmospheric Limb Sounding Using GNSS Radio Occultation Technique. PhD thesis, Niels Bohr Inst for Astron, Physics and Geophysics, Univ of Copenhagen

Syndergaard S (2000) On the Ionosphere Calibration in GPS Radio Occultation Measurements. Radio Sci 35:865–883

Vanhellemont F, Fussen D, Bingen C (2004) Global One-step Inversion of Satellite Occultation Measurements: A Practical Method. J Geophys Res 109(D18):9306, doi:10.1029/2003JD004168

Wickert J, Schmidt T, Beyerle G, König R, Reigber C, Jakowski N (2004) The radio occultation experiment aboard CHAMP: Operational data analysis and validation of vertical atmos. profiles. J Meteorol Soc Jpn 82:381–395

Modeling Errors of GOMOS Measurements: A Sensitivity Study

V. F. Sofieva[1], J. Tamminen[1], E. Kyrölä[1], and GOMOS CAL/VAL Team[2]

[1] Finnish Meteorological Institute, Earth Observation, Helsinki, Finland
viktoria.sofieva@fmi.fi

[2] Service d'Aronomie du CNRS, Verrieres le Buisson Cedex, France;
Finnish Meteorological Institute, Helsinki, Finland;
Institut d'Aeronomie Spatiale de Belgique, Brussels, Belgium;
ACRI-ST, Sophia Antipolis, France

Abstract. Error analysis and characterization of atmospheric profiles retrieved from GOMOS measurements is a challenging task, because of dependence of signal-to-noise ratio on stellar properties and various modeling errors and approximations. In addition, occultations suffer from scintillation caused by air density irregularities below 45 km. Neglecting the modeling errors results in underestimated error bars of retrieved profiles. In this paper, we present theoretical estimations of some modeling errors: uncertainties in cross-sections, dilution correction error, and ray tracing error. The uncertainty in atmospheric temperature is also briefly discussed. It is shown that inclusion of modeling errors provides more realistic error estimates of retrieved profiles and gives values of the χ^2-statistics close to the theoretical ones.

1 Introduction

The GOMOS (Global Ozone Monitoring by Occultations of Stars) instrument on board the Envisat satellite is the first operational instrument that uses stellar occultation technique to study the atmospheric composition from the troposphere to the mesosphere. The GOMOS products from the UV-Visible spectrometer are vertical profiles of ozone, neutral density, aerosols, NO_2, and NO_3. A global coverage, self-calibration, and a good vertical resolution are the strong features associated with the GOMOS measurements. The specific feature of stellar occultation measurements is the dependency of the signal-to-noise ratio on the stellar brightness and spectral class. This is illustrated in Fig. 1, where the signal-to-noise ratio at 30 km altitude is shown for various types of stars. Apart from the instrumental noise, errors arise from various approximations and modeling errors. The most important of them are:

- uncertainties in cross-sections,
- scintillation correction error,

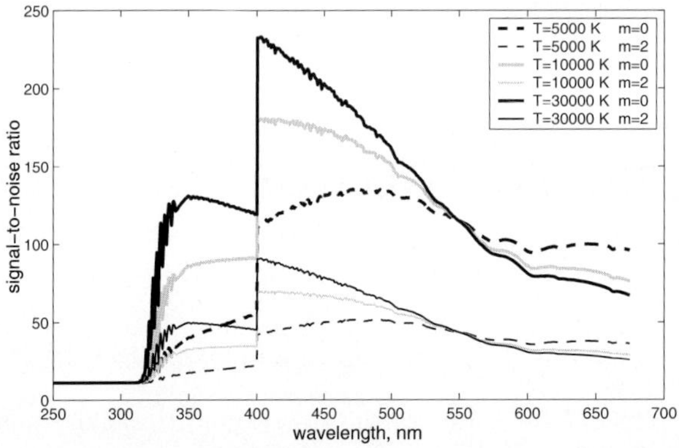

Fig. 1. Signal-to-noise ratio for some stellar classes (simulation) at 30 km; visual magnitudes m and effective temperatures T are specified in the legend.

- dilution correction error,
- uncertainty in the atmospheric temperature,
- error of the aerosol model.

This work is dedicated to sensitivity studies. We show how the modeling errors affect the accuracy of GOMOS profile retrievals. The theoretical estimation of modeling errors followed by estimation of retrieval errors via posterior covariance matrices is used in this work. In Section 2, we introduce GOMOS forward model and inversion needed for the theoretical estimations of the modeling errors. In Section 3, the principle of GOMOS error estimation is briefly described. Section 4 is dedicated to identification of problems in modeling error characterization and characterization of retrieval quality. In Section 5, we discuss the influence of modeling errors on accuracy of the retrieval. A discussion concludes the paper.

2 GOMOS Inversion

The benefit of the occultation principle is its self-calibrating measurement concept. The reference stellar spectrum is first measured when a star can be seen above the atmosphere. During the occultation, the measurements through the atmosphere provide spectra modified by absorption, scattering, and refraction.

The GOMOS processing of dark-limb occultations starts with various instrumental corrections. First, the dark current noise is subtracted from the reference and attenuated spectra. Second, the reference star spectrum is averaged from sufficiently many measurements above the atmosphere, thus giving the accurate estimate of the star spectrum S_{star}. Then the spectrum observed through the atmosphere S_{obs} is divided by the reference spectrum, yielding the atmospheric transmittance T_{atm}:

$$T_{\text{atm}} = \frac{S_{\text{obs}}}{S_{\text{star}}}. \tag{1}$$

Then the component due to refractive dilution and scintillation $T_{\text{r,scint}}$ is estimated and corrected (Dalaudier et al. 2001; GOMOS ESL 1999, 2005)

$$T_{\text{ext}} = \frac{T_{\text{atm}}}{T_{\text{r,scint}}}. \tag{2}$$

Finally, the transmittance due to absorption and scattering (extinction) T_{ext} is obtained. It can be described by the well-known Beer's law:

$$T_{\text{ext}} = e^{-\tau}, \tag{3}$$

where the optical depth τ at wavelength λ is given by

$$\tau(\lambda) = \sum_j \int_\ell \sigma_j(\lambda, \mathrm{T}(s))\rho_j(\mathbf{r}(s))\,\mathrm{ds}. \tag{4}$$

Here ρ_j's are constituent densities depending on the position $\mathbf{r}$ and σ_j's are the temperature-dependent absorption or scattering cross sections. The integration is performed along the optical path ℓ joining the instrument and the source.

In GOMOS data processing, the local spherical symmetry of the atmosphere is assumed. The inversion is split into two parts: the spectral inversion part and the vertical inversion part (Kyrölä et al. 2004). First, transmission data from every tangent height are inverted to horizontal column (line) densities for different constituents (spectral inversion). Second, for every constituent, the collection of line densities at successive tangent heights is converted to vertical density profiles (vertical inversion). The effective cross-section method (Sihvola 1994) allows this split of the inversion and provides a very efficient data processing. In this method, the optical depth τ is presented in the form

$$\tau(\lambda, \ell) = \sum_j \int \rho_j(s)\sigma_j(\lambda, \mathrm{T}(s))\,\mathrm{ds} = \sum_\mathrm{j} \sigma_\mathrm{j}^{\text{eff}}(\lambda, \ell)\mathrm{N_j}, \tag{5}$$

where N_j is the line density of the constituent j

$$N_j = \int_\ell \rho_j(z(s))\,\mathrm{ds} \tag{6}$$

and

$$\sigma_j^{\text{eff}}(\lambda, \ell) = \frac{\int_\ell \sigma_j(\lambda, \mathrm{T}(s))\rho_j(s)\,\mathrm{ds}}{\mathrm{N_j}} \tag{7}$$

is the effective cross-section of the constituent j. Here z denotes an altitude.

The spectral inversion part is given by (5) with the line densities N_j as unknowns. The vertical inversion part is given by (6) with local density $\rho_j(z)$ as the unknowns. The two parts are, however, coupled together by the unknown effective cross sections. This results in an iterative loop over spectral and vertical inversions.

3 Error Estimates of GOMOS Products

The estimation of line densities in the spectral inversion is based on the standard maximum likelihood method. Under assumption of Gaussian distribution of the measurement noise, it is equivalent to minimization of the normalized χ^2 statistics

$$\chi^2 = \frac{1}{n-k}(\mathbf{T}_{\text{mod}}(\mathbf{N}) - \mathbf{T}_{\text{ext}})^{\text{T}}\mathbf{C}^{-1}(\mathbf{T}_{\text{mod}}(\mathbf{N}) - \mathbf{T}_{\text{ext}}), \tag{8}$$

where $\mathbf{T}_{\text{mod}}$ is the vector of modeled transmittance and $\mathbf{C}$ is the covariance matrix of the transmission data, n is the number of measurements and k is the number of fitted parameters. The minimization is performed using the Levenberg-Marquardt algorithm (e.g., Press et al. 1992).

The covariance matrix $\mathbf{C}$ of the transmission data has two components. The first (and dominating) $\mathbf{C}_{\text{instr}}$ is due to noise in measurements (mainly a photon noise and a dark current of the CCD). If there is no operation destroying stochastic independence of data, this covariance matrix is diagonal. The second one $\mathbf{C}_{\text{mod}}$ comes from various modeling errors. If the data statistics and the model error are Gaussian, the total error is a sum of these two:

$$\mathbf{C} = \mathbf{C}_{\text{instr}} + \mathbf{C}_{\text{mod}}. \tag{9}$$

The Levenberg-Marquardt algorithm gives estimates of line densities and their covariance matrix. These error estimates serve as "measurement errors" for the second step – the vertical inversion. The error estimates of the final GOMOS products – local densities – are obtained with the standard Gaussian error propagation method, as the vertical inversion is linear.

4 Identification of Problems in Modeling Error Characterization

The χ^2 value together with residuals $R(\lambda) = T_{\text{mod}}(\lambda) - T_{\text{ext}}(\lambda)$ can be used as indicators of error estimates quality: if the modeling and instrumental error are Gaussian and well defined, then $\chi^2 \approx 1$ and residuals are white noise. Underestimated errors give values $\chi^2 > 1$.

The analysis of the GOMOS data has shown that the modeling errors significantly contribute to the total error budget for altitudes < 45 km: neglecting the modeling errors results in χ^2 values significantly exceeding 1.

The χ^2 values are larger for oblique (off orbital plane) occultations than for vertical (in orbital plane) ones. This is illustrated in Fig. 4, which will be discussed in more details in Sect. 6. Residuals have oscillating features in the case of oblique occultations (not shown here). These oscillating features are not correlated at adjacent altitudes and for different occultations. We believe that they are related to the presence of turbulence in the stratosphere: the GOMOS scintillation correction algorithm removes only the anisotropic component of the scintillation (caused by layered air density irregularities), while the isotropic component remains uncorrected. For vertical occultations, the scintillation correction is nearly complete, and residuals are close to white noise.

However, neglecting the modeling errors results in $\chi^2 > 1$ not only for oblique occultations, but also for vertical ones. This indicates the presence of other modeling errors, not related with the isotropic turbulence. These errors have no signature in residuals, therefore they can be considered as random. The possible error sources are errors in cross-sections, uncertainty in temperature for effective cross-sections computation, error of dilution correction and error of ray tracing. In this paper, we investigate sensitivity of the model to each of these error sources. We propose analytical estimates for these modeling errors and their propagation in inversion. The errors are considered as random, and the retrieval errors are estimated via posterior covariance matrices. The error of the aerosol model is not discussed in this paper, as it does not allow analytical description. The sensitivity of retrievals to the aerosol model has been studied in Tamminen (2004) using Markov Chain Monte Carlo method. The scintillation correction error will also be discussed in future publications.

5 Modeling Errors of GOMOS Measurements

5.1 Uncertainties in Cross-Sections

The uncertainties of the cross-sections are collected in Table 1.

Table 1. Uncertainties in cross-sections used in the simulation.

Cross-sections	Uncertainty	Reference and Comments
Ozone	2.5 %	c.f. (Malicet et al. 1995)
NO_2	3 %	c.f. (Merienne et al. 1995)
NO_3	6 %	c.f. (Sander 1986)
Rayleigh	2.5 %	uncertainty of depolarization factor

We assume that the uncertainties can be modeled as mutually independent Gaussian random variables with zero mean and standard deviations given in Table 1. We can present the optical depth as

$$\boldsymbol{\tau} = (\boldsymbol{\Sigma}_0 + \boldsymbol{\Sigma}_1)\mathbf{N} = \boldsymbol{\Sigma}_0\mathbf{N} + \boldsymbol{\Sigma}_1\mathbf{N}, \tag{10}$$

where $\boldsymbol{\Sigma}_0$ is the matrix of the exact cross-sections, $\boldsymbol{\Sigma}_1$ is the error in the cross-sections and $\mathbf{N}$ is the vector of line densities. The second term in (10), $\boldsymbol{\delta} = \boldsymbol{\Sigma}_1\mathbf{N}$, represents an additional, modeling error.

Assuming that the elements of the matrix $\boldsymbol{\Sigma}_1$ are Gaussian random variables with zero mean and covariance s_{ij}^2, the elements of the covariance matrix $\mathbf{C}_\delta$ corresponding to the error due to cross-section uncertainties can be estimated as

$$C_{\delta ii} = \sum_{j=1}^{N_{\mathrm{gas}}} s_{ij}^2 N_{0j}^2, \tag{11}$$

where N_{0j} are some a priori values of line densities, and N_{gas} is the number of retrieved constituents. $\mathbf{C}_\delta$ is assumed to be diagonal.

Provided that $\boldsymbol{\delta}$ is small ($\boldsymbol{\delta} \ll 1$), the spectral inversion problem with cross-sections uncertainty can be written as

$$\begin{aligned}\mathbf{T}_{\mathrm{ext}} &= \exp(-(\boldsymbol{\Sigma}_0\mathbf{N} + \boldsymbol{\delta})) + \varepsilon_{\mathrm{instr}} \\ &\approx \exp(-\boldsymbol{\Sigma}_0\mathbf{N})(1 - \boldsymbol{\delta}) + \varepsilon_{\mathrm{instr}} \\ &= \exp(-\boldsymbol{\Sigma}_0\mathbf{N}) + \varepsilon_{\mathrm{cross}} + \varepsilon_{\mathrm{instr}},\end{aligned} \tag{12}$$

where $\varepsilon_{\mathrm{cross}}$ and $\varepsilon_{\mathrm{instr}}$ correspond to the error due to uncertainty in cross-sections and due to instrumental noise, respectively. The error $\varepsilon_{\mathrm{cross}}$ is approximately a Gaussian random variable with zero mean and the standard deviation

$$\mathrm{std}_{\mathrm{cross}} = \sqrt{\mathbf{C}_\delta}\mathbf{T}_{\mathrm{ext}}^0. \tag{13}$$

Here $\mathbf{T}_{\mathrm{ext}}^0$ is the error-free transmittance and $\mathbf{C}_\delta$ is determined by (11).

We estimated the influence of the cross-section error on the accuracy of the line densities retrieval. Figure 2 shows the estimated error of reconstruction, for stars of magnitudes $m = 0$ and $m = 2$, in the the following cases:

- The total error includes the instrumental noise only (dashed lines).
- Both instrumental noise and uncertainties in the cross-sections are taken into account (solid lines).

The estimated error of the reconstruction is significantly larger if both instrumental and modeling errors are taken into account, especially for the star of magnitude 0. It is not surprising that the effect of cross-section uncertainties is more pronounced for bright stars: they have higher signal-to-noise ratio and, as a consequence, a larger contribution of modeling errors to the total error budget than dim ones.

5.2 Uncertainty in Temperature

The cross sections are temperature dependent. Since the atmospheric temperature is not known, the ECMWF temperature profile is used in the GOMOS

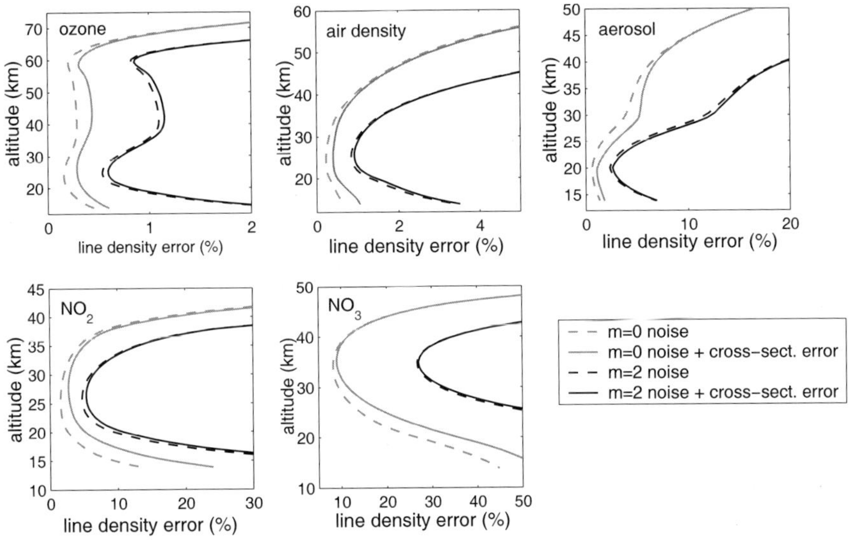

Fig. 2. Estimates of line density retrieval error. Dashed lines: only instrumental noise is accounted for; solid lines: both instrumental noise and uncertainties in cross sections are taken into account. The simulation is carried out for a star of an effective temperature of 10 000 K.

retrieval. This introduces some uncertainty to the retrieval. A sensitivity study in a bright star case shows that by assuming the temperature having Gaussian uncertainty with 2 K standard deviation the error estimates of ozone clearly increase at the altitude range 30 km to 55 km with the maximum around 40 km. The error estimate around the maximum is ~30 % larger compared to the case with no modelling error. At the other altitudes and for other constituents the impact was very small. The influence of the atmospheric temperature uncertainty on the accuracy of the retrieval is discussed in more detail in Tamminen (2004).

5.3 Error of Dilution Correction

The possible error of dilution correction is caused by uncertainty of air density data (ECMWF) used for ray tracing and subsequent dilution estimation. We assume that this uncertainty is Gaussian with zero mean and standard deviation given in Table 2 (GOMOS ESL 2005).

The dilution factor T_r is given by the formula (Hays and Roble 1968)

$$T_\mathrm{r} = \frac{1}{1 + L\frac{\mathrm{d}\alpha}{\mathrm{dp}}}, \tag{14}$$

where L is the distance from the tangent point to the satellite, p is the impact parameter and α is the refractive angle. The uncertainty in the dilution factor

Table 2. Relative uncertainties of the air density from external atmospheric model (ECMWF).

Altitude z [km]	Uncertainty $\delta\rho(z)$ [%]
0 – 25	2.5
25 – 35	$2.5 + 0.25(z - 25)$
35 – 45	5
45 – 55	$5 + 0.5(z - 45)$
55 – 120	$10 + 0.2(z - 55)$

δT_{r} is caused by the uncertainty in refractive angle $\delta\alpha$, which is, in turn, caused by the uncertainty in air density $\delta\rho$. Under assumption $\frac{\mathrm{d}}{\mathrm{dp}}(\delta\alpha) \ll \frac{\mathrm{d}\alpha}{\mathrm{dp}}$, this uncertainty can be estimated as follows:

$$\delta T_{\mathrm{r}} = -\frac{\delta(1 + L\frac{\mathrm{d}\alpha}{\mathrm{dp}})}{(1 + L\frac{\mathrm{d}\alpha}{\mathrm{dp}})^2} = -T_{\mathrm{r}}^2 L \frac{\mathrm{d}\alpha}{\mathrm{dp}}\frac{\delta\alpha}{\alpha} = -T_{\mathrm{r}}(1 - T_{\mathrm{r}})\frac{\delta\alpha}{\alpha}. \tag{15}$$

Here $\delta(\cdot)$ is used for denoting differentials. Finally, we get

$$\left|\frac{\delta T_{\mathrm{r}}}{T_{\mathrm{r}}}\right| \approx (1 - T_{\mathrm{r}})A\left|\frac{\delta\rho}{\rho}\right|, \tag{16}$$

where A is an amplification of error coefficient, which is ~ 0.5 (Sofieva and Kyrölä 2004). According to (16), in the upper atmosphere the dilution correction error vanishes even in the case of a large enough density uncertainty, as $T_{\mathrm{r}} \approx 1$. In the lower atmosphere the dilution correction error becomes more significant: it constitutes $\sim 2\,\%$ at 1 km altitude.

The dilution correction consists in elimination of this factor from the transmission data (2). The dilution correction error can be considered as an additional error term (the scintillation correction error is neglected here):

$$T_{\mathrm{ext}} = \frac{T_{\mathrm{atm}}}{T_{\mathrm{r}}^0 + \delta T_{\mathrm{r}}} \approx \frac{T_{\mathrm{atm}}}{T_{\mathrm{r}}^0}\left(1 - \frac{\delta T_{\mathrm{r}}}{T_{\mathrm{r}}^0}\right) = T_{\mathrm{ext}}^0 + \varepsilon_{\mathrm{dil}}, \tag{17}$$

where T_{r}^0 and T_{ext}^0 are the exact values of dilution and transmittance due to absorption and scattering, respectively. The additional error term $\varepsilon_{\mathrm{dil}}$ is assumed to be Gaussian with zero mean and the standard deviation

$$\mathrm{std}(\varepsilon_{\mathrm{dil}}) = T_{\mathrm{ext}}^0 \frac{\delta T_{\mathrm{r}}}{T_{\mathrm{r}}^0}. \tag{18}$$

The dilution correction error is significantly smaller than the instrumental error (even for bright stars). For the upper altitudes, it is negligible. Figure 3 compares the estimated errors of line densities, for a star of magnitude $m = 0$, assuming that (a) the total error includes the instrumental noise only (dashed lines) and (b) the instrumental noise and the dilution correction error are taken into account (solid lines). The incomplete dilution correction only slightly degrades the inversion accuracy.

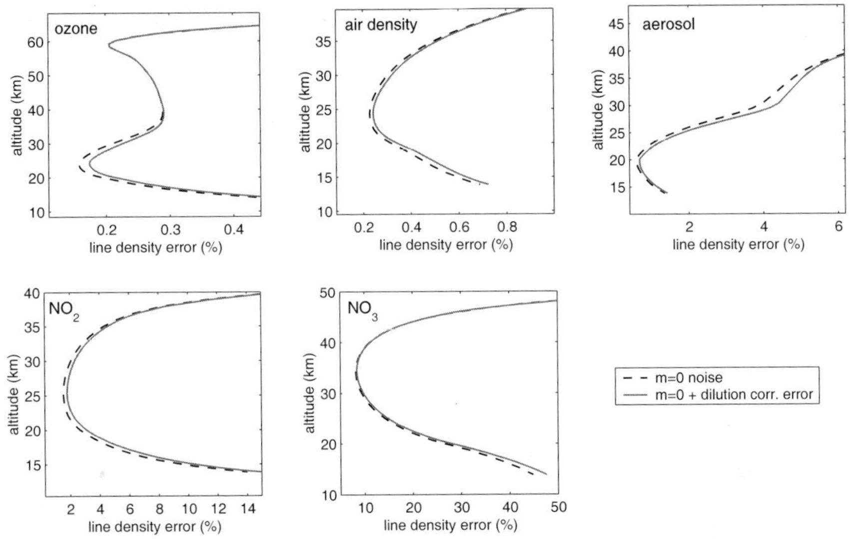

Fig. 3. Estimates of line density retrieval error. Dashed lines: only instrumental noise is accounted for; solid lines: both instrumental noise and dilution correction error are taken into account.

5.4 Error of Tangent Altitude Determination

The possible error of the tangent altitude determination is also caused by uncertainty in air density data (ECMWF) used for ray tracing. For the sake of simplicity, we assume that the atmosphere is spherically symmetric. In this case, the tangent altitude is the solution of the equation

$$n(r)r = p, \tag{19}$$

where $n(r)$ is the refractive index, r is the altitude measured from the Earth center and p is the impact parameter. From (19) we get

$$\left|\frac{\delta r_t}{r_t}\right| = \left|\frac{\delta n}{n}\right| = \frac{|\delta\nu|}{1+\nu}. \tag{20}$$

Here $\nu = n - 1$ is the refractivity of the air, connected with the air density via Edlen's formula (Edlen 1966). The error of tangent altitude determination is very small: it is only ∼15 m for the altitude 10 km and it is <5 m for altitudes above 20 km. This error cannot introduce any significant error in atmospheric profile reconstruction and can therefore be neglected.

6 Discussion and Summary

The presented analysis of modeling errors in the GOMOS inversion has shown that neglecting the modeling errors leads to significant underestimation of the

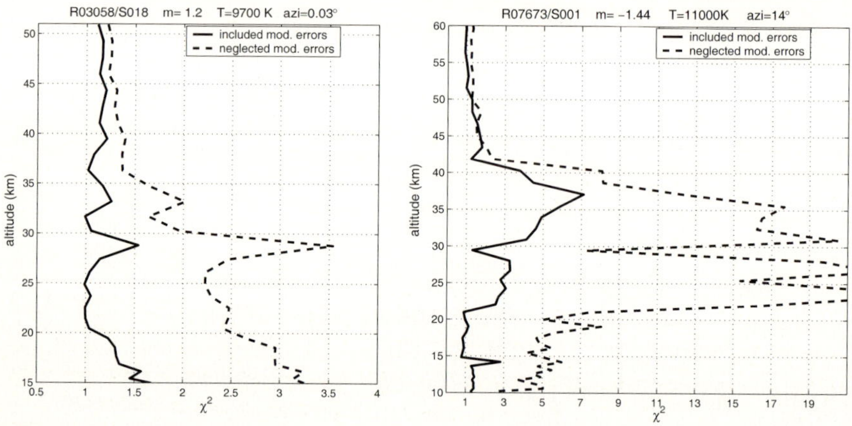

Fig. 4. χ^2 statistics in vertical (left) and oblique (right) occultations. Dashed lines: only instrumental noise is accounted for; solid lines: both instrumental noise and modeling errors are taken into account.

error of the retrieved profiles. This underestimation is more significant for bright stars, for which the contribution of modeling errors to the total error budget is larger than for dim ones. It was shown that the ray tracing errors are very small, while errors in cross sections, and, to a lesser extent, dilution correction error and uncertainty in atmospheric temperature are significant error sources below 45 km.

The accurate estimation of modeling errors provides realistic estimates of retrieval errors and gives the inversion χ^2 statistics close to the theoretical one. This is illustrated in Fig. 4: if modeling errors are neglected, χ^2 significantly exceeds the theoretical value 1. Inclusion of modeling errors in the inversion (in the considered case, cross-section uncertainties and dilution correction error were taken into account) makes values of χ^2 closer to 1. It is interesting to note that the χ^2 remains significantly larger than 1 at altitudes 30 km to 40 km (most sensitive to turbulence (Gurvich et al. 2005)) in the case of oblique occultations. These high values correspond to neglected scintillation correction error, which can be significant for oblique occultations.

However, it is difficult (even if possible in principle) to distinguish the contributions of uncertainties in cross-sections and scintillation correction error for oblique occultations, because they have similar influence almost in the same altitude range. Furthermore, the cross-section uncertainties also contain a systematic component, correlated in wavelength. Not excluded, that the uncertainties in cross-sections given in Table 1 and used in the simulation are overestimated, and they partially compensate for the scintillation correction error. Nevertheless, the χ^2 statistics close to 1 indicates that the modeling errors are correctly accounted for, and the error estimates of the retrieved products are close to reality.

The characterization of the scintillation correction error is a complicated task: it is very difficult to describe it analytically, because the model of stratospheric turbulence is three-dimensional. However, the total modeling error (without resolving the error sources) can be estimated a posteriori. This will be the subject of future publications.

References

Dalaudier F, Kan V, Gurvich AS (2001) Chromatic refraction with global ozone monitoring by occultation of stars. I. Description and scintillation correction. Applied Optics 40(6):866–877

Edlen B (1966) The refractive index of air. Metrologia 2(71)

GOMOS ESL (1999) GOMOS Algorithm Theoretical Basis Document, 1st edn. ESA

GOMOS ESL (2005) GOMOS Detailed Processing Model. ESA

Gurvich AS, Dalaudier F, Sofieva VF (2005) Study of stratospheric air density irregularities based on two-wavelength observation of stellar scintillation by Global Ozone Monitoring by Occultation of Stars (GOMOS) on Envisat. J Geophys Res 110(D11110), DOI 10.1029/2004JD005536

Hays RG, Roble PB (1968) Stellar spectra and atmospheric composition. J Atmos Sci 25:1141–1153

Kyrölä E, Tamminen J, Leppelmeier GW, Sofieva V, Hassinen S, Bertaux JL, Hauchecorne A, Dalaudier F, Cot C, Korablev O, d'Andon OF, Barrot G, Mangin A, Theodore B, Guirlet M, Etanchaud F, Snoij P, Koopman R, Saavedra L, Fraisse R, Fussen D, Vanhellemont F (2004) GOMOS on Envisat: An overview. Adv Space Res 33:1020–1028

Malicet J, Daumont D, Charbonnier J, Parisse C, Chakir A, Brion J (1995) Ozone UV Spectroscopy II: Absorption Cross-Sections and Temperature Dependence. J Atmos Chem 21:263–273

Merienne MF, Jenouvrier A, Coquart B (1995) The NO_2 absorption spectrum. I: Absorption cross-sections at ambient temperature in the 300–500 nm region. J Atmos Chem 20:281–297

Press WH, Teukolsky SA, Vetterling WT, Flannery BP (1992) Numerical Recipes in FORTRAN, The Art of Scientific Computing. Cambridge University Press, Cambridge

Sander SP (1986) Temperature dependence of NO_3 absorption spectrum. J Phys Chem 90:4135–4142

Sihvola E (1994) Coupling of spectral and vertical inversion in the analysis of stellar occultation data. Geophysical publications, no. 38, Finnish Meteorological Institute, Helsinki, Licentiate thesis at the University of Helsinki, Department of Theoretical Physics

Sofieva VF, Kyrölä E (2004) Abel integral inversion in occultation measurements. In: Kirchengast G, Foelsche U, Steiner A (eds) Occultations for Probing Atmosphere and Climate, Springer Verlag, pp 77–86

Tamminen J (2004) Validation of nonlinear inverse algorithms with Markov chain Monte Carlo method. J Geophys Res 109(D19):D19,303, DOI 10.1029/2004JD004927

Wave Optics Algorithms for the Processing of Radio Occultation Data

Asymptotic Wave Optics Methods in Inversion and Direct Modeling of Radio Occultations: Recent Achievements

M. E. Gorbunov[1] and K. B. Lauritsen[2]

[1] Institute for Atmospheric Physics, Pyzhevsky per. 3, Moscow 119017, Russia gorbunov@dkrz.de
[2] Danish Meteorological Institute, Lyngbyvej 100, Copenhagen DK-2100, Denmark

Abstract. We discuss the recent achievements in the application of asymptotic methods based on Fourier Integral Operators (FIOs) for inversion and direct modeling of radio occultations. We show that FIOs can be derived from the first principles: stationary phase principle and energy conservation. We discuss accurate and approximate solutions for the kernel of the FIOs. The approximations can be used for designing very efficient numerical algorithms, where the FIOs are reduced to a composition of multiplying with reference signals and Fourier transforms. Another application is inversion algorithms using an FIO that retrieves the geometric optical ray structure of wave fields. Asymptotic methods of forward modeling are based on inverse FIOs that map the geometric optical ray structure to wave fields. Such algorithms are very fast and significantly reduce numerical inaccuracies, which arise in computation of multiple diffractive integrals.

1 Introduction

Fourier Integral Operators (FIOs) are a very effective means of analysis of wave fields measured in radio occultation experiments (Gorbunov 2002a,b; Gorbunov and Lauritsen 2002; Jensen et al. 2002; Gorbunov 2003; Gorbunov and Kornblueh 2003; Jensen et al. 2003, 2004; Gorbunov et al. 2004; Gorbunov and Lauritsen 2004a). These operators generalize the standard construction of geometrical optics (GO) (Mishchenko et al. 1990). The basis of GO is the stationary phase principle which describes rays. Rays are curves in the phase space, where there is no multipath, because rays interfering at the same coordinate have different directions, or momenta. Multipath in the physical space is characterized by the projection type of the ray manifold. In the canonical transform method we change coordinates in the phase space so as to change the projection type of the ray manifold (Gorbunov 2002a; Gorbunov and Lauritsen 2004a). The new coordinates result in the same minimum action principle

(Kleinert 1995). FIOs generalize the concept of geometric optical canonical transform for wave fields.

The physical principles of stationary phase and energy conservation allow for establishing all the known phase functions of the FIO used in Canonical Transform (CT) method based on FIO of the 1st type, Full-Spectrum Inversion (FSI) method and Phase Matching (PM) (Jensen et al. 2003, 2004; Gorbunov and Lauritsen 2004a). The reduction of the FIOs to FT can be done by using some coordinate transform or approximation that linearizes the phase function (Gorbunov and Lauritsen 2004a). This significantly enhances the numerical efficiency.

By inverting the FIOs discussed above we can write the transform from the impact parameter representation to the representation of the physical coordinate (Gorbunov 2003). This results in fast asymptotic algorithm of forward modeling. If diffraction inside the medium is not negligible, e.g., in the presence of turbulence, direct modeling must use multiple phase screens. The final step from the last phase screen to the LEO orbit is performed by the computation of multiple diffractive integrals. This procedure is very inefficient numerically and, besides, it introduces computational inaccuracies. An alternative, very efficient solution is the FIO based on the corresponding linearized CT.

2 Basic Principles of Fourier Integral Operators

2.1 Basic Waveforms

FIOs are used for constructing asymptotic solutions of wave problems (Mishchenko et al. 1990), which generalize the standard geometric optics (GO). A FIO maps a geometric optical solution into an asymptotic solution of a wave problem. For radio occultations it is of primary importance that this construction can be inverted and the GO bending angle profiles can be retrieved from measurements of wave fields in multipath regions (Gorbunov 2002a,b). This also enhances the resolution beyond the Fresnel zone, which restricts the applicability of the standard geometric optical approach. Below we discuss the theory of FIOs using basic physical principles: Fermat's principle and energy conservation.

We discuss a 2D wave problem. This is a typical approximation for the atmosphere, where the vertical scale is significantly smaller than the horizontal scale. We introduce generic coordinates x, y in the occultation plane. Specific choice of the coordinate system can be different. It is only important that the x axis is the preferred direction of wave propagation. If we discuss a plane incident wave, then it is convenient to use Cartesian coordinates (x, y). For a spherical outgoing wave we can choose polar coordinates (r, θ).

A wave field $u(x, y)$ can be expanded with respect to basic wave forms. In the Cartesian coordinates it is represented as $u(x, y) = A(x, y) \exp(ik\Psi(x, y))$,

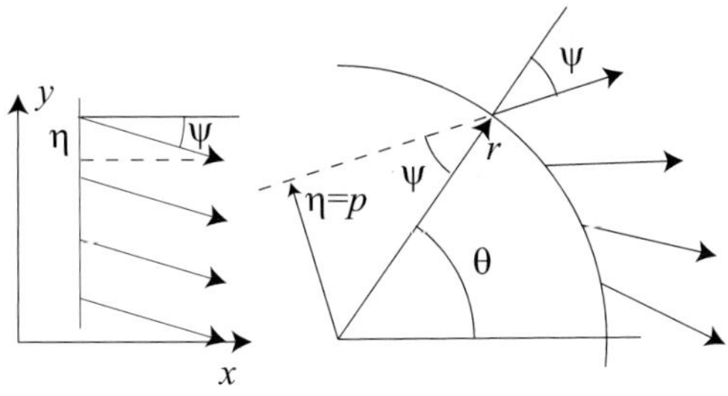

Fig. 1. Basic waveforms: (left) plane waves; (right) cylindrical harmonics.

where $A(x, y)$ is the amplitude, and $\Psi(x, y)$ is the eikonal. This form is convenient for description of the wave field in a single ray area, where the amplitude is a smooth function. For Cartesian coordinates, the most convenient choice of the basic wave forms will be plane waves (Figure 1):

$$u_\eta(x, y) = \tilde{u}(0, \eta) \exp\left[ik\left(\sqrt{1-\eta^2}x + \eta y\right)\right], \tag{1}$$

where $k = 2\pi/\lambda$ is the wavenumber, $\left(\sqrt{1-\eta^2}, \eta\right)$ is the unity ray direction vector, and $u(0, \eta)$ equals the Fourier transform of the wave field $u(0, y)$ in some source plane, $x = 0$, with respect to y:

$$\tilde{u}(0, \eta) = \sqrt{\frac{-ik}{2\pi}} \int u(0, y) \exp(-ik\eta y)\, \mathrm{d}y. \tag{2}$$

The wave vector of a plane wave equals $\mathbf{k} = (k_x, k_y) = \left(k\sqrt{1-\eta^2}, k\eta\right)$. Momentum equals $\eta = \sin\psi$. The element of optical path, $\mathrm{d}\Psi$, along the ray equals simply the element of the length, $\mathrm{d}s$ (Figure 1):

$$\mathrm{d}x = \sqrt{1-\eta^2}\mathrm{d}s, \quad \mathrm{d}y = \eta\,\mathrm{d}s, \tag{3}$$

$$\mathrm{d}\Psi = \eta\,\mathrm{d}y - H\,\mathrm{d}x = \sqrt{1-\eta^2}\,\mathrm{d}x + \eta\,\mathrm{d}y = \mathrm{d}s, \tag{4}$$

where H is the Hamilton function. Each plane wave corresponds to a family of parallel rays.

For polar coordinates r, θ, basic wave forms are cylindrical waves (Figure 1):

$$u_\eta(r, \theta) = \tilde{u}(r_0, \eta)\sqrt{\frac{r_0}{r}} \exp\left[ik\left(\sqrt{1-\frac{\eta^2}{r^2}}r + \eta\theta\right)\right], \tag{5}$$

and the momentum equals ray impact parameter, $\eta = r\sin\psi = p$.

Consider an arbitrary wave field $u(x, y) = A(x, y)\exp(ik\Psi(x, y))$. We can define ray direction, or momentum η, at every point using the momentum operator $\hat{\eta}$ as $\eta(y) = \hat{\eta}u(x, y) \approx \partial\Psi/\partial y$. This definition only works if there is a single ray, and the amplitude is a smooth function. In this case the derivative of

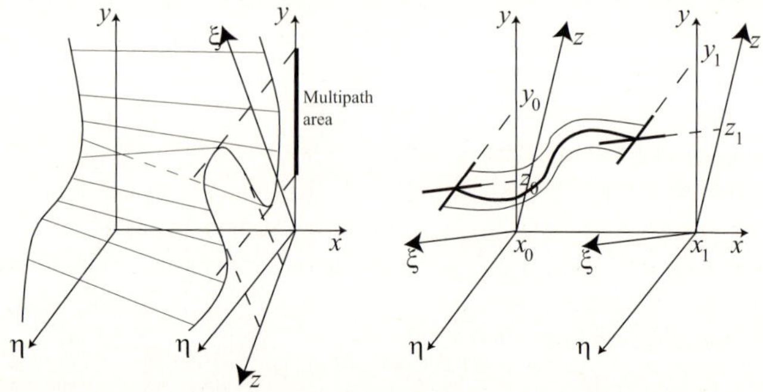

Fig. 2. (left) Schematic ray manifold in the phase space. The ray manifold evolves along the coordinate x. The type of its projection can be changed by choosing new coordinates (z, ξ) in the phase space. (right) Virtual variations of ray paths in different canonical coordinates in the phase space.

the amplitude can be neglected. Function $\eta(y)$ defines the ray manifold in the phase space with coordinates (y, η). The single-ray propagation corresponds to the situation when the ray manifold has an unique projection to the axis y. Multipath propagation means that $\eta(y)$ is a multi-valued function, and it cannot be expressed as $\partial\Psi/\partial y$.

From the geometric optical view point, the problem of multipath can be solved by another choice of coordinate and momentum than (y, η) in the phase space. We need to parameterize the phase space by a different coordinate and momentum (z, ξ) in such a way that the projection of the ray manifold to the new axis z is unique, and therefore, $\xi(z)$ is a single-valued function (Figure 2). For the new coordinates (z, ξ) we have to define the new Hamilton function and to find the corresponding transform $\hat{\Phi}$ of the wave field. The space of possible transforms is defined by two fundamental principles: 1) Fermat principle, 2) energy conservation.

2.2 Stationary Phase Principle and Canonical Transforms

The Fermat principle should hold both in the old and new coordinates (Arnold 1978; Kleinert 1995):

$$\delta\Psi = \delta \int_{(x_0,y_0)}^{(x_1,y_1)} \mathrm{d}\Psi = \delta \int_{(x_0,y_0)}^{(x_1,y_1)} \eta\,\mathrm{d}y - H\,\mathrm{d}x = 0, \tag{6}$$

$$\delta\Psi' = \delta \int_{(x_0,z_0)}^{(x_1,z_1)} \mathrm{d}\Psi' = \delta \int_{(x_0,z_0)}^{(x_1,z_1)} \xi\,\mathrm{d}z - H'\,\mathrm{d}x = 0, \tag{7}$$

where Ψ' and H' are the optical path and Hamilton function, respectively, in the new coordinates (z, ξ). Here the integrals are taken along the same physical ray parameterized by different coordinates. The ray connects points (y_0, η_0) and (y_1, η_1), or (z_0, ξ_0) and (z_1, ξ_1). The variations of ray paths have the following restrictions: $\delta y_{0,1} = 0$ and $\delta z_{0,1} = 0$ (Figure 2). The variations of momenta are not restricted (we allow arbitrary variations of the ray direction).

Because the new coordinate z is a function of (y, η), the boundary condition $\delta z = 0$ may imply some variation $\delta y \neq 0$, and vice versa. To establish the relation between $\mathrm{d}\Psi$ and $\mathrm{d}\Psi'$ we consider arbitrary variations of the optical paths form $\delta\Psi$ and $\delta\Psi'$ in the vicinity of a stationary path, not restricted with conditions $\delta y_{0,1} = 0$ and $\delta z_{0,1} = 0$:

$$\delta\Psi = \eta\, \delta y|_{x_0}^{x_1}, \quad \delta\Psi' = \xi\, \delta z|_{x_0}^{x_1}. \tag{8}$$

We require that $\mathrm{d}\Psi' - \mathrm{d}\Psi = \xi\, \mathrm{d}z - \eta\, \mathrm{d}y - (H' - H)\, \mathrm{d}x$ should be equal to a full differential $\mathrm{d}S$ (Arnold 1978; Kleinert 1995). For arbitrary variations of an arbitrary trajectory in the phase space we have then the following relation:

$$\delta\Psi' - \delta\Psi = \delta \int_{x_0}^{x_1} \mathrm{d}S = \delta S(y, z)|_{x_0}^{x_1} = \xi\, \delta z - \eta\, \delta y|_{x_0}^{x_1}. \tag{9}$$

Therefore, if for some path $\delta\Psi = 0$ with the boundary condition $\delta y_{0,1} = 0$, then for the same path $\delta\Psi' = 0$ with boundary condition $\delta z_{0,1} = 0$. If we consider a cross section of the phase space frozen at some fixed x, then we can write the reduced equation:

$$\mathrm{d}S = \xi\, \mathrm{d}z - \eta\, \mathrm{d}y, \quad \frac{\partial S}{\partial z} = \xi, \quad \frac{\partial S}{\partial y} = -\eta \tag{10}$$

A transform from (y, η) to (z, ξ) such that $\xi\, \mathrm{d}z - \eta\, \mathrm{d}y$ equals a full differential $\mathrm{d}S$ is termed *canonical*, $S(z, y)$ being its *generating function* (Arnold 1978).

2.3 Fourier Integral Operator of 2nd Type

Consider now a complex integral transform on the wave field $u(y)$:

$$v(z) = \hat{\Phi}_2 u(z) = \sqrt{\frac{-ik}{2\pi}} \int a_2(z, y) \exp\left(ikS_2(z, y)\right) u(y)\mathrm{d}y. \tag{11}$$

We will refer to this operator as a FIO of the 2nd type (Gorbunov and Lauritsen 2002, 2004a). The transformed wave field $v(z)$ can be written in the form $A'(z) \exp\left(ik\Psi'(z)\right)$. Our aim is to find a transform of the wave field that implements a canonical transform, i.e. given the momentum $\partial\Psi(y)/\partial y = \eta$ of the wave field $u(y)$, the momentum $\partial\Psi'(z)/\partial z$ of the transformed wave field $v(z)$ should be equal to ξ.

2.4 Energy Conservation

Another important requirement is that this transform conserves the energy of the wave field (Egorov 1985; Egorov et al. 1999): $\int v\, v^* \mathrm{d}z = \int u\, u^* \mathrm{d}y$. From this it follows that the conjugated operator $\hat{\Phi}_2^*$ should be equal to the inverse operator $\hat{\Phi}_2^{-1}$, because by definition

$$\int \hat{\Phi}_2 u \left(\hat{\Phi}_2 u\right)^* \mathrm{d}z = \int u \left(\hat{\Phi}_2^* \hat{\Phi}_2 u\right)^* \mathrm{d}z, \tag{12}$$

$$\hat{\Phi}_2^* v(y) = \hat{\Phi}_2^{-1} v(z) = \sqrt{\frac{ik}{2\pi}} \int a_2(z,y) \exp\left(-ikS_2(z,y)\right) v(z) \mathrm{d}z. \tag{13}$$

Substituting $u(y) = \delta(y - y_0)$ and considering $\hat{\Phi} u(z)$ we have $\Psi'(z) = S_2(z, y_0)$. From here it follows that $\partial S_2/\partial z = \xi$. Substituting $v(z) = \delta(z - z_0)$ and considering $\hat{\Phi}^{-1} v(z)$ we have $\Psi(y) = -S_2(z_0, y)$. From here it follows that $\partial S_2/\partial y = -\eta$. Thus we see that $S_2(z, y)$ equals the generating function of the canonical transform $S(z, y)$. The derivation of the amplitude function a_2 is discussed in Gorbunov et al. (2004).

The simplest form of a Fourier Integral Operator is the Fourier transform:

$$S_2(z,y) = -zy; \quad \xi = \frac{\partial S_2}{\partial z} = -y; \quad -\eta = \frac{\partial S_2}{\partial y} = -z, \tag{14}$$

which implements a $\pi/2$ rotation of the phase plane $(y, \eta) \to (z = \eta, \xi = -y)$.

2.5 Fourier Integral Operator of 1st Type

Another, 1st, type of FIO can be expressed as a composition of two FIOs of the 2nd type, one of which is the Fourier transform (Egorov 1985):

$$v(z) = \hat{\Phi}_1 u(z) = \sqrt{\frac{ik}{2\pi}} \int a_1(z,\eta) \exp\left(ikS_1(z,\eta)\right) \tilde{u}(\eta) \mathrm{d}\eta, \tag{15}$$

where the equation for the generating function S_2 is similar to that for S_1:

$$\mathrm{d}S_1 = \xi\, \mathrm{d}z + y\, \mathrm{d}\eta, \quad \frac{\partial S_1}{\partial z} = \xi, \quad \frac{\partial S_1}{\partial \eta} = y. \tag{16}$$

3 Processing Radio Occultations

3.1 Phase Function: Accurate and Approximate Solutions

The application of the technique of FIOs for processing radio occultation data uses the fact that impact parameters uniquely characterize rays in a spherically symmetric atmosphere (Gorbunov 2002a,b; Gorbunov and Lauritsen 2002; Jensen et al. 2002; Gorbunov 2003; Gorbunov and Kornblueh

2003; Jensen et al. 2003, 2004; Gorbunov et al. 2004; Gorbunov and Lauritsen 2004a). Equations (10,16) can be directly applied for the derivation of the phase functions (Gorbunov and Lauritsen 2004a):

$$S_2(p, y) = -\int \eta(p, y)\mathrm{d}y, \quad S_1(p, \eta) = -\int y(p, \eta)\mathrm{d}\eta. \tag{17}$$

Consider the expression for the derivative of the optical path of a radio occultation signal:

$$\dot{\Psi} = \eta(p, t) \equiv \dot{\theta}p + \frac{\dot{r}_\mathrm{L}}{r_\mathrm{L}}\sqrt{r_\mathrm{L}^2 - p^2} + \frac{\dot{r}_\mathrm{G}}{r_\mathrm{G}}\sqrt{r_\mathrm{G}^2 - p^2}, \tag{18}$$

where r_G and r_L are the distances from the Earth's center to the GPS and LEO satellite, respectively, and θ is the angle between the GPS and LEO radius vectors. This allows for the derivation of the exact phase function (Jensen et al. 2004):

$$\begin{aligned} S_2(p, t) &= -\int \left(p\mathrm{d}\theta + \frac{\mathrm{d}r_\mathrm{G}}{r_\mathrm{G}}\sqrt{r_\mathrm{G}^2 - p^2} + \frac{\mathrm{d}r_\mathrm{L}}{r_\mathrm{L}}\sqrt{r_\mathrm{L}^2 - p^2}\right) = \\ &= -p\theta - \sqrt{r_\mathrm{G}^2 - p^2} + p\arccos\frac{p}{r_\mathrm{G}} - \sqrt{r_\mathrm{L}^2 - p^2} + p\arccos\frac{p}{r_\mathrm{L}}. \end{aligned} \tag{19}$$

The corresponding momentum equals minus refraction angle:

$$\xi(p) = \frac{\partial\Psi'(p)}{\partial p} = \frac{\partial S_2(p, t = t_s(p))}{\partial p} = -\theta + \arccos\frac{p}{r_\mathrm{G}} + \arccos\frac{p}{r_\mathrm{L}} = -\epsilon(p), \tag{20}$$

where $t_s(p)$ is the moment of time, when the ray with impact parameter p was observed. The exact solution for the corresponding amplitude function a_2 was obtained in Gorbunov et al. (2004).

The precise phase function is inconvenient for numerical implementation, because the corresponding operator cannot be reduced to the Fourier transform. To reduce this operator to the Fourier transform, it is necessary to use some approximation. The simplest approximation is that implemented in the Full-Spectrum Inversion method (Jensen et al. 2003):

$$S_2(p, \theta) = -p\theta - \int \left(\frac{\dot{r}_\mathrm{G}}{r_\mathrm{G}}\sqrt{r_\mathrm{G}^2 - p_\mathrm{m}^2} + \frac{\dot{r}_\mathrm{L}}{r_\mathrm{L}}\sqrt{r_\mathrm{L}^2 - p_\mathrm{m}^2}\right) d\theta \equiv -p\theta + \int F(\theta)d\theta, \tag{21}$$

where $p_\mathrm{m} = p_\mathrm{m}(\theta)$ is the a priori model of impact parameter variation. The corresponding momentum $\xi(p)$ equals minus satellite-to-satellite angle, $\partial S_2/\partial p = -\theta$.

More accurate approximation is based on the linearization of the canonical transform from (t, η) to (p, ξ) in the vicinity of the model $[p_0(t), \eta_0(t)]$ (Gorbunov and Lauritsen 2004a,b):

$$\tilde{p}(t,\eta) = p_0 + \frac{\partial p_0}{\partial \eta}(\eta - \eta_0) = f(t) + \frac{\partial p_0}{\partial \eta}\eta, \tag{22}$$

$$f(t) = p_0 - \left(\dot{\theta} - \frac{\dot{r}_G}{r_G}\frac{p_0}{\sqrt{r_G^2 - p_0^2}} - \frac{\dot{r}_L}{r_L}\frac{p_0}{\sqrt{r_L^2 - p_0^2}}\right)^{-1}\eta_0. \tag{23}$$

For the generation of the model, we use geometric optical modeling to obtain multivalued functions $[p(t), \eta(t)]$ corresponding to multipath propagation. From them we obtain nearest smoothed single-valued ray structure model $[p_0(t),\ \eta_0(t)]$. Instead of (t, η) we introduce scaled coordinate and frequency (Υ, σ):

$$\mathrm{d}\Upsilon = \left(\frac{\partial p_0}{\partial \eta}\right)^{-1}\mathrm{d}t = \mathrm{d}\theta - \frac{\mathrm{d}r_G}{r_G}\frac{p_0}{\sqrt{r_G^2 - p_0^2}} - \frac{\mathrm{d}r_L}{r_L}\frac{p_0}{\sqrt{r_L^2 - p_0^2}}, \tag{24}$$

$$\sigma = \frac{\partial p_0}{\partial \eta}\eta, \tag{25}$$

where we expressed $\partial p_0/\partial\eta$ using (18). This scaling allows for the derivation of the linear canonical transform and its generating function:

$$\tilde{p} = f(\Upsilon) + \sigma, \quad \xi = -\Upsilon, \tag{26}$$

$$S_2(\tilde{p}, \Upsilon) = -\int \sigma(\tilde{p}, \Upsilon)\mathrm{d}\Upsilon = -\tilde{p}\Upsilon + \int f(\Upsilon)\,\mathrm{d}\Upsilon. \tag{27}$$

The FIO with this phase function is very similar to FSI, but it uses more precise approximation, and the definition of scaled coordinate Υ instead of θ takes into account the deviation of the trajectory from a circle. The derivation of the amplitude function for this approximation is discussed in Gorbunov et al. (2004).

Figure 3 shows the profiles of temperature, humidity, real refractivity and specific absorption for GPS frequencies from a high-resolution radiosonde data. This profile was used for modeling a spherically symmetrical atmosphere. A radio occultation experiment was simulated using multiple phase screen technique, and the simulated data were processed by CT method based on the FIO with the phase function (27). The simulation was performed for GPS frequencies. The results of processing the simulated data are shown in Figure 3. The complicated profile of bending angle is retrieved with a good accuracy.

4 Forward Modeling

4.1 Wave Propagation in Atmosphere

Fourier integral operators can also be used for asymptotic direct modeling (Gorbunov 2003; Gorbunov and Lauritsen 2004a). The FIOs $\hat{\Phi}_{1,2}$ can be easily inverted: $\hat{\Phi}_{1,2}^{-1} = \hat{\Phi}_{1,2}^{*}$. If we use the representation of approximate impact

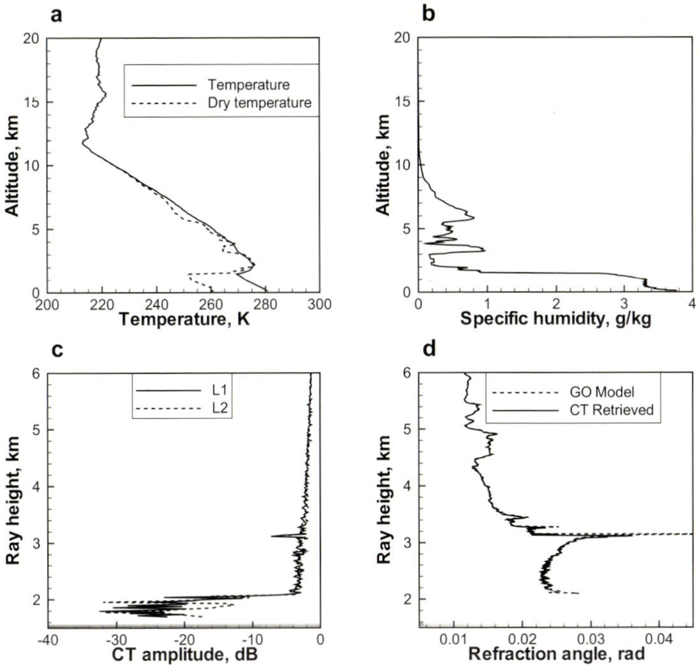

Fig. 3. Simulated occultation event for a high resolution radiosonde data: (a) temperature, T, and dry temperature, T_{dry}, (b) specific humidity, q, (c) CT amplitudes for the two channels, and (d) refraction angles, computed by the GO model and retrieved by the CT method.

parameter $\tilde{p}$, then the direct model is especially efficient. Given a 3D atmospheric model, we first perform geometric optical modeling, and iteratively find the trajectory point $\Upsilon_s(\tilde{p})$, where the ray with the impact parameter $p(\tilde{p})$ is observed. The wave function in the $\tilde{p}$-representation is then equal to $w(\tilde{p}) = A'(\tilde{p}) \exp\left(-ik \int \Upsilon_s(\tilde{p}) \mathrm{d}\tilde{p}\right)$, where the amplitude $A'(\tilde{p})$ equals a normalizing constant in the light zone and 0 in the geometric optical shadow. This function is then mapped into the Υ-representation by the inverse FIO (Gorbunov and Lauritsen 2004a):

$$u(\Upsilon) = \sqrt{\frac{ik}{2\pi}} \exp\left(-ik \int_0^{\Upsilon} f(\Upsilon')\, \mathrm{d}\Upsilon'\right) \int \exp(ik\tilde{p}\Upsilon)\, a_2(\tilde{p}, \Upsilon_s(\tilde{p}))\, w(\tilde{p})\, \mathrm{d}\tilde{p}, \tag{28}$$

For modeling atmospheric absorption, the amplitude $A'(\tilde{p})$ must also be multiplied by a factor of $\exp\left(-k \int n'' \mathrm{d}s\right)$, where n'' is the imaginary part of refractive index, and the integral is taken along the ray with the impact parameter $p(\tilde{p})$. A similar direct modeling algorithm can be constructed by inverting the operator used in the FSI method.

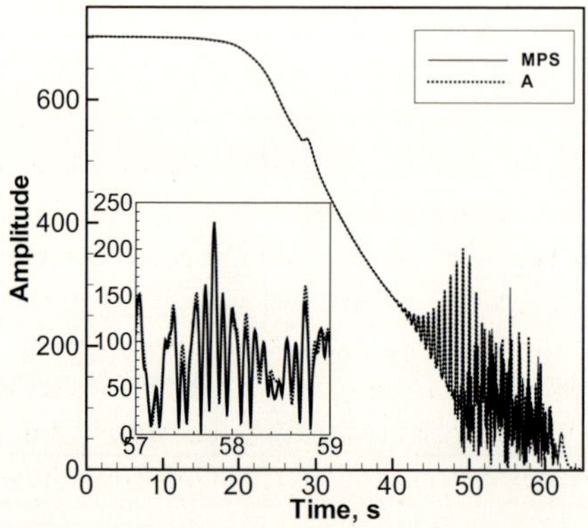

Fig. 4. Validation of asymptotic direct modeling. Amplitude of simulated radio occultation signal as function of time: (1) MPS simulation (solid line) and (2) asymptotic simulation based on the FIO2 (A, dotted line).

For the validation of the asymptotic direct modeling we performed numerical simulations with a simple spherically-symmetrical phantom (refractive index field model). The phantom represents an exponential model with a quasi-periodical perturbation:

$$n(z) = 1 + N_0 \exp\left(-\frac{z}{H}\right)\left[1 + \alpha \cos\left(\frac{2\pi z}{h}\right)\exp\left(-\frac{z^2}{L^2}\right)\right], \tag{29}$$

where z is the height above the Earth's surface, $N_0 = 300 \times 10^{-6}$ is the characteristic refractivity at the Earth's surface (300 N-units), $H = 7.5$ km is the characteristic vertical scale of refractivity field, $\alpha = 0.003$ is the relative magnitude of the perturbation, $h = 0.3$ km is the period of the perturbation, $L = 3.0$ km is the characteristic height of the perturbation area. This phantom was smoothly combined with the MSIS climatological model above 20 km. We simulated radio occultation signals using multiple phase screens (MPS) and the asymptotic solution (A) for the frequency 9.7 GHz, which is intended to be used in LEO-LEO occultations. The results of the comparison of the amplitude of the simulated wave field for these two modeling techniques are presented in Figure 4. The peculiarity of the amplitude around 28.5 s is due to the transfer from MSIS to the test phantom. Between 40 s and 47.5 s the amplitude indicates large-scale oscillations reproducing the oscillations of the refractivity profile. In this area there is no multipath propagation. After 47.5 s we notice increasing small-scale scintillations due to emergence of multipath propagation. The occultation fragment from 57 s to 59 s with strong multipath scintillations is enlarged and shown separately. Figure 4 illustrates a good agreement of both these simulation techniques.

4.2 Wave Propagation in Vacuum

Asymptotic forward modeling has the following applicability limitation: it cannot be used for modeling effects of diffraction inside the atmosphere. Accurate account of diffraction on small-scale atmospheric structures is necessary e.g., for modeling wave propagation in a turbulent atmosphere. In this case it is necessary to apply multiple phase screen technique. Previously, the step from the last phase screen to the LEO orbit is performed by the computation of multiple diffractive integrals. The computation of diffractive integrals is not only very ineffective numerically, it is also a source of computational inaccuracies. However, using the technique of FIOs it is possible to construct an asymptotic solution of wave propagation in a vacuum from a straight phase screen to an arbitrary observation curve that can be reduced to the Fourier transform.

Consider the wave field $u_0(y)$ in the phase screen plane, and the observation curve $X(t), Y(t)$ (Figure 5). The accurate solution is obtained from the plane wave expansion of the source field (1):

$$u(t) = \sqrt{\frac{ik}{2\pi}} \int \exp\left(ikX(t)\sqrt{1-\eta^2} + ikY(t)\eta\right) \tilde{u}_0(\eta)\mathrm{d}\eta. \tag{30}$$

This equation is another form of the diffraction integral of Fresnel, Kirchhoff and Helmholtz, represented as a FIO of the first type with the following phase function:

$$S_1(t,\eta) = X(t)\sqrt{1-\eta^2} + Y(t)\eta. \tag{31}$$

This operator transforms the wave field from the representation (y, η) in the source plane to the representation (t, σ) on the observation curve. We can write the standard differential equation for the phase function:

$$\mathrm{d}S_1 = \sigma \mathrm{d}t + y\mathrm{d}\eta, \qquad \frac{\partial^2 S_1}{\partial t\, \partial \eta} = \frac{\partial \sigma}{\partial \eta} = \frac{\partial y}{\partial t} \tag{32}$$

which corresponds to the following equations describing straight rays:

$$\sigma = \frac{\partial S_1}{\partial t} = \dot{X}(t)\sqrt{1-\eta^2} + \dot{Y}(t)\eta, \tag{33}$$

$$y = \frac{\partial S_1}{\partial \eta} = Y(t) - X(t)\frac{\eta}{\sqrt{1-\eta^2}}. \tag{34}$$

Operator (30) provides an accurate solution. However, it cannot be reduced to the Fourier transform in general case. For a vertical observation trajectory, $X(t) = \text{const}$, we can parameterize the observation trajectory with coordinate Y, and the operator will turn into a Fourier transform.

For a general case, we will construct an approximation based on the linearization of the canonical transform $(y, \eta) \to (t, \sigma)$. For this we introduce smooth model of the ray structure $[t_0(\eta), \sigma_0(\eta), y_0(\eta), \eta_0(t)]$. Then the canonical transform can be linearized:

$$t = t_0 + \frac{\partial t_0}{\partial y}(y - y_0) = t_0 - \frac{\partial t_0}{\partial y} y_0 + \frac{\partial t_0}{\partial y} y = h(\eta) + z, \tag{35}$$

$$\sigma = \sigma_0 + \frac{\partial \sigma_0}{\partial \eta}(\eta - \eta_0) = \sigma_0 - \xi_0 + \xi = g(t) + \xi, \tag{36}$$

where $g(t) = \sigma_0(t) - \xi_0(t)$ and $h(\eta) = t_0(\eta) - \frac{\partial t_0}{\partial y} y_0(\eta)$. Instead of (y, η), we introduced scaled coordinate and momentum using (32):

$$z = \frac{\partial t_0}{\partial y} y \quad \mathrm{d}\xi = \left(\frac{\partial t_0}{\partial y}\right)^{-1} \mathrm{d}\eta. \tag{37}$$

This allows for the derivation of the phase function:

$$\mathrm{d}S = \sigma \mathrm{d}t + z\mathrm{d}\xi = (g(t) + \xi)\mathrm{d}t + (t - h(\xi))\mathrm{d}\xi, \tag{38}$$

$$S(t, \xi) = \int g(t)\mathrm{d}t - \int h(\xi)\mathrm{d}\xi + t\xi. \tag{39}$$

The approximate FIO is represented as a composition of multiplication with the first reference signal $a(t_0(\xi), \xi) \exp\left(-ik \int f(\xi)\mathrm{d}\xi\right)$, Fourier transform, and multiplication with the second reference signal $\exp\left(ik \int g(t)\mathrm{d}t\right)$:

$$u(t) = \sqrt{\frac{ik}{2\pi}} \exp\left(ik \int g(t)\mathrm{d}t\right) \times$$
$$\times \int a(t_0(\xi), \xi) \exp(ikt\xi) \exp\left(-ik \int h(\xi)\mathrm{d}\xi\right) \tilde{u}_0(\eta(\xi))\mathrm{d}\xi. \tag{40}$$

We used the same radiosonde profile as above for validation of the algorithm of wave propagation based on FIO (40). Figure 5 shows a good agreement of the amplitude of the wave field computed by the FIO and that computed by the standard algorithm based on Fresnel integrals. The occultation data obtained by Fresnel integral break at 62 s, while the data obtained by the FIO still continue. This is linked with the difficulty of the identification of the stationary point of the Fresnel integral near the border of shadow zone, where the signal becomes weak. In this example, we observe a sharp spike of refraction angle (Figure 3d) producing a very weak signal, which is difficult to compute by Fresnel integral. However, such signals are not a problem for the FIO technique.

5 Conclusions

FIOs provide a generalization of geometric optical canonical formalism for wave optics. This allows for using FIOs for solving problem of disentangling multipath by finding an unique projection of ray manifold. Another, equivalent

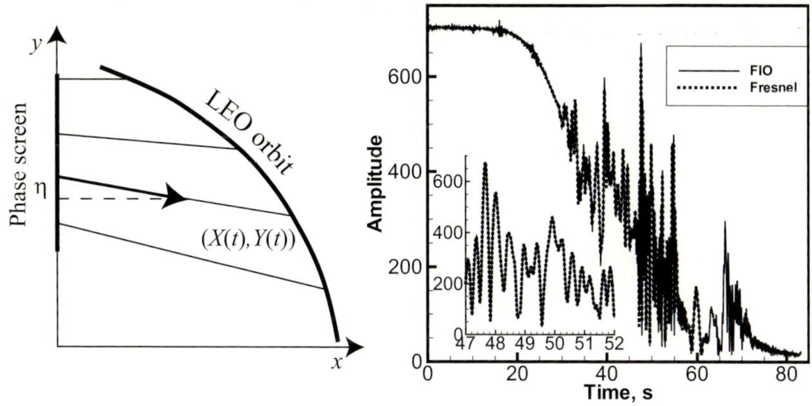

Fig. 5. (left) Geometry of wave propagation from phase screen to LEO orbit. (right) Validation of FIO-based algorithm of wave propagation from phase screen to LEO orbit. Amplitude of simulated radio occultation signal as function of time: (1) FIO algorithm (solid line) and (2) Fresnel integrals (dotted line).

view of FIOs is based on signal processing approach and frequency matching principle: at the stationary point of the oscillating integral the frequency of the signal is matched by the frequency of the oscillating kernel. This principle allows for sorting signal components with different instantaneous frequencies.

The practical importance of FIOs for inversion and forward modeling of radio occultation data cannot be underestimated: 1) FIOs provide high accuracy and vertical resolution in the retrieval of refraction angles. 2) FIOs allow for the retrieval of transmission due to atmospheric absorption. 3) Using reasonable approximations, it is possible to reduce the FIOs to a composition of multiplication with a reference signal and Fourier transforms. This is important for achieving high numerical efficiency of inversion algorithms. 4) FIOs can be very effectively used in the forward modeling of radio occultation signals.

Acknowledgements. The Authors are grateful to A. S. Gurvich, S. V. Sokolovskiy, V. I. Klyatskin, I. G. Yakushkin (Institute for Atmospheric Physics, Russian Academy of Sciences, Moscow), A. S. Jensen (Danish Meteorological Institute, Copenhagen), G. Kirchengast (Institute for Geophysics, Astrophysics, and Meteorology, University of Graz, Austria) for useful scientific discussions.

References

Arnold VI (1978) Mathematical Methods of Classical Mechanics. Springer-Verlag, New York

Egorov YV (1985) Lectures on Partial Differential Equations. Additional Chapters. (In Russian) Moscow State University Press, Moscow

Egorov YV, Komech AI, Shubin MA (1999) Elements of the Modern Theory of Partial Differential Equations. Springer-Verlag, Berlin

Gorbunov ME (2002a) Canonical transform method for processing GPS radio occultation data in lower troposphere. Radio Sci 37, 9–1–9–10, doi:10.1029/2000RS002592

Gorbunov ME (2002b) Radio-holographic analysis of Microlab-1 radio occultation data in the lower troposphere. J Geophys Res – Atm 107, 7–1–7–10, doi:10.1029/2001JD000889

Gorbunov ME (2003) An asymptotic method of modeling radio occultations. J Atm Sol-Terrestr Phys 65:1361–1367

Gorbunov ME, Kornblueh L (2003) Analysis and validation of Challenging Minisatellite Payload (CHAMP) radio occultation data. J Geophys Res 108, 4584, doi:10.1029/2002JD003175

Gorbunov ME, Lauritsen KB (2002) Canonical transform methods for radio occultation data. Scientific Report 02-10, Danish Meteorological Institute, Copenhagen, Denmark, http://www.dmi.dk/dmi/Sr02-10.pdf

Gorbunov ME, Lauritsen KB (2004a) Analysis of wave fields by Fourier Integral Operators and its application for radio occultations. Radio Sci 39, RS4010, doi:10.1029/2003RS002971

Gorbunov ME, Lauritsen KB (2004b) Canonical transform methods for radio occultation data. In Kirchengast G, Foelsche U, Steiner AK (eds) Occultations for Probing Atmosphere and Climate. Springer, Berlin, pp 61–68

Gorbunov ME, Benzon H-H, Jensen AS, Lohmann MS, Nielsen AS (2004) Comparative analysis of radio occultation processing approaches based on Fourier integral operators. Radio Sci accepted

Jensen AS, Benzon H-H, Lohmann MS (2002) A new high resolution method for processing radio occultation data. Scientific Report 02-06, Danish Meteorological Institute, Copenhagen, Denmark

Jensen AS, Lohmann MS, Benzon H-H, Nielsen AS (2003) Full spectrum inversion of radio occultation signals. Radio Sci 38, 6–1–6–15, doi:10.1029/2002RS002763

Jensen AS, Lohmann MS, Nielsen AS, Benzon H-H (2004) Geometrical optics phase matching of radio occultation signals. Radio Sci 39, RS3009, doi:10.1029/2003RS002899

Kleinert H (1995) Path Integrals in Quantum Mechanics, Statistics und Polymer Physics, World Scientific, Singapore

Mishchenko AS, Shatalov VE, Sternin BY (1990) Lagrangian manifolds and the Maslov operator, Springer-Verlag, Berlin

Processing Radio Occultation Data by Full Spectrum Inversion Techniques: An Overview and Recent Developments

A. S. Jensen[1], H.-H. Benzon[1], M. S. Lohmann[2], and A. S. Nielsen[1]

[1] Danish Meteorological Institute, Research and Development Department, Lyngbyvej 100, 2100 Copenhagen, Denmark
arne_skov_jensen@hotmail.com
[2] University Corporation for Atmosphere Research, Box 3000, Boulder, CO 80307, USA

Abstract. The *Full Spectrum Inversion* (FSI) method was developed in the beginning of 2002 in an effort to solve the multipath problem in radio occultation measurements. The physical ideas, which lead to the method, were that the occultation path could be considered as a synthetic aperture and the radio occultation Doppler frequency in a single path was a monotonic function of time. In star occultations, at optical wavelengths, the multipath problem is easily solved by having a lens in front of an array detector separating the beams in space. The lens is performing a spatial Fourier transform i.e., a plane wave is focused into a point displaced from the optical axis an amount given by the direction (the spatial frequency) of the plane wave. The analogy to this space processing method in time, is to have a "time lens", which can separate multiple temporal frequencies occurring at the same time: Obviously this is what a temporal Fourier transform does. These ideas were implemented in 2001 and tested successfully on simulations of radio occultation signals, which had circular satellite orbits. However, for non-circular orbits the plain Fourier method turned out to give a not fully correct result and the work on the FSI emerged realizing that some preprocessing steps were necessary in order to eliminate the impact of non-radial orbits. This involves pre-calculation of phases, which, multiplied on the occultation signal, reduces the impact of the non-circular orbits on the resulting Fourier transform of the preprocessed signal. In 2003 the phasematching method was developed, where the impact of the non-circular orbits was totally solved, but with the cost that the processing could not be implemented with a fast Fourier transform. Both the FSI and the phasematching methods will be discussed in detail in this paper. The present development on the FSI method includes its practical implementation and making the method robust for mass processing of radio occultation signals. Filtering of signals in the FSI method is important and different filtering methods will be discussed in this paper.

1 Introduction

A short overview of the basic physical and mathematical tools needed for a description of radio occultation signals is given in the following sections. Most of the formulas and theoretical descriptions can be found in Jensen et al. (2003) and Jensen et al. (2004b), though the presentation here hopefully represents a more pedagogical view. In Sect. 2 the general phasematching scheme is described. In Sect. 3.1 an implementation of the FSI method is shown and discussed. The presented implementation of the FSI is the one used by the authors of this paper; other implementation schemes are surely possible. This is an ongoing development and research topic. In Sect. 3.2 a new method for filtering and a conversion of the time of arrival uncertainties to Doppler frequency uncertainties is proposed. This part is important and gives estimates of the spatial resolution of the refractivity or the bending angle for an actual occultation signal. In Sect. 3.3 the effect of thermal noise on the arrival times are calculated. These results are not used here but just noted for eventual future use. The idea of using both the amplitude and phase of the received radio occultation signal in a radio holographic approach has also been investigated by other authors in the papers Pavelyev (1998); Hocke et al. (1999); Beyerle and Hocke (2001); Igarashi et al. (2000, 2001); Beyerle et al. (2002); Pavelyev et al. (2002).

1.1 The Radio Occultation Signal

The received radio signal at the LEO satellite is a narrow banded signal with a carrier frequency determined by the transmitting GPS satellite. The electromagnetic field is formed by the refractive structure of the atmosphere. The GPS and the LEO satellite are moving during the occultation (see Fig. 1). The path of the LEO satellite can be thought of as a synthetic aperture. In principle, the resulting field should be computed from Maxwell equations using an actual refractivity structure. An analytical solution has not been found and it can probably not be done. Instead, a geometrical optical description is applied in order to analyze possible signal processing methods.

1.2 Geometrical Optical Description

The geometrical optical description implies a ray description instead of a field description with the loss of diffraction effects. A light ray is a path, where the center of the electromagnetic energy flows and with a direction equal to the normal of the electromagnetic fields wave front. As a special case, the analytical expression for the phase, amplitude and Doppler frequency can be found, if it can be assumed that the refractivity structure of the atmosphere is spherical symmetric. By applying the formula of Bouguer (Snell's law for a spherical stratified media) a simple analytical description can be obtained. The formula of Bouguer can be generalized

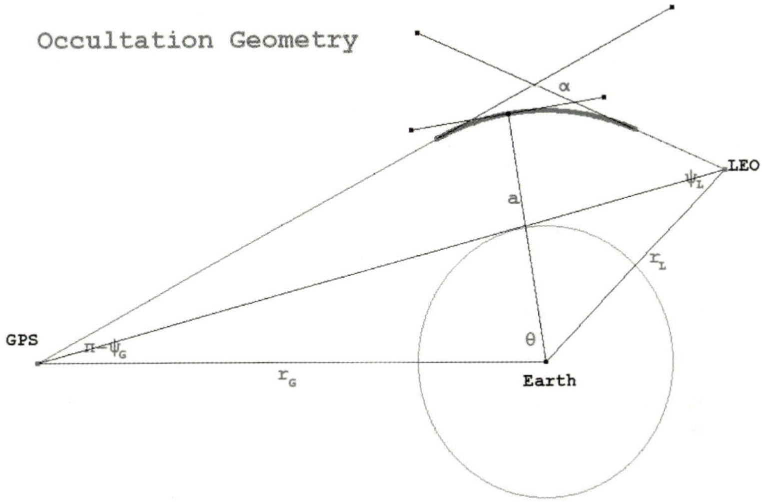

Fig. 1. The drawing shows the radio occultation geometry. Here α is the bending angle and r_G, r_L are the distances to the GPS and LEO satellite, respectively. Θ and a are the angle between the satellites and the impact parameter. Ψ_L is the angle between the ray and the position vector to the GPS satellite while Ψ_R is the angle between the ray and the position vector to the LEO satellite.

to cover a non-spherical media, but in this case the inversion cannot be done without pre-knowledge of the structure of the refractivity.

The validity of the optical geometrical description of radio occultation signals in the Earth's atmosphere is due to the weak refractivity and the small (in most cases) refractivity gradients. However, in the case of a strong refractivity gradient it must be expected that the geometrical description will break down. The theory needed for the geometrical optical description can be found in Fjeldbo (1964) and Born and Wolf (1999). A geometric optics description of the refraction attenuation for radio occultations valid for the 3-D case has been introduced in the paper Pavelyev and Kucherjavenkov (1978).

Depending on the structure of the refractivity, several rays can be present at the same time. In this multipath situation, the LEO satellite receives a signal, which is a sum of sub signals with different Doppler frequencies. This complicates the signal processing.

1.3 The Phase and the Doppler Frequency

The phase of a single ray is given by

$$\psi = k\int_{s_G}^{s_L} n\mathrm{d}s = k\left[\sqrt{r_L^2 - a^2} + \sqrt{r_G^2 - a^2} + a\alpha + \int_a^{\infty}\alpha(x)\mathrm{d}x\right], \tag{1}$$

where a = impact parameter, α = bending angle and r_L, r_G the LEO and the GPS radii.

It is seen that the phase has a memory term, the integral of the bending angle. This means that the phase of the occultation signal at all heights always includes an ionosphere contribution.

The Doppler frequency for a single ray is given by

$$\omega(t) = \dot{\psi} = k\left(\frac{\dot{r}_G}{r_G}\sqrt{r_G^2 - a^2} + \frac{\dot{r}_L}{r_L}\sqrt{r_L^2 - a^2} + a\dot{\theta}\right), \tag{2}$$

where the dots denote time derivates.

2 Processing of Occultation Signals

Optimal signal processing means utilization of all available pre-knowledge of the physical nature of the signal. A radio occultation signal has the following characteristics:

- The signal is locally narrow banded i.e., by subtracting the vacuum phase the frequency bandwidth is down to 10 Hz.
- By using the synthetic aperture the frequency resolution can be optimized.
- The instantaneous Doppler frequency is a decreasing/increasing function of time. In the case of multipath, several frequencies can be present at the same time.

If the signal is processed by an array of frequency filters and these give a signal when a frequency is present, the recording of these temporal events will display the Doppler frequency as function of time. This means that a Fourier transform of the signal will resolve multipath and give an optimal resolution.

The Fourier transform can be considered as a phasematching, which selects frequencies in the signal. The selection of the frequencies can be resolved by using the *stationary phase method.* If there are multiple stationary points i.e., the same frequency is present at several time points we have a situation, which could be called temporal multipath. The temporal multipath can be caused by the non-radial path of the satellites and by horizontal gradients. Especially, the plain Fourier transform would give errors in the Doppler frequency curve when the satellite paths are non-radial.

In order to remove the temporal multipath a more general phasematching than the Fourier transform method can be found.

2.1 Phasematching

The occultation signal $A(t)e^{i\psi(t)}$ is multiplied with the phasor, $e^{-i\psi_0(c,t)}$, where $\psi_0(c,t)$ is a phase depending on the parameter c. The result is then integrated in time (or another variable, which is a mapping of the time space):

$$
\begin{aligned}
u(c) &= \int_0^T A(t)\, e^{i(\psi(t)-\psi_0(c,t))} dt \\
&\cong \sqrt{\frac{2\pi i}{\ddot{\psi}(t_1)-\ddot{\psi}_0(c,t_1)}} A(t_1)\, e^{i[\psi(t_1)-\psi_0(c,t_1)]}\Bigg|_{\dot{\psi}(t_1)=\dot{\psi}_0(c,t_1)} .
\end{aligned}
\tag{3}
$$

The condition $\dot{\psi}(t_1)=\dot{\psi}_0(c,t_1)$ denotes a stationary point in time. If we, with the choice of $\psi_0(c,t)$ have accomplished that there only exists one stationary point at any time, the function $u(c)$ can explicit be represented by the approximation above.

The condition for absence of temporal multipath is

$$
\ddot{\psi}(t)-\ddot{\psi}_0(c,t) \neq 0 \quad \text{for} \quad 0<t<T. \tag{4}
$$

In the design of the phase $\psi_0(c,t)$ this condition should be fulfilled. However, this depends on the measured parameter and is therefore not possible in all cases. The search for the right matching phase has resulted in the development of the *Fourier*, the *FSI* and the *Phasematching* methods.

2.2 The Generic Methods of Phasematching

The phase of $u(c)$, $\psi(t_1)-\psi_0(c,t_1)$, is a function of the parameter c, and by differentiating the phase with respect to c, an auxiliary function on a parametrical form can be found:

$$
\left[c, \frac{\partial(\psi-\psi_o)}{\partial c}\right]_{\dot{\psi}(t_1)=\dot{\psi}_0(c,t_1)}, \tag{5}
$$

which expresses a functional relation between the parameter c and the partial derivative of the matching phase.

I: *The Fourier method* $\psi_0(c,t) = ct$

$$
\left[c, \frac{\partial(\psi-\psi_o)}{\partial c}\right]_{\dot{\psi}(t_1)=\dot{\psi}_0(c,t_1)} = [c, -t_1], \tag{6}
$$

where c is related to the impact parameter through the Doppler equation:

$$
c = k\left(a\dot{\theta} + \frac{\dot{r}_L}{r_L}\sqrt{r_L^2-a^2} + \frac{\dot{r}_G}{r_G}\sqrt{r_G^2-a^2}\right). \tag{7}
$$

In the case where the radial velocities are zero, the impact parameter is directly proportional to Doppler frequency c.

II. *The FSI method*

$$\left[c, \frac{\partial(\psi - \psi_o)}{\partial c}\right]_{\dot{\psi}(t_1)=\dot{\psi}_0(c,t_1)} = [c, -t_1]. \tag{8}$$

Method A, with the matching phase given by:

$$\psi_0(c, c_0, t) = k\left(\sqrt{r_L^2 - c_0^2} + \sqrt{r_G^2 - c_0^2} - c_0\left(\tan\left(\frac{\sqrt{r_L^2 - c_0^2}}{c_0}\right) + \tan\left(\frac{\sqrt{r_G^2 - c_0^2}}{c_0}\right)\right) + c\theta\right). \tag{9}$$

The impact parameter is found from the Doppler equation:

$$a\dot{\theta} + \frac{\dot{r}_L}{r_L}\sqrt{r_L^2 - a^2} + \frac{\dot{r}_G}{r_G}\sqrt{r_G^2 - a^2} = c\dot{\theta} + \frac{\dot{r}_L}{r_L}\sqrt{r_L^2 - c_0^2} + \frac{\dot{r}_G}{r_G}\sqrt{r_G^2 - c_0^2}. \tag{10}$$

It is seen that the impact of the radial velocities are reduced with a proper choice of c_0.

Method B, with the matching phase given by:

$$\psi_0(c, t) = k\left(ct + \int_0^t \left(\frac{\dot{r}_L}{r_L}\sqrt{r_L^2 - c_0^2(t)} + \frac{\dot{r}_G}{r_G}\sqrt{r_G^2 - c_0^2(t)}\right) dt\right), \tag{11}$$

where $c_0(t)$ is a model of the impact parameter. A special case of this is $c_0(t) = a_0(t)n_0(a_0)$, where $a_0(t)$ is the vacuum impact parameter and n_0 is the model of the refractivity.

Again, the impact parameter is found from the Doppler equation:

$$a\dot{\theta} + \frac{\dot{r}_L}{r_L}\sqrt{r_L^2 - a^2} + \frac{\dot{r}_G}{r_G}\sqrt{r_G^2 - a^2} = c\dot{\theta} + \frac{\dot{r}_L}{r_L}\sqrt{r_L^2 - c_0^2(t)} + \frac{\dot{r}_G}{r_G}\sqrt{r_G^2 - c_0^2(t)}. \tag{12}$$

It is seen that the impact of the radial velocities are reduced overall with a proper choice of the model of c_0. Other variations of the FSI method exist (Gorbunov and Lauritsen 2004).

III. *The Phasematching method*

$$\left[c, \frac{\partial(\psi - \psi_o)}{\partial c}\right]_{\dot{\psi}(t_1)=\dot{\psi}_0(c,t_1)} = [c, \beta], \tag{13}$$

with the matching phase given by

$$\psi_o(c, t) = k\left(\sqrt{r_L^2 - c^2} + \sqrt{r_G^2 - c^2} + c\beta\right) \tag{14}$$

with $\beta = \theta - a\tan\left(\frac{\sqrt{r_L^2 - c^2}}{c}\right) - a\tan\left(\frac{\sqrt{r_G^2 - c^2}}{c}\right)$.

Again, the impact parameter is found from the Doppler equation:

$$a\dot{\theta} + \frac{\dot{r}_L}{r}\sqrt{r_L^2 - a^2} + \frac{\dot{r}_G}{r_G}\sqrt{r_G^2 - a^2} = c\dot{\theta} + \frac{\dot{r}_L}{r}\sqrt{r_L^2 - c^2} + \frac{\dot{r}_G}{r_G}\sqrt{r_G^2 - c^2}, \tag{15}$$

which gives $c = a$.

The phasematching method solves the problem with the radial velocities perfectly. It gives directly the bending-angle as a function of the impact parameter. However the phasematching function is non-linear and the processing cannot be implemented with a fast Fourier transform.

3 FSI Implementation Problems

The FSI method is fairly easy to implement. Tests on simulated noise free occultation signals give good results. However, real signals are more difficult to process. They are noisy and they occasionally contain detection errors. In the following, the different steps in the processing chain will be described and discussed.

3.1 Implementation of the FSI Processing

In this section, technical issues regarding the implementation of the FSI method are addressed. The occultation signal is supposed to be given as amplitude and phase (the excess phase) sampled with 50 Hz.

1. The signal phase (the excess phase) is modified:
   - The vacuum phase is added, so the real signal phase is established.
   - The FSI phasematching phase is subtracted from the phase.
2. The amplitude is (optional) modified.
3. The center frequency of the now modified signal is found and a new signal is formed with a center frequency of zero.
4. The phase and amplitude of the new signal are upsampled 20–30 times depending on the bandwidth of the new signal.
5. The Fourier transform is performed with a FFT.
6. The phase difference (the difference between two frequency neighbor points) of the FFT phase is computed.

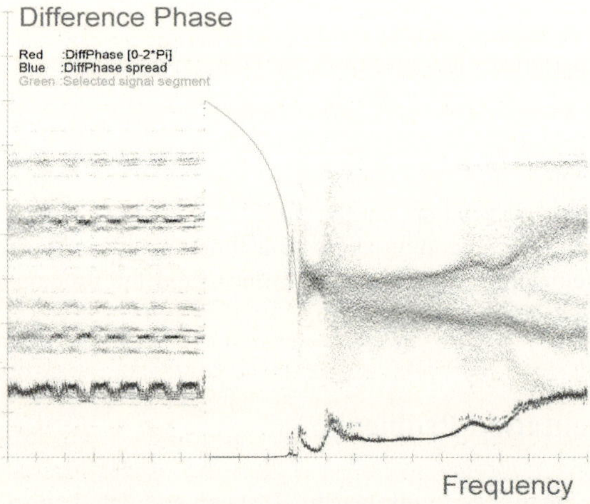

Fig. 2. The difference of the Fourier phase displayed as function of frequency. The red (and green) curve shows the difference phase. The blue curve shows the local spread of the phase difference. The green curve shows the selected signal as a result of the discrimination. It is seen that the difference phase spread is very small under the green curve. The difference phase is proportional to the differential of the phase or to the time of arrival of a ray. The noise structure is special in this simulated case, for real CHAMP signals the noise looks more symmetrical. The frequency increases to the left. See Jensen et al. (2006) this issue, page 116, for a color version of this figure).

The phase difference spectrum, which is proportional to the spectrum of the "arrival" times of the rays, consists of a signal area with noise plus a pure noise area (see Fig. 2). The task now is to separate the signal from the pure noise. Two methods for performing this task have been investigated: Discrimination of the FSI spectrum amplitude and discrimination of the phase difference shown in Fig. 2. The experience with the amplitude discrimination was that it was difficult to get a robust and reliable method, which preserved an optimum amount of information of the signal. The discrimination of the phase difference spectrum is done by computing a local mean spread (the spread is calculated within a window) as function of the frequency. This local spread function is shown in Fig. 2 in blue. It is clearly seen that the local phase difference spread is high in areas with noise and low in areas where the signal is present. With a proper threshold value, the areas with noise can be separated from the signal. If the threshold is too large, too much noise from the two ends of the signal will be taken into account in the further signal processing and if it is too low the signal will be truncated. The selection of the threshold value is therefore always a compromise. Another complication is that the "optimum" threshold varies for each occultation. The solution to this problem has here been to compute the mean sum of the above mentioned local spreads and choose the threshold as a fraction of this sum. This has, until now, shown to be a very reliable and robust method for establishing the threshold for a large number of real occultation data. With the threshold the start and end of the occultation can be found i.e., the information bearing part of the signal can be isolated. The signal does still contain receiver noise and numerical noise from the finite discrete Fourier transform. This noise can partly be filtered out on the cost of a reduced spatial resolution.

Another observation from the blue curve in Fig. 2 is that the spread in the signal is very small at the start of the occultation (the left side of the figure) whereas the spread is increasing at the end of the occultation. This can be used for further discrimination of noise embedded in the signal. By applying a varying threshold (here implemented as a linear function) single points in the signal can be taken out. This gives, however, only minor changes in the signal. The philosophy behind this is that the local spread of the signal can be considered as an error measure.

7. From the selected phase difference spectrum, the arrival time and successively the bending angle as function of the impact parameter is found.

These seven steps described do not involve any filtering. The errors in the signal are due to receiver noise and numerical noise. This could be the raw product delivered by the FSI method. Further averaging and extrapolation of the bending angle, reducing the intrinsic resolution of the FSI method would then be up to the user.

The filtering problem is complicated. A relatively simple method is to use an expansion in Chebyshev polynomials (Goodwin 1961), which both gives an interpolation and a smoothing of the functions. This is convenient, but the essential documentation of the final spatial resolution is missing. A deeper analysis of the problem will be given below.

3.2 Filtering and Spatial Resolution

In the time of arrival spectrum (the phase difference spectrum Fig. 2) fluctuations in the arrival times cause large fluctuations in the bending angle i.e., for real data fluctuations up to 1 mrad have been seen unless smoothing and averaging was done on the data. Averaging of data is a legal way to process data as long as the variance of the data is reported. In this case the problem is that we want to know the uncertainty in the frequency or spatial domain and not in the time or bending angle domain. For real data sampled with 50 Hz, the time of arrival of the rays will have a basic uncertainty of 20 ms, and the upsampling of the signal to 1000 Hz or larger will only give a pseudo time of arrival resolution of 1 ms or better. The thermal noise in the original signal is also upsampled, which means that the noise is correlated in the pseudo sample intervals. The discrete nature of the FFT and the transform of the noise into a multiplicative phase noise in the spectrum, give both contributions to the time of arrival noise. The deterministic variation of the time of arrival is given by the persistence time (Jensen et al. 2004a) i.e., the time interval in which the Doppler frequency is approximately constant. Usually the persistence time is larger than the 20 ms sampling interval (this is the basis for choosing 50 Hz sampling in radio occultation) and one must expect that only a finite number of discrete frequencies are present in a sample. If there is a distribution of frequencies this must be due to thermal noise, sample noise and processing noise as mentioned above.

Now, from this analysis a filtering scheme can be proposed. If the time of arrival function $t(\omega)$ is considered in N intervals $t_i < t < t_i+\Delta t$, where $t_0 = 0$ and $t_{N\text{-}1} = T$ (the recording time of the occultation) then we can define a function $\xi_i(\omega)$ with the following properties:

$$\xi_i(\omega) = \begin{cases} 1 & \text{if} \quad t_i < t(\omega) < t_i + \Delta t \\ 0 & \text{else.} \end{cases} \tag{16}$$

It is natural to chose Δt as the sampling time for the actual occultation.

In order to illustrate the possible behavior of $\xi_i(\omega)$ in a time interval i the following graphic is shown: The vertical bars indicate that $\xi_i(\omega) = 1$

1. || →*frequency*

2. ||| || →*frequency*

3. ||||||| →*frequency*

4. ||| || || →*frequency*

In example 1, two frequencies occur in the time interval i. The frequencies are close together and the spread is small. In 2, clusters of frequencies are seen, which can be interpreted in two ways. Either the spread in frequencies is large or we have a multipath situation. In 3, clusters of frequencies are seen. The spread is probably large. In 4, three distinct clusters of frequencies are seen, which indicates a clear multipath situation. Using the function $\xi_i(\omega)$ the local mean and spread with respect to ω can be computed and the arrival time function $t(\omega)$ can then be transformed to a discrete function or a table $[t_i, <\omega; >\sigma_\omega]$ with the time and the mean local Doppler frequency as the variables and with the local mean spread as a parameter. This table can then be used to calculate the bending angle as function of the impact parameter and by transforming the local mean spread of the Doppler frequency into an uncertainty in the impact parameter (σ_a) i.e.,

$$[t_i, <\omega>; \sigma_\omega] \rightarrow [\alpha_i, <a>; \sigma_a]. \tag{17}$$

The local means and spreads can only be calculated in the case of possible multipath if certain constraints are imposed. This is seen from example 2 and 3 above. We have to make a constraint saying that we only accept a separation between clusters, which is larger than a certain $\Delta\omega(\Delta a)$ before it can be accepted as a multipath. In order to compute the local mean(s) and spread(s) $\zeta_i(\omega)$ is correlated with a square shaped function $s(\omega)$ of the width $\Delta\omega$ (Eq. 18).

$$\Gamma_i(\omega) = \int_{\omega_{\min}}^{\omega_{\max}} s(\omega - \omega', \Delta\omega)\xi_i(\omega')\mathrm{d}\omega' \tag{18}$$

The peaks of the correlation function $\Gamma_i(\omega)$ give the local means and the local spread can be defined in various ways: As the half width of the peak, the radius of curvature of the peak or the spread of the peak. The limits of the correlation integral can be reduced so the processing speed can be improved.

In Gorbunov et al. (2005) the uncertainty of the Doppler frequency is evaluated by using running Fourier transforms. This method is questionable though it will produce an estimate of the Doppler uncertainty. The question is how this measure is related to the "real" Doppler spread in the FSI function $t(\omega)$. First, the time length of this local Fourier transform gives an intrinsic uncertainty in the spread around the peak maxima in the Fourier amplitude spectrum. Second, the impact of the noise will be different from the noise impact in the FSI spectrum. Third, the numerical noise due to the short length of the discrete Fourier transform will also give a contribution to the spread or uncertainty in the Doppler frequency. In the best case the running Fourier method will give a very conservative measure of the uncertainty.

3.3 The Phase Noise in the FSI Spectrum

In Appendix A the impact of thermal noise on the arrival time is calculated. The main result here is the variance of the time of arrival (Eq. A17). It yields

$$\langle(\partial t)^2\rangle \cong T^2 \frac{1}{\pi\Gamma}\left(\left(\frac{t_1}{T}\right)^2 + \frac{1}{4\pi}\frac{1}{\Delta f T}\right), \tag{19}$$

where $\Gamma = A_0^2/E_0\Delta f$ is the square of the signal-to-noise ratio in the Fourier space (∂t and t_1 are introduced in the introduction of the appendix). The product $E_0\Delta f$ is the noise energy in the frequency band and ΔfT is the time-bandwidth product. In the FSI method this last quantity will usually be a large quantity. The equation is an approximation, which is valid in the limit of large signal-to-noise ratios. It is seen that the relative variance is constant for large time of arrivals.

4 Conclusion

An overview of the present development of the FSI method has been given. The method proposed in Sect. 3.2 converts the time of arrival uncertainties into uncertainties in the Doppler frequency. The method has the advantage that it operates directly on the measured arrival time as function of the Doppler frequency taking the available pre-information into account. The analysis shows that an improvement in the spatial resolution of the refractivity profile can be achieved if the sampling rate is increased. The result from the phase noise analysis (Sect. 3.3) can theoretically be used to extend the method if the arrival time spread (see Eq. 19) is used to define varying time intervals. However, this will demand longer computation times and will only be exact in case of simulations.

5 References

Beyerle G, Hocke K (2001) Observation and simulation of direct and reflected GPS signals in radio occultation experiments. Geophys Res Lett 28(9):1895–1898

Beyerle G, Hocke K, Wickert J, Schmidt T, Marquardt C, Reigber C (2002) GPS radio occultations with CHAMP: A radio holographic analysis of GPS signal propagation in the troposphere and surface reflections. J Geophys Res 107(D24), 4802, doi:10.1029/2001JD001402

Born M, Wolf E (1999) Principles of Optics. Cambridge University Press, England

Fjeldbo G (1964) Bistatic radar methods for studying planetary ionospheres and surfaces. PhD thesis, Department of Electrical Engineering, Stanford University, California

Goodwin ET (1961) Modern Computing Methods. Notes on Applied Science No 16, London, Her Majesty's Stationary Office

Gorbunov ME, Lauritsen KB (2004) Analysis of wave fields by Fourier integral operators and their application for radio occultations. Radio Sci 39(4), RS4010, doi:10.1029/2003RS002971

Gorbunov ME, Lauritsen KB, Rhodin A, Tomassini M, Kornblueh L (2005) Analysis of CHAMP radio occultation data for Earth's atmosphere. Izvestiya RAN, Atm and Ocean Phys, submitted 2005

Hocke K, Pavelyev A, Yakovlev O, Barthes L, Jakowski N (1999) Radio occultation data analysis by radio holographic method. J Atmos Solar-Terr Phys 61:1169–1177

Igarashi K, Pavelyev A, Hocke K, Pavelyev D, Wickert J (2001) Observation of wave structures in the upper atmosphere by means of radio holographic analysis of the RO data. Adv Space Res 27(6-7):1321–1327

Igarashi K, Pavelyev A, Hocke K, Pavelyev D, Kucherjavenkov IA, Matugov S, Zakharov A, Yakovlev O (2000) Radio holographic principle for observing natural processes in the atmosphere and retrieving meteorological parameters from radio occultation data. Earth Planets Space 52(11):893–899

Jensen AS, Benzon H-H, Lohmann MS (2002) A new high resolution method for processing radio occultation data. Scientific Report 02-06, ISSN No 0905-3263, ISBN No. 87-7478-458-7, Danish Meteorological Institute, Copenhagen

Jensen AS, Lohmann MS, Benzon H-H, Nielsen AS (2003) Full spectrum inversion of radio occultation signals. Radio Sci 38(3), 1040, doi:10.1029/2002RS002763

Jensen AS, Lohmann MS, Benzon H-H, Nielsen AS (2004a) Fourier analysis of GNSS-LEO radio occultation signals, resolution and limitations. In: Kirchengast G, Foelsche U, Steiner AK (eds) Occultations for Probing Atmosphere and Climate. Springer-Verlag, Berlin Heidelberg, pp 53–60

Jensen AS, Lohmann MS, Nielsen AS, Benzon H-H (2004b) Geometrical optics phase matching of radio occultation signals. Radio Sci 39(3), RS3009, doi:10.1029/2003RS002899

Jensen AS, Marquardt C, Benzon H-H, Lohmann MS, Nielsen AS (2006) Evaluation of the Processing of Radio Occultation Signals by Reconstruction of the Real Signals. This issue

Pavelyev AG (1998) On the possibility of radio holographic investigation on communication link satellite-to-satellite. J Comm Techn and Electron 43(8):126–131

Pavelyev AG, Kucherjavenkov AI (1978) Refraction attenuation in the planetary atmospheres. Radio Eng and Electron Phys 23(7):13–19

Pavelyev A, Igarashi K, Reigber C, Hocke K, Wickert J, Beyerle G, Matyugov S, Kucherjavenkov A, Pavelyev D, Yakovlev O (2002) First application of radioholographic method to wave observations in the upper atmosphere. Radio Sci 37(3), 1043, doi:10.1029/2000RS002501

Appendix A: The Impact of Thermal Noise in the FSI Processing Method

In this appendix, the effect of thermal additive noise on the retrieval of the refractive index from occultation measurements will be analyzed. It is assumed that the processing of the signal is done with the FSI method (Jensen et al. 2003). The outline for this analysis is:

- A description of the Fourier transform of band-pass filtered noise.
- A calculation of the mean value and variance of the frequency derivate of the phase of the Fourier transform.

The FSI method gives as output the Doppler frequency as function of time. From the Doppler frequency, the impact parameter a, and the bending angle α can be derived. The Doppler frequency is the frequency of the Fourier transform and the time is found by differentiating the phase $\varphi(\omega)$ of the Fourier transform i.e., $t_1 = -\mathrm{d}\varphi(\omega)\mathrm{d}\omega$ (see Jensen et al. 2002). The uncertainty in the time determination ∂t, gives an uncertainty in the impact parameter ∂a. The relation is simply given by $\partial a = (\mathrm{d}a/\mathrm{d}t)\delta t$. The mean value of δt and its variance are calculated in Appendix I.2 for additive band-pass filtered noise.

The effect of the Fourier transformed noise, are the issues in Appendix A.1. This section can be skipped, if the reader accepts the application of the results in Appendix A.2.

A.1 Bandpass Filtered Noise

In order to describe the impact of the noise we will first define and describe the properties of the band pass filtered noise, which are the result of the pre-processing of the signal. The noise is passing in the band pass $\pm\frac{1}{2}\Delta\omega$ around the center frequencies q and $-q$. The noise contribution around q is denoted by N^+ and around $-q$, N^-.

$$N^{+}(t)=\frac{1}{2\pi}\int_{q-\frac{1}{2}\Delta\omega}^{q+\frac{1}{2}\Delta\omega} n(\omega)\mathrm{e}^{i\omega t}\mathrm{d}t=\mathrm{e}^{iqt}\int_{-\frac{1}{2}\Delta\omega}^{\frac{1}{2}\Delta\omega} n(\omega+q)\mathrm{e}^{i\omega t}\mathrm{d}\omega$$
$$N^{-}(t)=\frac{1}{2\pi}\int_{-q-\frac{1}{2}\Delta\omega}^{-q+\frac{1}{2}\Delta\omega} n(\omega)\mathrm{e}^{i\omega t}\mathrm{d}t=\mathrm{e}^{-iqt}\int_{-\frac{1}{2}\Delta\omega}^{\frac{1}{2}\Delta\omega} n(\omega-q)\mathrm{e}^{i\omega t}\mathrm{d}\omega \tag{A1}$$

The total noise $N(t) = N^{+}+N^{-}$ can be expressed as a sum of cosine and sine functions, as shown below:

$$N(t) = N^{+} + N^{-} = N_C(t)\cos(qt) - N_S(t)\sin(qt),$$

where

$$N_C(t)=\frac{1}{2\pi}\int_{-\frac{1}{2}\Delta\omega}^{\frac{1}{2}\Delta\omega}(n(\omega+q)+n(\omega-q))\mathrm{e}^{i\omega t}\mathrm{d}\omega, \tag{A2}$$
$$N_S(t)=-\frac{i}{2\pi}\int_{-\frac{1}{2}\Delta\omega}^{\frac{1}{2}\Delta\omega}(n(\omega+q)-n(\omega-q))\mathrm{e}^{i\omega t}\mathrm{d}\omega.$$

The pair (N_C,N_S) can be considered as a pair of noise signals in quadrature.

The noise function $K = N_C+iN_S$ is a function of time. In the following, the noise considerations are done in the Fourier space and for that reason; the properties of the finite Fourier transform of K will be analyzed. The mean value of K and $\mathrm{d}K/\mathrm{d}\omega$ and their variances will be calculated, with the intension of using the result to determine the mean and variance of the impact parameter.

$$\begin{aligned}K(\omega)&=\frac{1}{T}\int_{-T/2}^{T/2}K(t)\mathrm{e}^{-i\omega t}\mathrm{d}t\\&=\frac{1}{\pi}\int_{-T/2}^{T/2}\mathrm{e}^{-i\omega t}\int_{-\frac{1}{2}\Delta\omega}^{\frac{1}{2}\Delta\omega}n(\omega'+q)\mathrm{e}^{i\omega' t}\mathrm{d}\omega'\mathrm{d}t\\&=\frac{T}{\pi}\int_{-\frac{1}{2}\Delta\omega}^{\frac{1}{2}\Delta\omega}n(\omega'+q)\frac{\sin\frac{1}{2}(\omega'-\omega)T}{\frac{1}{2}(\omega'-\omega)T}\mathrm{d}\omega',\end{aligned} \tag{A3}$$

where T is the time length of the signal.

It is seen that

$$\langle K(\omega)\rangle = 0\,. \tag{A4}$$

The variance of the spectral noise becomes:

$$\left\langle |K(\omega)|^2\right\rangle \cong \frac{1}{\pi^2}E_0\Delta\omega=\frac{2}{\pi}E_0\Delta f\,, \tag{A5}$$

where E_0 is the noise energy per frequency.

The derivate of the Fourier noise with respect to ω yields:

$$\frac{\mathrm{d}K(\omega)}{\mathrm{d}\omega}=\frac{1}{\pi}\int\limits_{-\frac{1}{2}\Delta\omega}^{\frac{1}{2}\Delta\omega} n(\omega'+q)\frac{\mathrm{d}}{\mathrm{d}\omega}\left(\frac{\sin\frac{1}{2}(\omega'-\omega)T}{\frac{1}{2}(\omega'-\omega)T}\right)\mathrm{d}\omega' . \tag{A6}$$

It is easily seen that the mean value of the noise derivative is zero. The variance of this yields:

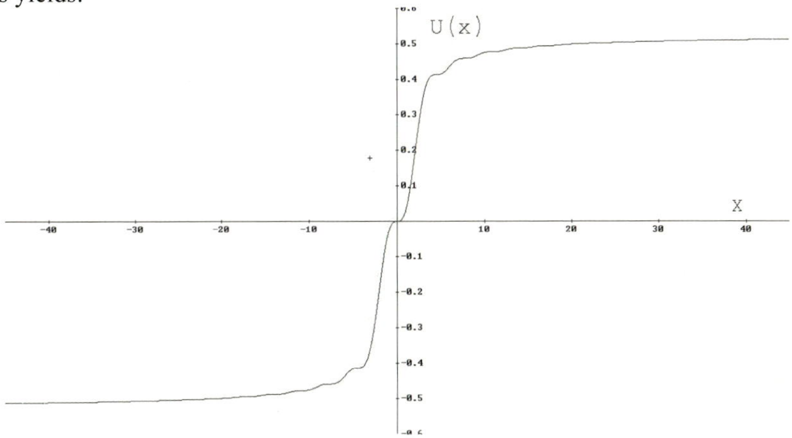

Fig. A1. $U(x)$ is defined in Eq. A8.

$$\left\langle\left|\frac{\mathrm{d}K(\omega)}{\mathrm{d}\omega}\right|^2\right\rangle=\frac{E_0}{\pi^2}\int\limits_{-\frac{1}{2}\Delta\omega}^{\frac{1}{2}\Delta\omega}\left(\frac{\mathrm{d}}{\mathrm{d}\omega}\left(\frac{\sin\frac{1}{2}(\omega'-\omega)T}{\frac{1}{2}(\omega'-\omega)T}\right)\right)^2\mathrm{d}\omega'$$

$$=\frac{E_0T^2}{4\pi^2}\int\limits_{-\frac{1}{2}\Delta\omega}^{\frac{1}{2}\Delta\omega}\left(\frac{\sin\frac{1}{2}(\omega'-\omega)T-\frac{1}{2}(\omega'-\omega)T\cos\frac{1}{2}(\omega'-\omega)T)^2}{(\frac{1}{2}(\omega'-\omega)T)^2}\right)\mathrm{d}\omega' \tag{A7}$$

$$=\frac{E_0T}{2\pi^2}\int\limits_{\frac{1}{2}(T\omega-\frac{1}{2}T\Delta\omega)}^{\frac{1}{2}(T\omega+\frac{1}{2}T\Delta\omega)}\left[\frac{\sin(x)-x\cos(x)}{x^2}\right]^2\mathrm{d}x.$$

The resulting integral above cannot be expressed explicitly as function of ω and $\Delta\omega$, but the result can be described approximated. To do this the function $U(x)$, defined below, has been calculated analytically and the result is displayed on the graph below.

$$U(x)=\int\limits_0^x\left[\frac{\sin(t)-t\cos(t)}{t^2}\right]^2\mathrm{d}t \tag{A8}$$

It is seen that the function is a two-valued function with a narrow transition area around zero (Fig. A1). The transition area is as seen from the graph approximately restricted by $|x|\leq 10$. In the calculation of the noise term, due to the derivate of the noise, the difference between $U(\frac{1}{2}\omega T+\frac{1}{4}\Delta\omega T)-U(\frac{1}{2}\omega T-\frac{1}{4}\Delta\omega T)$ has to be approximated. First it is seen that if $|\omega|\geq\frac{1}{2}\Delta\omega$ the contribution will be approximately zero. This is also to be expected from a physical point of view, due to bandpass nature of the noise. If $|\omega| \leq ½\Delta\omega$, the contribution will approximately be 1, by near the narrow areas where the transition occurs. However since these points always will give a contribution smaller than 1, we do make a conservative estimate by setting these contributions equal to 1.

$$\left\langle \left| \frac{\mathrm{d}K(\omega)}{\mathrm{d}\omega} \right|^2 \right\rangle \cong \begin{cases} \dfrac{E_0 T}{2\pi^2} & |\omega| < \frac{1}{2}\Delta\omega \\ 0 & |\omega| > \frac{1}{2}\Delta\omega \end{cases} \tag{A9}$$

A.2 Mean Value and Variance of the Fourier Phase Frequency Derivate

The output of the FSI method, the Fourier transform of the signal plus noise, can be written as:

$$F(\omega) = |F(\omega)| \mathrm{e}^{i\varphi(\omega)} = A_0 \mathrm{e}^{i\varphi_0(\omega)} + K(\omega), \tag{A10}$$

where A_0 is the amplitude of the Fourier transform. A_0 is a constant or a weak function of the frequency in case of absorption in the atmosphere, so it is safe to consider it to be a constant. $\varphi(\omega)$ is the phase of the Fourier transform and $\varphi_0(\omega)$ is the Fourier phase of the noise free radio occultation signal.

The Fourier phase is given by:

$$\varphi(\omega) = A\tan\left(\frac{A_0 \sin(\varphi_0) + \mathrm{Im}(K)}{A_0 \cos(\varphi_0) + \mathrm{Re}(K)} \right). \tag{A11}$$

The derivative of the detected Fourier phase yields:

$$\begin{aligned} \frac{\mathrm{d}\varphi}{\mathrm{d}\omega} = & \frac{\left(A_0^2 + A_0 \,\mathrm{Im}(K)\sin(\varphi_0) + A_0 \,\mathrm{Re}(K)\cos(\varphi_0)\right)\dfrac{\mathrm{d}\varphi_0}{\mathrm{d}\omega}}{|F(\omega)|^2} + \\ & + \frac{A_0\left(\cos(\varphi_0)\dfrac{\mathrm{d}\,\mathrm{Im}(K)}{\mathrm{d}\omega} - \sin(\varphi_0)\dfrac{\mathrm{d}\,\mathrm{Re}(K)}{\mathrm{d}\omega} \right)}{|F(\omega)|^2} + \\ & + \frac{\mathrm{Re}(N)\dfrac{\mathrm{d}\,\mathrm{Im}(K)}{\mathrm{d}\omega} - \mathrm{Re}(N)\dfrac{\mathrm{d}\,\mathrm{Im}(K)}{\mathrm{d}\omega}}{|F(\omega)|^2}. \end{aligned} \tag{A12}$$

Expanding the equation above to $(A_0)^{-1}$ yields:

$$\begin{aligned} \frac{\mathrm{d}\varphi}{\mathrm{d}\omega} \cong & \left(1 - \frac{1}{A_0}\left(\mathrm{Im}(K)\sin(\varphi_0) + \mathrm{Re}(K)\cos(\varphi_0)\right) \right)\frac{\mathrm{d}\varphi_0}{\mathrm{d}\omega} + \\ & + \frac{1}{A_0}\left(\cos(\varphi_0)\frac{\mathrm{d}\,\mathrm{Im}(K)}{\mathrm{d}\omega} - \sin(\varphi_0)\frac{\mathrm{d}\,\mathrm{Re}(K)}{\mathrm{d}\omega} \right) + \ldots\ldots \end{aligned} \tag{A13}$$

From this the mean value of the detected time can be calculated

$$\left\langle \frac{\mathrm{d}\varphi}{\mathrm{d}\omega} \right\rangle \cong \frac{\mathrm{d}\varphi_0}{\mathrm{d}\omega}. \tag{A14}$$

It is seen that the time $t_1 = -\mathrm{d}\varphi(\omega)/\mathrm{d}\omega$ is bias free. This is important, in that respect that a bias will not be present in the retrieval of the refractivity.

The calculation of the variance of the detected time is done by squaring Eq. A13 with a subsequent averaging and keeping terms to second order in the amplitude. Mean values of cross terms i.e., products where the real and imaginary part of the Fourier noise is involved are set to zero.

$$\begin{aligned}\left(\frac{\mathrm{d}\varphi}{\mathrm{d}\omega} - \frac{\mathrm{d}\varphi_0}{\mathrm{d}\omega}\right)^2 \cong\ & \frac{1}{A_0^2}\left(\mathrm{Im}(K)\sin(\varphi_0) + \mathrm{Re}(K)\cos(\varphi_0)\right)^2 \left(\frac{\mathrm{d}\varphi_0}{\mathrm{d}\omega}\right)^2 \\ & + \frac{2}{A_0^2}\left(\cos(\varphi_0)\frac{\mathrm{d}\,\mathrm{Im}(K)}{\mathrm{d}\omega} - \sin(\varphi_0)\frac{\mathrm{d}\,\mathrm{Re}(K)}{\mathrm{d}\omega}\right)\cdot \\ & \cdot\left(\mathrm{Im}(K)\sin(\varphi_0) + \mathrm{Re}(K)\cos(\varphi_0)\right)\frac{\mathrm{d}\varphi_0}{\mathrm{d}\omega} + \\ & + \frac{1}{A_0^2}\left(\cos(\varphi_0)\frac{\mathrm{d}\,\mathrm{Im}(K)}{\mathrm{d}\omega} - \sin(\varphi_0)\frac{\mathrm{d}\,\mathrm{Re}(K)}{\mathrm{d}\omega}\right)^2.\end{aligned} \tag{A15}$$

The mean value of Eq. A15 is obtained by using the results in Sect. 1.1 (Eqs. A5, A9).

$$\begin{aligned}\left\langle \left(\frac{\mathrm{d}\varphi}{\mathrm{d}\omega} - \frac{\mathrm{d}\varphi_0}{\mathrm{d}\omega}\right)^2 \right\rangle &\cong \frac{1}{2A_0^2}\left(\left\langle |K|^2 \right\rangle \left(\frac{\mathrm{d}\varphi_0}{\mathrm{d}\omega}\right)^2 + \left\langle \left|\frac{\mathrm{d}K}{\mathrm{d}\omega}\right|^2 \right\rangle\right) \\ &= T^2 \frac{E_0 \Delta f}{\pi A_0^2}\left(\left(\frac{1}{T}\frac{\mathrm{d}\varphi_0}{\mathrm{d}\omega}\right)^2 + \frac{1}{4\pi}\frac{1}{\Delta f T}\right).\end{aligned} \tag{A16}$$

Eq. A16 can be rewritten to yield:

$$\left\langle (\partial t)^2 \right\rangle \cong T^2 \frac{1}{\pi\Gamma}\left(\left(\frac{t_1}{T}\right)^2 + \frac{1}{4\pi}\frac{1}{\Delta f T}\right), \tag{A17}$$

where $\Gamma = A_0^2 / E_0 \Delta f$ is the square of the signal-to-noise ratio in the Fourier space (∂t and t_1 are introduced in the introduction). The product $E_0\Delta f$ is the noise energy in the frequency band, and $\Delta f T$ is the time-bandwidth product. In the FSI method this last quantity will usual be a large quantity.

The found results can now be used to find the mean and the variance of the impact parameter. The relation between the uncertainty in the impact parameter and the time is given by: $\partial a = (\mathrm{d}a/\mathrm{d}t)\delta t$.

By using Eq. A14 we obtain:

$$\langle \partial a \rangle = \frac{\mathrm{d}a}{\mathrm{d}t}\langle \partial t \rangle = 0, \tag{A18}$$

i.e., the measuring of impact parameter is bias free in the FSI method.

The variance of the impact parameter yields:

$$\langle(\partial a)^2\rangle = \left(\frac{\mathrm{d}a}{\mathrm{d}t}\right)^2 \langle(\partial t)^2\rangle \cong T^2 \frac{1}{\pi\Gamma}\left[\left(\frac{t_1}{T}\right)^2 + \frac{1}{4\pi}\frac{1}{\Delta f T}\right]\left(\frac{\mathrm{d}a}{\mathrm{d}t}\right)^2. \tag{A19}$$

This is an interesting result, which shows that the uncertainty of the impact parameter increases when the detection time increases. This means that the advantage of having a long synthetic aperture is limited by the thermal noise and that for a given strength of the noise there exists an optimal detection time. The uncertainty in the impact parameter due to the detection time is given by $2\pi/(k\dot{\theta}T)$, where k is the wave number and $\dot{\theta}$ the relative angular velocity of the pair of satellites.

The effect of the thermal noise propagating in the retrieval chain is, as seen in the previous chapter, a very complex issue. The practical use of the results will mostly be related to Eq. A19 which can be used to find an optimal time aperture in the processing of the occultation signal. In the error analysis of the Abel transform it is seen that the errors coming from various sources are affected by the variation of the refractive index and that especially a large gradient in the refractive index amplifies the final errors. The effect of these errors can only be described in a statistical framework i.e., in an extensive simulation with "all" possible refractive index profiles weighted with the probability of their existence.

Evaluation of the Processing of Radio Occultation Signals by Reconstruction of the Real Signals

A. S. Jensen[1], C. Marquardt[2], H.-H. Benzon[1], M. S. Lohmann[3], and A. S. Nielsen[1]

[1] Danish Meteorological Institute, Research and Development Department, Lyngbyvej 100, 2100 Copenhagen, Denmark
arne_skov_jensen@hotmail.com
[2] Eumetsat, Am Kavalleriesand 31, D-64295 Darmstadt, Germany
[3] University Corporation for Atmosphere Research, Box 3000, Boulder, CO 80307, USA

Abstract. The processing of radio occultation signals has the purpose of extracting the bending angle as function of the impact parameter so that a following Abel transform gives the refractivity profile of the atmosphere. The radio occultation signal is the detected electromagnetic field from a remote transmitter (a GPS satellite). Various processing techniques can be used to retrieve the bending angle. The Full Spectrum Inversion (FSI) method relies on geometrical optics and will produce the actual bending angle as function of the impact parameter, if it can be assumed that the refractivity is spherically symmetric around the Earth and the geometrical optical description of the propagating electromagnetic field is correct. The inverse problem, construction of the electromagnetic field from knowledge of the refractivity profile (or the bending angle) and the path of the receiving and transmitting antennas can also be solved in the framework of geometrical optics. This opens a way to evaluate the quality of actual occultation measurements both with respect to the computational algorithms and the assumptions of spherical symmetry. When the refractivity profile has been measured from an occultation signal, this profile together with information about the satellites' path can be used to generate a simulated signal, which can be compared with the original signal. Processing of real data is shown here and the results are discussed.

1 Signal Processing and Information Measures

Optimal signal processing is usually a statistical concept. Knowing the structure of the signal and noise sources one can in many cases define an information measure, which can be used to optimize the signal processing, for instance, by using the maximum likelihood method. Information or quality measures for an individual signal, which is the aim for the following discussions, depends on the reliability of the processing method and of the underlying assumptions for the method i.e., how well the assumed structure of the signal is present.

Retrieving the wanted information from a signal usually involves filtering, which reduces the noise or unwanted information in the signal. Filtering means that certain filter parameter values have to be set. In order to optimize the signal-to-noise ratio and without reducing the relevant information, these parameters are usually set as a compromise. If the filter parameters for a generic group of signals can be fixed and still give optimum results, the processing method can be said to be robust and the method will be useful for automatic signal processing. However, if this is not the case, the filter parameters have to be adjusted to individual signals in order to avoid information loss.

In the case of radio occultation signals no exact practical model exists and one is left with different physical models and assumptions. Theoretically Maxwell equations can give an exact solution but this is a very comprehensive computation work. A simple and practical approach is to model the signal using geometrical optics together with some assumptions about the atmosphere. A strict verification of this is difficult and is probably also impossible due to diffraction effects, which inevitable will be present in the signal. Also other effects such as scintillations, reflections, horizontal refractivity gradients are difficult or impossible to model and to detect. Horizontal gradients, for instance, can be modeled but not detected in the framework of geometrical optics. When the impact parameter (see below) is found from the Doppler frequency, it is assumed that the refractivity profile is spherically symmetric.

Since the outcome of the processing of the radio occultation signal is the refractivity or bending angle, a way to test whether the processing method and the assumptions have been correct is to reconstruct the radio occultation signal and compare it with the original signal (in general, signal processing, signal reconstruction is not possible).

Depending on the processing methods for the reconstruction, the results of this comparison will be an evaluation of the processing method, or a mixed outcome with respect to the evaluation of the processing method and the assumptions. The result from the evaluation is an error measure or an information measure, here found by computing the correlation coefficients for the phase and amplitude of the real occultation signal and the reconstructed signal. A similar approach has been used in Pavelyev (2003). The reconstruction of the signal is here done using a geometric optics approach, other techniques exist. These include the multiple phase screens technique and asymptotic modeling. See the references Gorbunov (2003) and Gorbunov and Lauritsen (2004).

Before the details about how to reconstruct the radio occultation signal, we will give a brief description of the FSI (Jensen et al. 2003) and of the inverse problem.

1.1 The FSI Method

The processing of the occultation signal is performed by the FSI method. In this method the complex radio signal, multiplied with a phasor depending on the satellite path, is Fourier transformed giving a complex spectrum. Ideally the amplitude of the spectrum is a constant (when no absorption is present in the atmosphere). By differentiation of the spectral phase with respect to the frequency, the time (arrival time of a ray) of the occurrence of the frequency is revealed. After a trivial correction, (due to an applied phase term in the FSI) this gives the Doppler frequency as function of time, from which the impact parameter a, can be found as a function of time. The functional form of the Doppler frequency as a function of the impact parameter assumes that the atmosphere is spherically symmetric. If there are horizontal gradients present in the atmosphere, the formula is not correct and the impact parameter derived from the equation will have an error.

$$\omega(t) = k\left[\dot{r}_{\mathrm{G}}\sqrt{1-\left(\frac{a}{r_{\mathrm{G}}}\right)^2} + \dot{r}_{\mathrm{L}}\sqrt{1-\left(\frac{a}{r_{\mathrm{L}}}\right)^2} + a\dot{\theta}\right] \tag{1}$$

where r_{L}, r_{G} are the LEO and the GPS radii, respectively, and θ the angle between the satellites. The dots denote time differentiation.

Knowing the impact parameter and time, the bending angle α as function of the impact parameter is found from the angular equation below.

$$\alpha = \theta - \pi + a\sin\left(\frac{a}{r_{\mathrm{G}}}\right) + a\sin\left(\frac{a}{r_{\mathrm{L}}}\right) \tag{2}$$

With the knowledge of $\alpha(a)$ the refractivity profile can be found by performing an Abel transform on $\alpha(a)$.

The details in the implementation of the FSI processing method is described in (Jensen et al. 2002, 2004b, 2006). In Fig. 1 the difference phase (the difference between two frequency neighbor points) is shown. The curve illustrates the time of arrivals as function of frequency. This function is then used to find the impact parameter and the bending angle (Eqs. (1) and (2)).

The problems, which concern the reconstruction of the signal, are the smoothing and the interpolation of the bending angle. The filtering methods used here are based on expanding the functions as Chebyshev polynomials (Goodwin 1961), both of which give an interpolation and a smoothing of the functions. This is convenient, but the essential documentation of the final spatial resolution is missing. However, when the issue is a signal reconstruction, a visual inspection of the result should be sufficient.

The inverse problem, constructing the signal from the knowledge of the refractivity profile or the bending angle, can in principle be solved by reversing the chain described above. However, in praxis it is not fully correct to use the Fourier transform for the inversion, due to the discrete nature of the data. The discrete Fourier transform produces oscillations (Gibbs phenomena). Instead, the phase and amplitude of the single rays as function of time are calculated whereby the occultation signal can be (re-)constructed. The phase can be reconstructed in two ways: A reconstruction of the phase of the FSI spectrum and thereby deriving the phase in the time domain or a reconstruction from the geometrical optical expression of the phase in a spherically symmetric atmosphere. Comparison of the reconstructed phase to the real phase in the first approach will reveal processing errors whereas the second approach will reveal both processing errors and errors in the assumptions about the spherical symmetry. So in the later case, the difference between the two phases should in principle reveal broken assumptions about spherical symmetry.

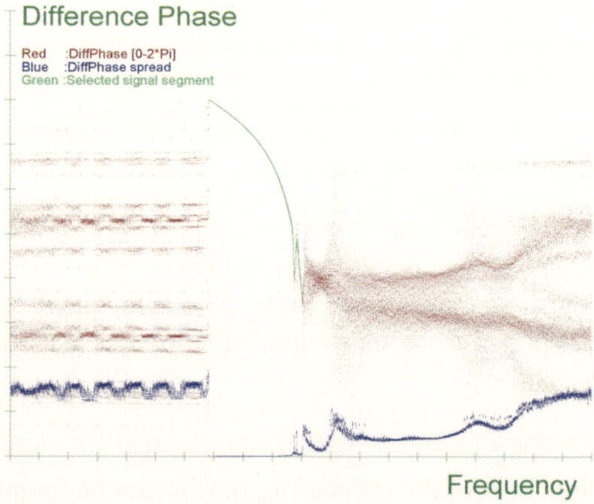

Fig. 1. The difference of the Fourier phase displayed as function of frequency. The red (and green) curve shows the difference phase. The blue curve shows the local spread of the phase difference. The green curve shows the selected signal as a result of the discrimination. It is seen that spread in the difference phase is very small under the green curve. The difference phase is proportional to the differential of the phase or to the time of arrival of a ray. The noise structure is special in this simulated case, for real CHAMP signals the noise looks more symmetrical.

2 Signal Reconstruction

Reconstruction of signals from the results of the processing chain above is not always possible. In case of radio occultations the signal can be constructed if it can be assumed that the propagation of the electromagnetic field can be described with geometrical optics. Knowing the arrival time of the rays as function of the Doppler frequency the reconstruction can be done. From a comparison between the real signal and the reconstructed signal information about the efficiency of the signal processing algorithms can be found. Another way to reconstruct the signal is to apply the assumption of spherical symmetry of the refractivity profile in the atmosphere. If this condition is not fulfilled for the actual atmosphere the reconstructed signal will be different from the real signal, especially the reconstructed signal phase should exhibit deviations form the real signal phase in areas where the atmosphere is non-spherically symmetric. In this paper the reconstruction is performed with the assumption of a spherically symmetric refraction profile. The examples shown here are from randomly selected CHAMP data, and no attempt has been made to reveal the presence of non-spherical conditions. The goal is to make an initial investigation and to verify the geometrical optical assumption.

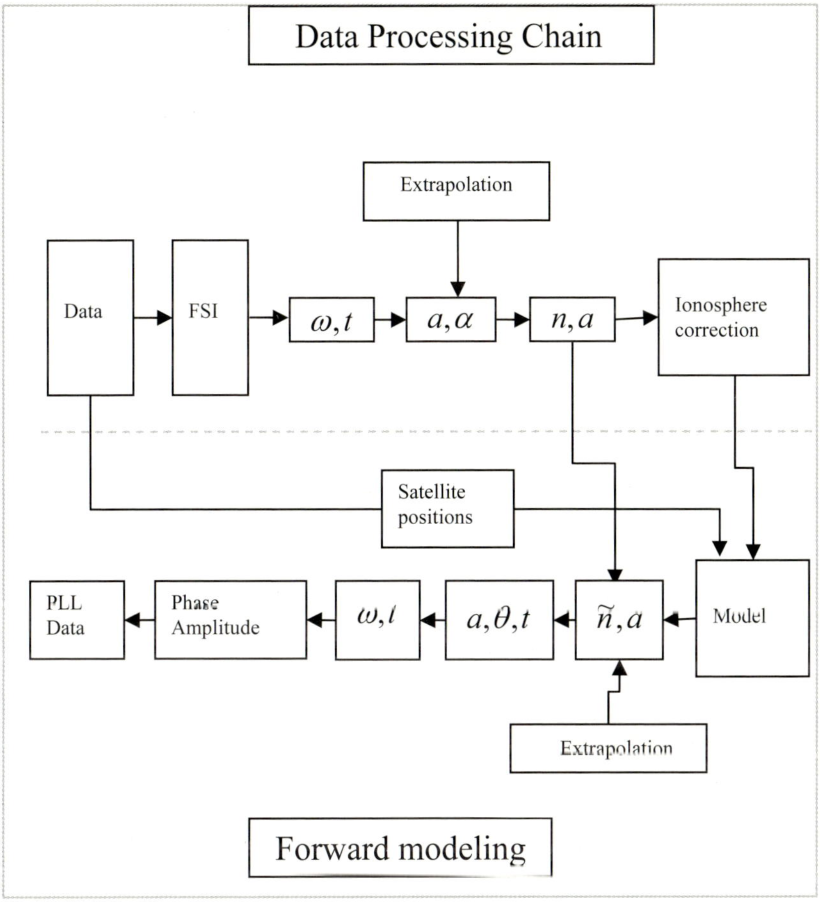

Fig. 2. Block diagram of the processing chain and the forward modeling showing various possibilities for signal reconstruction and data assimilation.

In general, reconstruction is done when the refractivity profile of the bending angle is used for assimilation purpose. Average values of temperature and water vapor from a numerical weather model give an estimate of the refractivity profile or the bending angle. This profile is iteratively compared with a measured profile from an occultation measurement, until the best estimates of temperature and water vapor have been achieved. In Fig. 2 a sketch of the data processing chain is shown together with a scheme for forward modeling or reconstruction of the signal. The figure illustrates the possibilities of signal reconstruction and assimilation at various levels: refractivity, bending angle and Doppler frequency. If a model of the ionosphere is present the assimilation and the signal reconstruction can be done simultaneous. In the following section the principles of the forward modeling approach are presented.

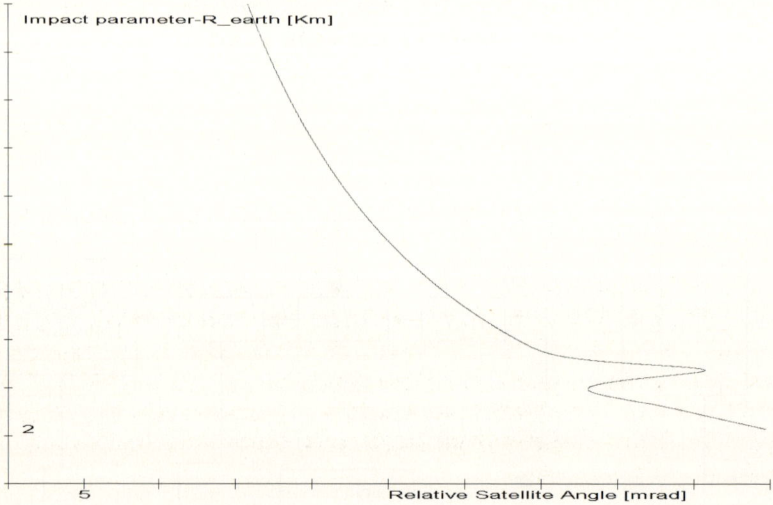

Fig. 3. The impact parameter as a function of the satellite angle for satellite radii at time *t* (Eq. (3)). The impact parameter(s) at the time *t* is found by the crossings of vertical line at $\theta(t)$ with the curve. By using Eqs. (2), (4) and (5) the field can then be reconstructed.

2.1 Geometrical Optical Forward Propagator

The output from the FSI processing method and the Abel transform is the refractivity as function of the impact parameter. With a set of satellite radii, the angle between the satellites θ can be calculated from Eq. (3) (which can be derived from Bouguers' formula, (Born and Wolf, 1999)) as function of the impact parameter. This is displayed in Fig. 3.

$$\theta = a\int_{r_o}^{r_G} \frac{\mathrm{d}r}{r\sqrt{n^2r^2 - a^2}} + a\int_{r_o}^{r_L} \frac{\mathrm{d}r}{r\sqrt{n^2r^2 - a^2}} \tag{3}$$
$$= \int_a^{r_G} \frac{a}{\sqrt{u^2 - a^2}}\left(\frac{1}{u} - \frac{\mathrm{d}\ln(n)}{\mathrm{d}u}\right)\mathrm{d}u + \int_a^{r_L} \frac{a}{\sqrt{u^2 - a^2}}\left(\frac{1}{u} - \frac{\mathrm{d}\ln(n)}{\mathrm{d}u}\right)\mathrm{d}u$$

where a is the impact parameter, n the refractive index and r_L, r_G the radii of the LEO and GPS satellite, respectively.

With a set of satellite path data r_L,r_G, θ, t we can find the associated impact parameter(s), which with the use of Eq. (2) gives the bending angle(s). Each solution is equivalent to a ray and for each ray we can find the phase ψ (Eq. (4)) and the amplitude A (Eq. (5)). Both formulas can be found in Jensen et al. (2004a).

$$\psi = k\left[\sqrt{r_L^2 - a^2} + \sqrt{r_G^2 - a^2} + a\alpha + \int_a^\infty \alpha(x)\mathrm{d}x\right], \tag{4}$$

$$A = \left(\frac{P}{2\pi}\frac{a}{r_G r_L \sin(\theta)\sqrt{r_G^2 - a^2}\sqrt{r_L^2 - a^2}\dfrac{\partial\theta}{\partial a}}\right)^{1/2} \tag{5}$$

where P is the emitted power from the GPS satellite and $\partial\theta/\partial a$ is the partial derivative of θ with respect to the impact parameter. This quantity can be found by differentiation of the curve in Fig. 3 where r_L,r_G are constants. A formula for the amplitude can also be found in Eshleman et al. (1980).

By adding the complex field of the rays the occultation signal can be found as function of time. In a similar way the signal reconstruction can be done from the knowledge of the bending angle.

2.2 Implementation of the Signal Reconstruction

The bending angle profile from an occultation processed by the FSI method is interpolated using Chebyshev series. The use of the Chebyshev series has certain programming and computational advances. Practical, n Chebyshev polynomials are used for calculating n interpolation coefficients from which an arbitrary point on the bending angle profile can be calculated. If the profile was exactly a polynomial of degree n, the interpolation would give the exact result. In other cases the use of the interpolation coefficients will result in a smoothed version of the bending angle. After an extrapolation of the bending angle, the Abel transform is used for computing the refractivity profile. Experience with this interpolation technique shows that if n is chosen to low, oscillations occur in the results. In Fig. 4 this effect is vaguely seen on the curves. The implementation of the signal reconstruction is then done accordingly to the scheme laid out in Sect. 2.1.

2.3 Results

Results from the reconstruction process are shown in Figs. 4 to 9. In each case the amplitude and phase from the original signal and the reconstructed forward propagated signal are compared visually and by calculation the correlation coefficients. The correlation coefficients are computed for the whole signal.

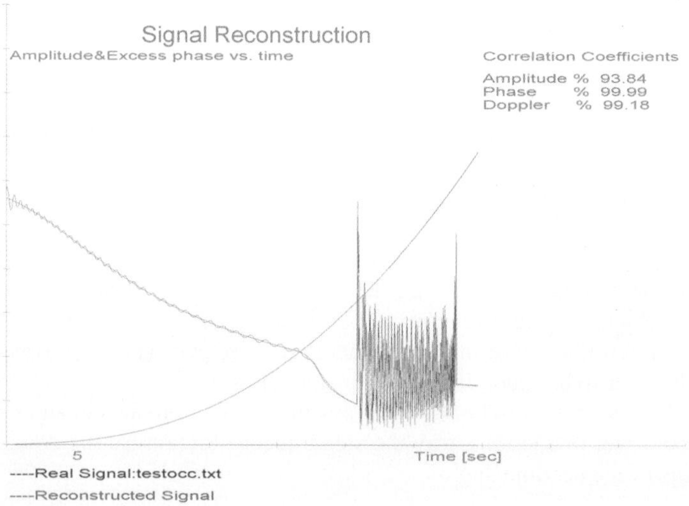

Fig. 4. Reconstruction of a simulated radio occultation signal.

A local measure would give more information, but this is not done in this paper. In the case of real data, the fly wheeling time slots are indicated on the graphs.

In the first case (Fig. 4) the signal is generated with a simulated refractivity profile, which has a clear multipath zone. For the simulated signal the satellite paths have constant radii. The radio occultation signal is then generated with the method outlined in Sect. 2.1 with a small amount of additive noise. After the use of the FSI method, the resulting refractivity profile is then again used to generate a signal. The results are shown in Fig. 4. Ideally the amplitude and the phase should be identical, but as seen, the amplitude correlation coefficient does only reach 94 %, whereas the phase and the Doppler coefficients are approximately 100 %. In the beginning of the occultation and up to the start of the multipath area, the reconstruction is close to perfect. In the multipath area it is seen from the graph that the signals are very look alike, but they are not totally overlapping. The reason for this must be numerical computational problems, numerical accuracy or algorithms, which are not optimal. Figure 1 (which is the result from FSI processing of the test signal) illustrates the problem: In the multipath area there are two extreme points, which indicate the temporal start and end of the multipath; if these points are smoothed the exact start and end of the multipath will have an offset relative to the real points. This will affect the whole multipath area as seen in Fig. 4 and results in a lower correlation coefficient for the amplitude. In general, this will be a problem for real occultation events, but for the near deterministic case (shown in Fig. 4) it should be possible to create a perfect reconstruction. The problem comes from the interpolation and smoothing of the bending angle. A lot of effort have been done in order to solve this problem, but until now with no results. The missing perfect match for the amplitude should be taken into account when the results for the reconstruction of real signals are evaluated.

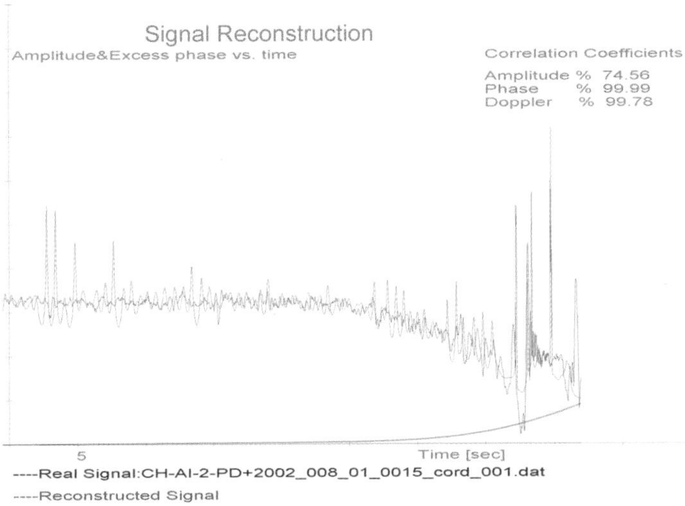

Fig. 5. Reconstruction of CHAMP data. The results show a high degree of similarities especially in the envelope of the signals. The large fluctuations in the reconstructed amplitude could be damped if the bending angle had been smoothed more.

In the reconstruction of real occultation signals it is assumed that the signal can be described with geometrical optics and that the spherical symmetrical assumption is fulfilled. The reconstruction does not take the actual detection process with the phase locked loop into account i.e., it is assumed that the detection process is ideal. In some cases (see Figs. 6, 7 and 8) the fly wheeling flag has been set during the detection process indicating that the phase lock loop has not been properly working at all times.

The origin of the errors or lack of correlation between the original signal and the reconstructed signal can be listed in four groups:

1. Geometrical optical assumption is not fulfilled or only partly fulfilled
2. The spherical symmetrical assumption is not fully fulfilled
3. The detection errors in the signal
4. Numerical errors due to noise, programming errors and interpolation and smoothing of the data

The errors of type 1 are difficult to recognize in the present investigation. Most of the reconstructed signals look very similar to the original signals and derivations can be explained as one of the other errors listed above. However, a conclusion that the geometrical optical assumption is fulfilled 100 %, cannot be drawn from this investigation.

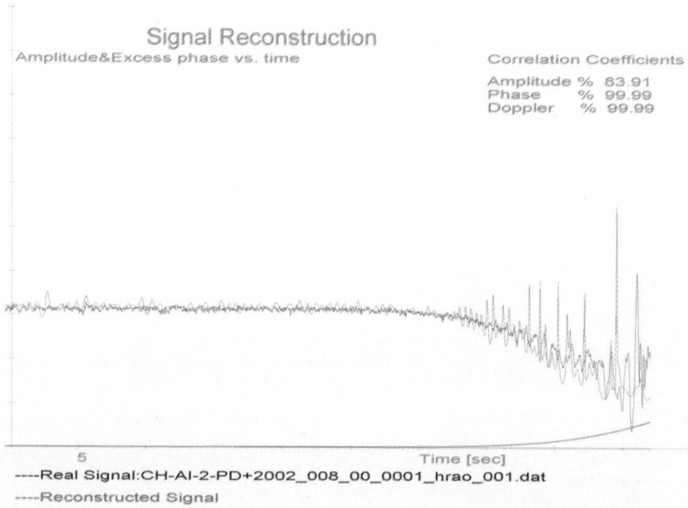

Fig. 6. Reconstruction of CHAMP data. The results show a high degree of similarities especially in the envelope of the signals. The large fluctuations in reconstructed amplitude could be damped if the bending angle had been smoothed more. With respect to correlation coefficients, this is the best of the examples shown.

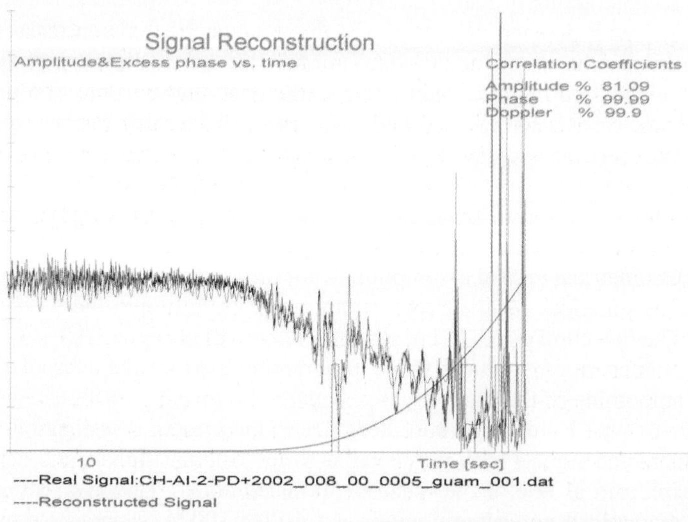

Fig. 7. Reconstruction of CHAMP data. The results show a high degree of similarities especially in the envelope of the signals. The large fluctuations in reconstructed amplitude could be damped if the bending angle had been smoothed more. Fly wheeling is within the boxes.

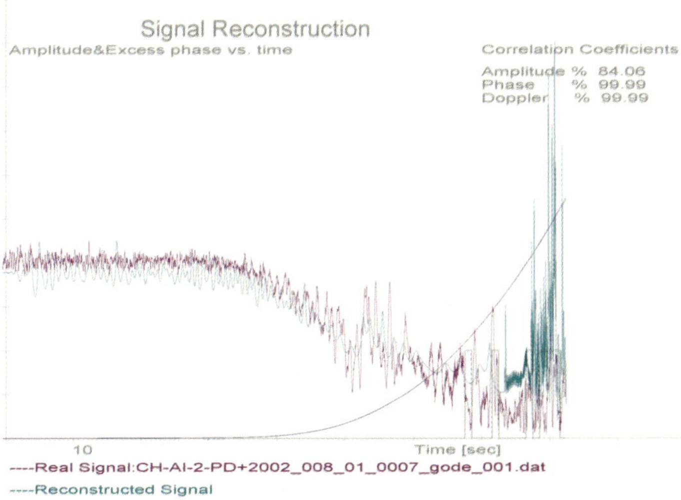

Fig. 8. Reconstruction of CHAMP data. The results show a high degree of similarities especially in the envelope of the signals. The large fluctuations in reconstructed amplitude could be damped if the bending angle had been smoothed more. Fly wheeling is within the boxes.

Errors of type 2 should be detectable in the reconstructed phase because spherical symmetry has been assumed in the reconstruction. In the present examples (Figs. 5, 6, 7, 8, 9 and 10) the phase correlation is always close to 100 %. If this has to be revealed data for selected locations have to be investigated.

Errors of type 3 are seen in Figs. 7, 8 and 9 where fly wheeling is present. The errors seen here are detectable, but they are not more significant than other errors. This could lead to the conclusion that the fly wheeling gives a margin improvement in the results or that detection errors also are present outside the fly wheeling areas.

In all examples the errors of type 4 are present. By changing the smoothing of the bending angle the appearance of the amplitude will change and the correlation coefficient will also change slightly.

In general, it is seen that the match between the signal and the reconstructed signal is best in the start of the occultation. In the last couple of seconds, where the rays are close to the surface of the Earth and the signal to noise ratio is low, the correlation is low. If this part of the signal was excluded much higher correlation coefficients would be obtained.

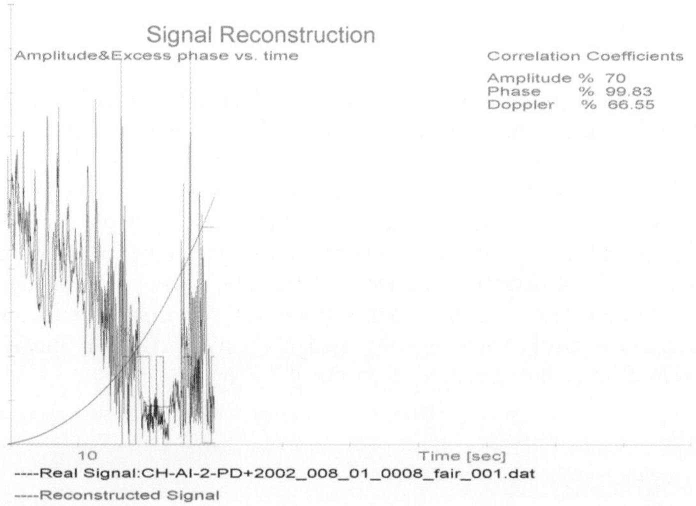

Fig. 9. Reconstruction of CHAMP data. The results show a high degree of similarities especially in the envelope of the signals. The large fluctuations in reconstructed amplitude could be damped if the bending angle had been smoothed more. Fly wheeling is within the boxes. In this the Doppler correlation coefficient is low, which can not be explained unless the signal can be assumed to be corrupted.

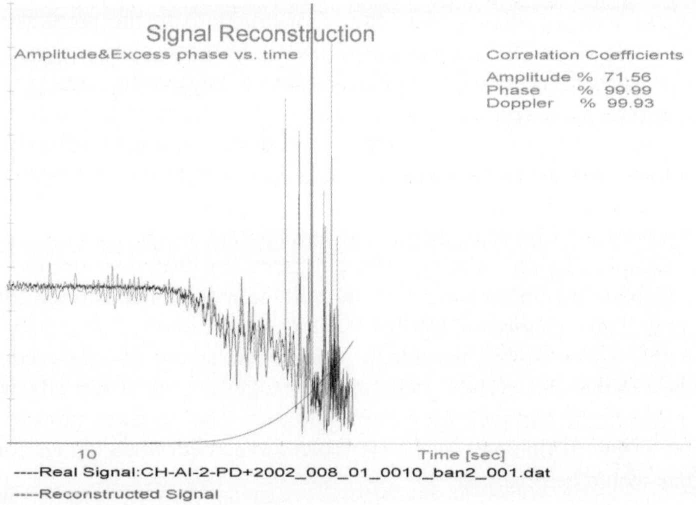

Fig. 10. Reconstruction of CHAMP data. The results show a high degree of similarities especially in the envelope of the signals. The large fluctuations in reconstructed amplitude could be damped if the bending angle had been smoothed more.

3 Conclusions

The results from the signal reconstruction technique investigated here show the possibilities for an evaluation of the quality of the occultation signals, both with respect to the computational algorithms and the assumptions of spherical symmetry. The interpretation of the results is however not simple due to the mixing of errors in the reconstruction. The result from the simulated refractive index profile shows the difficulties with the used algorithms and the smoothing and interpolation technique. This demands further investigation. Ideally, the reconstruction should only reveal errors due to the non-symmetrical atmosphere, noise and detection errors. These errors could then be used when the occultation measurements are assimilated into a numerical weather model.

References

Born M, Wolf E (1999) Principles of Optics. Cambridge University Press, England

Eshleman VR, Muhleman DO, Nicholson PD, Steffes PG (1980) Comment on Absorbing Regions in the Atmosphere of Venus as Measured by Radio Occultation. Icarus 44:793–803

Goodwin ET (1961) Modern Computing Methods. Notes on Applied Science No. 16, London, Her Majesty's Stationary Office

Gorbunov ME (2003) An asymptotic method of modeling radio occultations. J Atmos Solar-Terr Phys 65(16–18):1361–1367

Gorbunov ME, Lauritsen KB (2004) Analysis of wave fields by Fourier integral operators and their application for radio occultations. Radio Sci 39(4), RS4010, doi:10.1029/2003RS002971

Jensen, AS, Benzon IIII, Lohmann MS (2002) A New High Resolution Method for Processing Radio Occultation Data. Scientific Report 02-06, ISSN no. 0905-3263, ISBN no 87-7478-458-7, Danish Meteorological Institute, Copenhagen

Jensen AS, Lohmann MS, Benzon HH, Nielsen AS (2003) Full Spectrum Inversion of Radio Occultation Signals. Radio Sci 38(3):1040, doi:10.1029/2002RS002763

Jensen AS, Lohmann MS, Nielsen AS, Benzon HH (2004a) Geometrical Optics Phase Matching of Radio Occultation Signals. Radio Sci 39, 3009, doi:10.1029/2003rs002899

Jensen, AS, Lohmann MS, Benzon HH, Nielsen AS (2004b) Fourier Analysis of GNSS-LEO Radio Occultation Signals, Resolution and Limitations. In: Kirchengast G, Foelsche U, Steiner AK (eds) Occultations for Probing Atmosphere and Climate. Springer, Berlin, pp 53–60

Jensen AS, Benzon HH, Lohmann MS, Nielsen AS (2006) Processing Radio Occultation Data by Full Spectrum Inversion Techniques: An Overview and Recent Developments. This issue

Pavelyev A, Tsuda T, Igarashi K, Liou YA, Hocke K (2003) Wave structures in the electron density profile in the ionospheric D- and E-layers observed by radio holography analysis of the GPS/MET radio occultation data. Journal of Atmospheric and Solar-Terrestrial Physics 65:59–70

Radio Holographic Filtering of Noisy Radio Occultations

M. E. Gorbunov[1] and K. B. Lauritsen[2]

[1] Institute for Atmospheric Physics, Pyzhevsky per. 3, Moscow 119017, Russia gorbunov@dkrz.de

[2] Danish Meteorological Institute, Lyngbyvej 100, DK-2100 Copenhagen, Denmark

Abstract. We investigate algorithms for filtering noisy radio occultation data with atmospheric multipath behavior in order to improve the accuracy of the inversion based on canonical transform/full spectrum methods. The noise filtering procedure uses the compression of the signal spectrum by multiplying it with a reference signal, Fourier filtering of the narrow-banded signal, and decompression of its spectrum. We study filtering in the time domain and in the impact parameter domain. Our results show that the inversion methods are able to handle additive white noise and high resolution bending angle profiles are obtained. For comparison we also present results where the inversion of multipath behavior is based on using phase data only.

1 Introduction

Recently radio occultation data with atmospheric multipath behavior have attracted much attention. It has been shown that canonical transform/full spectrum inversion (CT, FSI, CT2) methods based on Fourier integral operators can effectively unfold multipath behavior [1, 2, 3, 6]. Furthermore, this approach gives a high accuracy for the retrieved bending angle profiles. Some investigations of the CT/FSI inversion in the presence of white noise were performed in [5].

In the present work, we investigate the effect of noise on the inversion of radio occultations by CT/FSI canonical transform methods. We introduce radio-holographic filtering methods that use reference signals for compression of the spectrum of radio occultation measurements and investigate how this affects the resolution. We find that it is possible to obtain high resolution unfolding of multipath behavior even in the presence of relatively strong noise. For comparison, we also study the case where the inversion is based on the phase data without using the amplitude. It is still possible to unfold the multipath behavior without using amplitude data but it results in reduced accuracy for the bending angle profiles. By applying a model for the amplitude (or using a strongly smoothed version of the measured, noisy amplitude) one can obtain an approximately constant CT/FSI amplitude. Our investigations are based on simulated radio occultation data with multipath behavior.

For a spherical symmetric atmosphere the impact parameter, p, uniquely defines a ray [1, 2] (with horizontal gradients the assumption will be fulfilled to a good approximation). Thus, the procedure for unfolding atmospheric multipath behavior in radio occultations consists in mapping the measured field $u(t) = A(t)\exp(i\phi(t))$ to the p-representation, $w(p)$, using methods based on Fourier integral operators (FIO) and canonical transforms (FSI [6], CT2 [3, 4]). The resulting operator can be written as the composition of multiplication with a reference signal $\exp\left(ik\int f(Y)\mathrm{d}Y\right)$, the Fourier transform, and an amplitude factor:

$$\Phi_2 u(\tilde{p}) = \sqrt{\frac{-ik}{2\pi}} a_2(\tilde{p}, Y_\mathrm{s}(\tilde{p})) \int \exp\left(-ik\tilde{p}Y\right) \exp\left(ik\int_0^Y f(Y')\,\mathrm{d}Y'\right) u(Y)\,\mathrm{d}Y. \quad (1)$$

The coordinate Y depends on time, $Y = Y(t)$, and the stationary phase point $Y_\mathrm{s}(\tilde{p})$ equals $-\tilde{\xi}$, where the momentum $\tilde{\xi}$ is the derivative of the eikonal of the integral term in (1): $w(\tilde{p}) \equiv \Phi_2 u(\tilde{p}) = B(\tilde{p})\exp(ik\Psi(\tilde{p}))$, $\tilde{\xi} \equiv \mathrm{d}\Psi(\tilde{p})/\mathrm{d}\tilde{p}$. This operator maps the wave field to the representation of the approximate impact parameter $\tilde{p}$. Knowing $\tilde{p}$, the exact p can be obtained (practically, the difference between them is small and can be neglected). The factor $a_2(\tilde{p}, Y_\mathrm{s}(\tilde{p}))$ is the amplitude function of the FIO operator and is needed in order to ensure energy conservation (see e.g., [3, 4]). The bending (or refraction) angle, $\epsilon(p)$, is obtained from the derivative of the phase of $w(p)$ in combination with the geometric formula relating the various angles in the occultation geometry.

2 Inversion of Data Without Noise

In Fig. 1 we show the amplitude of a simulated field with a simple atmospheric multipath behavior that we used in our investigations. The simulations have been performed by a combined multiple phase screen method and asymptotic modeling [3].

In Figs. 2–4 we show the retrieved bending angle profiles $\epsilon(p)$ with CT and GO (geometric optics) algorithms (without doing noise filtering). The left panels also show the (sliding) spectra in the (p, ϵ)-space (where ray height is p minus the Earth local curvature radius). The right panels show the CT amplitudes as function of ray height. One observes that high resolution bending angle profiles can be obtained even in the case where the amplitude is not used. With the strongly smoothed amplitude, it is observed that the CT amplitude is not constant in the multipath region (in the p-representation). This reflects the fact that the strong smoothed version of the simulated amplitude does not respect the conservation of energy.

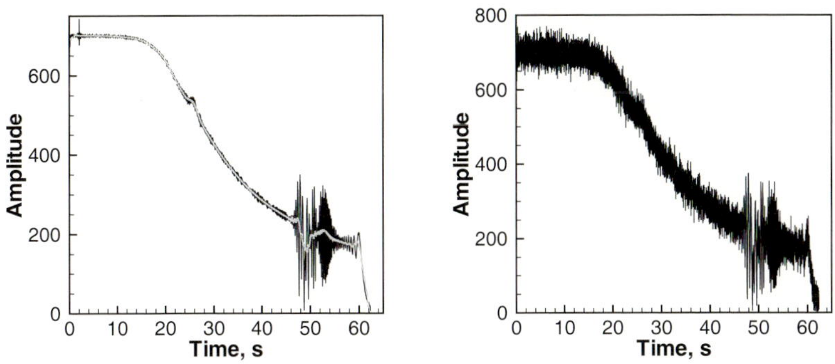

Fig. 1. Amplitude of simulated field without noise with multipath behavior (left). The superimposed line (light gray) is a strongly smoothed version of the amplitude. The simulated field with additive Gaussian white noise with a strength of 30 [5] (right).

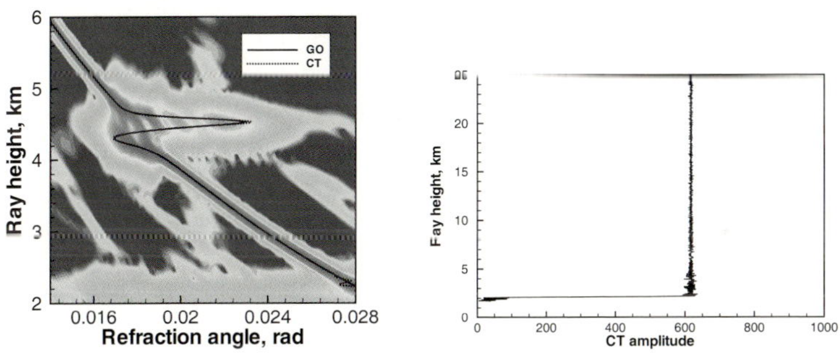

Fig. 2. Bending angle and (p, ϵ)-spectrum (left) and CT amplitude as function of ray height (right). The results are obtained directly from the simulated amplitude.

3 Phase Models

We construct smooth models for the phase variations, $\phi_{\mathrm{m}}(t)$ and $\Psi_{\mathrm{m}}(p)$, based on the measured signal and on the retrieved bending angle profile without doing any noise filtering. The measured signal $u(t)$ and the transformed signal $w(p)$ are then multiplied by the phase model reference signals as follows:

$$u(t) \to u_{\mathrm{m}}(t) \equiv u(t) \exp(-i\phi_{\mathrm{m}}(t)), \tag{2}$$

where the phase model $\phi_{\mathrm{m}}(t)$ is obtained by smoothing the phase of the measured wave field $u(t)$.

Analogously, for the $w(p)$ field:

$$w(p) \to w_{\mathrm{m}}(p) \equiv w(p) \exp(-ik\Psi_{\mathrm{m}}(p)), \tag{3}$$

where the phase model $\Psi_{\mathrm{m}}(p)$ is obtained by smoothing the phase of $w(p)$.

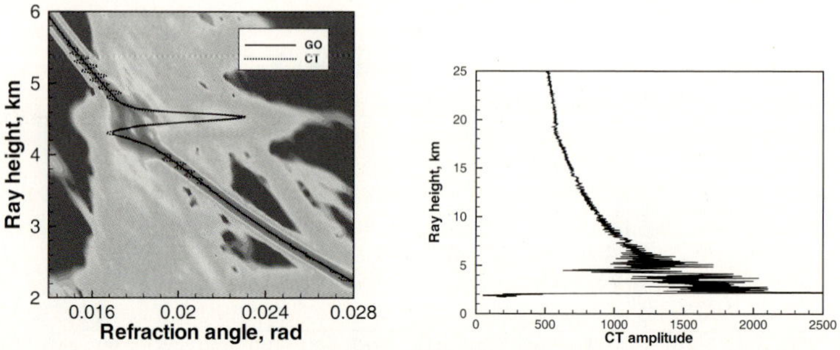

Fig. 3. Bending angle and (p, ϵ)-spectrum (left) and CT amplitude as function of ray height (right). The results are obtained by using a constant amplitude ($A \equiv 1$).

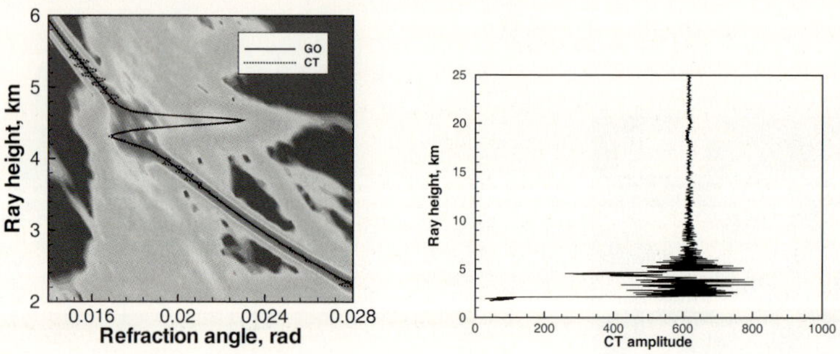

Fig. 4. Bending angle and (p, ϵ)-spectrum (left) and CT amplitude as function of ray height (right). The results are obtained by using a strongly smoothed version of the simulated amplitude.

4 Radio Holographic Noise Filtering

The filtering of the noisy signals are done in three steps: first, the signals are multiplied by reference signals. This step ensures that the spectrum will be narrow-banded. Next, the modified signals are Fourier transformed, a Gaussian weighting function applied, and inverse Fourier transformed. Finally, the reference signals are removed. In schematized form, the steps are as follows for the noise filtering in the t-domain:

$$
\begin{aligned}
u(t) &\rightarrow u_{\rm m}(t) \qquad [\text{add } \phi_{\rm m}(t)] && (4)\\
&\rightarrow \tilde{u}_{\rm m}(\omega) \qquad [F] && (5)\\
&\rightarrow \tilde{u}_{\rm m}^{F_t}(\omega) \equiv \tilde{u}_{\rm m}(\omega)\exp(-\omega^2/2\sigma_\omega^2) \qquad [\text{Gaussian filtering}] && (6)\\
&\rightarrow u_{\rm m}^{F_t}(\omega) \qquad [F^{-1}] && (7)\\
&\rightarrow u^{F_t}(t) \qquad [\text{remove } \phi_{\rm m}(t)] && (8)
\end{aligned}
$$

Next, apply Φ_2 in order to obtain the field in the p-representation, $w^{F_t}(p)$, and finally, obtain the bending angle $\epsilon^{F_t}(p)$ from the derivative of the phase of $w^{F_t}(p)$.

For the noise filtering in the p-domain it reads:

$$w(p) \to w_{\mathrm{m}}(p) \qquad [\text{add } \Psi_{\mathrm{m}}(p)] \tag{9}$$

$$\to \tilde{w}_{\mathrm{m}}(\xi) \qquad [F] \tag{10}$$

$$\to \tilde{w}_{\mathrm{m}}^{F_p}(\xi) \equiv \tilde{w}_{\mathrm{m}}(\xi)\exp(-\xi^2/2\sigma_\xi^2) \qquad [\text{Gaussian filtering}] \tag{11}$$

$$\to w_{\mathrm{m}}^{F_p}(\xi) \qquad [F^{-1}] \tag{12}$$

$$\to w^{F_p}(p) \qquad [\text{remove } \Psi_{\mathrm{m}}(p)] \tag{13}$$

Next, obtain the bending angle $\epsilon^{F_p}(p)$ from the derivative of the phase of $w^{F_p}(p)$.

5 Inversion of Noisy Data

The results of our noise filtering procedure are shown in Figs. 5–7 where we show the retrieved bending angle profiles, the (p,ϵ)-spectrum for the noisy simulation, and the CT/FSI amplitudes. The results have been obtained by using a smoothing window of a size corresponding to about 30 m. The phase model $\phi_{\mathrm{m}}(t)$ has been obtained by smoothing at the scale 2 s (corresponding to about 4 km in the vertical scale in the sliding spectra). The width of the filter was chosen to be $\sigma_\omega = 200\ \mathrm{s}^{-1}$, corresponding to impact parameter intervals of $\Delta p \approx 4$ km ($\omega \approx kp\Omega$, with $k \approx 33\ \mathrm{m}^{-1}$, and the derivative of the satellite-to-satellite angle θ being approximately $\Omega = d\theta/dt \approx 0.0015$ rad/s for the simulation we have investigated). For $\Psi_{\mathrm{m}}(p)$, the smoothing scale was chosen to correspond to the radio-holographic filter width, thus $\sigma_\xi = 0.005$ rad.

Figure 6 shows that the noise filtering has removed some energy from the signal since the CT amplitude is no longer (approximately) constant. The sharp decrease in amplitude seen around 5 km signals that the original multipath behavior (in the t-space) has been slightly corrupted by the Gaussian filtering (in the ω-space). This can be traced to the fact that the width of the multipath spectrum has been larger than the used noise filtering width σ_ω. Phrased differently, the unfolding of the multipath behavior is not complete in the ω-space, therefore the spectrum will be slightly broader in the multipath region than outside it (i.e., in the single path regions) and accordingly the weighting with a Gaussian can slightly cut off the spectrum width.

On the contrary, the filtering in the ξ-space, using the impact parameter reference model $\Psi_{\mathrm{m}}(p)$, has not corrupted the multipath spectrum (Fig. 7). This is also to be expected, since the multipath behavior will be completely unfolded in the p-representation. Thus, the field $\tilde{w}(\xi)$ will be more narrow-banded than $\tilde{u}(\omega)$ and therefore the noise filtering of $\tilde{w}(\xi)$ will not interfere with the multipath behavior. In this sense, the filtering in the ξ-space is optimal.

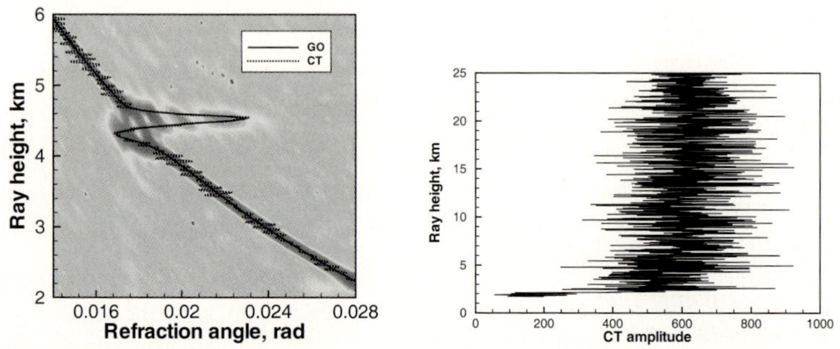

Fig. 5. Bending angle and (p, ϵ)-spectrum (left) and CT amplitude as function of ray height (right). The bending angle $\epsilon(p)$ and CT amplitude are obtained without any noise filtering.

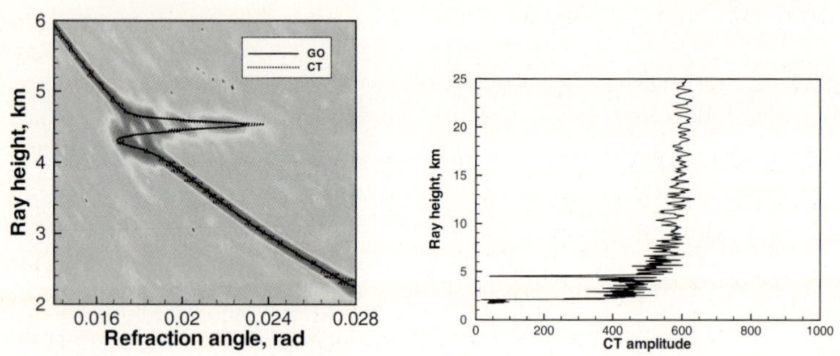

Fig. 6. Bending angle and (p, ϵ)-spectrum (left) and CT amplitude as function of ray height (right). The bending angle $\epsilon^{F_t}(p)$ and CT amplitude are obtained with the $\phi_{\mathrm{m}}(t)$ phase model.

We have carried out simulations for different strengths of the noise level. We find that the $\phi_{\mathrm{m}}(t)$ filtering breaks down for strengths about 100 whereas the $\Psi_{\mathrm{m}}(p)$ filtering breaks down for strengths about 50. Thus, the filtering in the ω-domain seems to be more robust (but note the fact that the spectrum in this space is broader and thus the filtering may slightly disturb the multipath structure). One way to interpret this is that the (small) suppression of some of the multipath structures seems to have created a slightly more robust signal.

6 Conclusions

We have investigated the effect of noise on CT/FSI retrieval methods. To this end, we have introduced a filtering approach based on phase models. The noise reduction is performed in either the ω-space (conjugate to the time domain) or the ξ-space (conjugate to the impact parameter domain). We find that

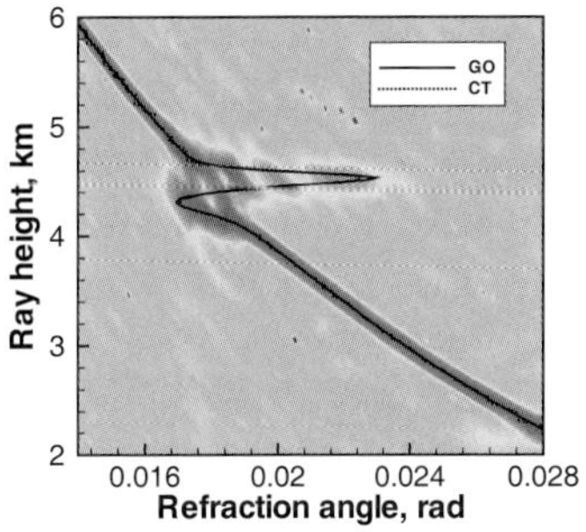

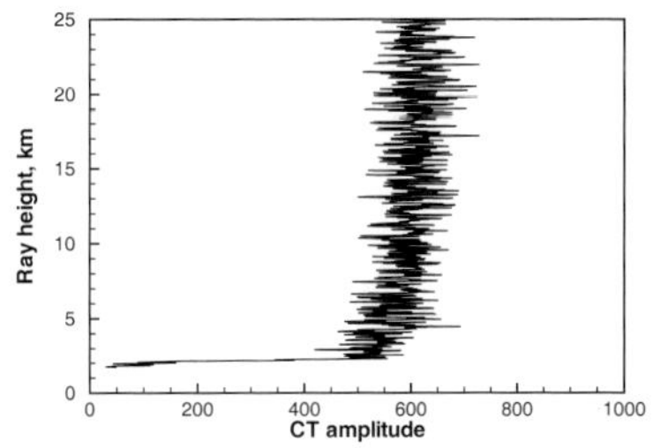

Fig. 7. Bending angle and (p, ϵ)-spectrum (left) and CT amplitude as function of ray height (right). The bending angle $\epsilon^{F_p}(p)$ and CT amplitude are obtained with the $\Psi_{\mathrm{m}}(p)$ phase model.

high resolution bending angle profiles can be obtained, even in the presence of relatively strong noise levels. Thus, FIO-based retrieval methods are robust with respect to additive wide band noise. However, the noise filtering breaks down for very high noise levels.

We also investigated the case where the inversion is performed using the phase signal alone (and a constant or strongly smoothed amplitude signal). In such cases, the CT/FSI amplitude is no longer constant, signaling that the inversion did not respect the conservation of energy. On the other hand, by imposing a constant CT/FSI amplitude, and mapping such a signal back to the time domain, one can construct a model for the measured amplitude that can reveal the structure of the multipath behavior in the signal.

Acknowledgements. This work has been supported by the EUMETSAT GRAS Meteorology SAF project.

References

[1] Gorbunov ME (2002) Radio-holographic analysis of Microlab-1 radio occultation data in the lower troposphere. J of Geophys Res 107(D12), 4156, doi:10.1029/2001JD000889;

[2] Gorbunov ME (2002) Canonical transform method for processing GPS radio occultation data in the lower troposphere. Radio Science 37(5), 1076, doi:10.1029/2000RS002592

[3] Gorbunov ME, Lauritsen KB (2004) Analysis of wave fields by Fourier integral operators and their application for radio occultations. Radio Science 39(4), RS4010, doi:10.1029/2003RS002971

[4] Gorbunov ME, Lauritsen KB (2005) Canonical Transform Methods for Analysis of Radio Occultations. In: Reigber C, Luehr H, Schwintzer P, and Wickert J

(eds) Earth Observation with CHAMP: Results from Three Years in Orbit. Springer Verlag, Berlin, pp 519–524

[5] Gorbunov ME, Benzon HH, Jensen AS, Lohmann MS, Nielsen AS (2004) Comparative Analysis of Radio Occultation Processing Approaches Based on Fourier Integral Operators. Radio Science 39(6), RS6004, doi:10.1029/2003RS002916

[6] Jensen AS, Lohmann MS, Benzon HH, and Nielsen AS (2003) Full spectrum inversion of radio occultation signals. Radio Science 38(3), 1040, doi:10.1029/2002RS002763

Future GNSS Occultation Missions and the LEO-LEO Occultation Concept

Preparing for COSMIC: Inversion and Analysis of Ionospheric Data Products

S. Syndergaard[1], W. S. Schreiner[1], C. Rocken[1], D. C. Hunt[1], and K. F. Dymond[2]

[1] COSMIC Project Office, University Corporation for Atmospheric Research, Boulder, CO, U.S.A.
ssy@ucar.edu

[2] Thermospheric and Ionospheric Research and Application Group, Naval Research Laboratory, Washington, DC, U.S.A.

Abstract. The Constellation Observing System for Meteorology, Ionosphere, and Climate (COSMIC) is scheduled for launch in 2006. COSMIC will consist of six low earth orbiting satellites in planes separated by 24° to provide global atmospheric and ionospheric observations. One of the goals is to demonstrate near real-time processing of data products for numerical weather prediction and space weather applications. Each COSMIC satellite will carry three payloads: (1) a Global Positioning System (GPS) occultation receiver with two high-gain limb viewing antennas and two antennas for precision orbit determination, (2) a Tiny Ionospheric Photometer (TIP) for monitoring the electron density via nadir radiance measurements along the sub-satellite track, and (3) a Tri-Band Beacon (TBB) transmitter for ionospheric tomography and scintillation studies. The data from all these payloads will be processed at the COSMIC Data Analysis and Archival Center (CDAAC). Here we give an overview of the ionospheric data products from COSMIC and focus on the plans and preliminary simulation studies for analyzing the ionospheric occultation data and combining them with ground-based GPS, TIP, and TBB observations.

1 Introduction

The Constellation Observing System for Meteorology, Ionosphere, and Climate (COSMIC) is a joint Taiwan–U.S. mission with the goal to launch six low earth orbiting (LEO) micro-satellites in early 2006. All six satellites will be launched together on a Minotaur rocket. After the launch vehicle reaches the injection orbit, the satellites will be released one by one. During the following thirteen months, the satellites will slowly be distributed to their final configuration in six different orbital planes at 750 km to 800 km altitude (all orbits will be circular), with 72° inclination and 24° separation. During this deployment phase the satellites will be fully operational. The expected life-time of the mission is about five years.

Each COSMIC satellite will carry three payloads to study the Earth's neutral atmosphere and ionosphere: A Global Positioning System (GPS) receiver, connected to four antennas (two limb viewing antennas for neutral atmospheric radio occultation sounding and two antennas for precise orbit determination and ionospheric monitoring), will provide data for atmospheric and ionospheric research, weather prediction, and climate change studies. A Tiny Ionospheric Photometer (TIP) will measure the ultraviolet emission due to recombination of oxygen ions and electrons in the ionosphere along the sub-satellite track on the Earth's night-side. Finally, a Tri-Band Beacon (TBB) will transmit radio signals on three frequencies (150 MHz, 400 MHz, and 1067 MHz) which will be received by chains of receivers on the ground with the main goal to determine the line-of-sight total electron content (TEC) and ionospheric scintillation levels. A collection of papers with a detailed description of COSMIC and its potential science applications can be found in [12].

One key objective of COSMIC is to demonstrate the value of the radio occultation (RO) data for weather forecasting and inclusion in space weather models. Thus, COSMIC "real-time" data products will be available to researchers and leading numerical weather prediction centers worldwide within less than 150 minutes of data collection. In this paper we focus on the plans and ongoing preparations at the COSMIC Data Analysis and Archival Center (CDAAC) in Boulder, Colorado, for analyzing the ionospheric data anticipated from COSMIC.

2 Ionospheric Data Products

At the time of writing, CDAAC considers to provide the following baseline ionospheric data products from COSMIC:

GPS receiver:

- High-resolution (1 Hz) absolute TEC to all GPS satellites in view at all times (useful for global ionospheric tomography and assimilation into space weather models).
- Occultation TEC and derived electron density profiles.
- Scintillation parameters for the GPS transmitter–LEO receiver links.

Tiny Ionospheric Photometer:

- Nadir intensity on the night-side (along the sub-satellite track) from radiative recombination emission at 1356 Å(135.6 nm).
- Derived F-layer peak density and critical frequency (foF2).
- Location and intensity of ionospheric anomalous structures such as the Auroral oval.

Tri-Band Beacon:

- Phase and amplitude of radio signals at 150, 400, and 1067 MHz transmitted from the COSMIC satellites and received by chains of ground receivers.

- TEC between the COSMIC satellites and the ground receivers.
- Scintillation parameters for the LEO transmitter–ground receiver links.

In addition to providing these baseline ionospheric products, CDAAC will also work to combine different data types to provide improved products for ionospheric research. For example, it is well known that the accuracy of RO-derived electron density profiles is limited by horizontal ionospheric gradients. The TIP on each satellite, as well as the signals received on ground from the TBB transmitters, will provide valuable information about the ionospheric horizontal gradients in the vicinity of the occultations. Thus, the ionospheric occultation data are complementary to the data from the TIP and TBB instruments and it is anticipated that the different observations can be combined to improve derived electron density profiles and to estimate two-dimensional (2D) electron density structure in the plane of occultation.

3 Ionospheric Profiles from Occultations

During the first months after launch, the COSMIC satellites will gradually be lifted into their final orbits. Thus, at the beginning of the mission, most ionospheric occultations will start at a relatively low altitude (450 km to 500 km), similar to the altitude of the German CHAMP satellite at the beginning of its mission. As practice, CDAAC has therefore begun the processing of a subset of the CHAMP ionospheric RO data. Figure 1 shows a few examples of derived electron density profiles from CHAMP differential (L1−L2) phase observations, using the not always valid assumption of local spherical symmetry (presumably giving rise to large errors below the F-layer).

The electron density at the orbit altitude was obtained from the observed TEC near the orbit altitude using a novel approach that will be described in more detail in a forthcoming paper. Disregarding horizontal gradients, it can be shown that the TEC for tangent radius, r, just below the orbit altitude is related to the electron density, N_e, at the orbit altitude, as

$$\mathrm{TEC}(r) - \mathrm{TEC}(r_\mathrm{orb}) \approx \sqrt{2r_\mathrm{orb}}N_\mathrm{e}(r_\mathrm{orb})\sqrt{r_\mathrm{orb} - r}\ , \qquad (1)$$

where r_orb is the radius at the orbit altitude. Essentially, the electron density at orbit altitude at the beginning of an occultation was derived from the occultation data by fitting a square root function to the uppermost few kilometers (about 10 km) of TEC observations. Equation (1) was derived under the assumption of a circular satellite orbit and constant electron density along the orbit track. The latter assumption may cause a significant error in the estimate of the electron density using (1). In Fig. 1 the asterisks indicate the in situ electron density provided by the Planar Langmuir Probe on board CHAMP. The comparisons to the uppermost points of the electron density profiles indicate errors (almost 20 % in one case) in the derived electron density at the top of the profiles. This is presumably due to horizontal gradients along the

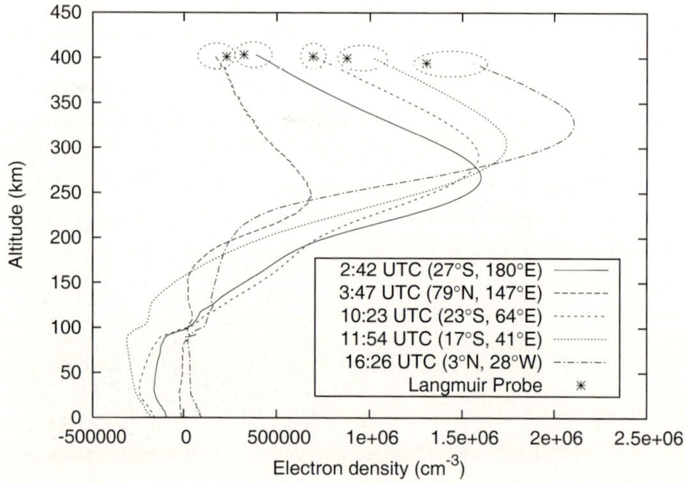

Fig. 1. Examples of retrieved electron density profiles from CHAMP occultations on October 12, 2003. Corresponding electron density measured by the CHAMP Langmuir Probe is indicated by an asterisk at the top of each profile.

orbit track, not accounted for in (1). An alternative approach [11], uses an adaptive electron density model of the topside ionosphere and plasmasphere in the inversion of CHAMP ionospheric occultation observations.

The profiles in Fig. 1 have been processed from so-called calibrated TEC [14], an approach to estimate the occultation TEC below the orbit. For the processing of CHAMP data, the calibration method was modified using the estimated electron density at the satellite orbit and assuming exponential decay of the electron density above the orbit (CHAMP does not collect positive elevation angle data necessary to apply the calibration as described in [14]).

4 Combining TIP and Occultation Data

The TIP will provide nadir observations of radiative recombination emission at 1356 Å, with a temporal resolution of several seconds. These observations will give information about the horizontal ionospheric gradients along the sub-satellite track, and can be used in conjunction with the GPS occultation data to estimate the 2D electron density structure in the occultation plane (assuming that the occultation plane is near coincident with the orbit plane). Figure 2 shows the setup for a simulation experiment, using the IRI-90 ionosphere, where the occultation takes place in a region of large horizontal gradients. Synthetic data were obtained as integrated electron density (occultation data) and integrated squared electron density (radiation data). These data were then inverted using weighted least squares according to assumed error

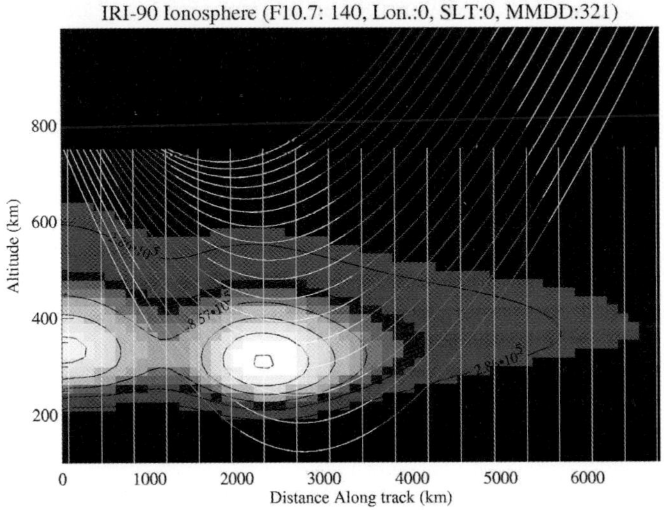

Fig. 2. Simulations of GPS occultation measurements (curved lines across the image) and TIP measurements (vertical lines) through the IRI-90 ionosphere.

covariances (see [4] for more details). The reconstruction algorithm was based on a parameterization of the vertical structure assumed to be a generalized Chapman profile, with the parameters being the height and density at the F-layer peak, as well as three parameters describing an altitude dependent scale height. Fifty-six parameters were used to parameterize the horizontal variation via the F-layer peak density.

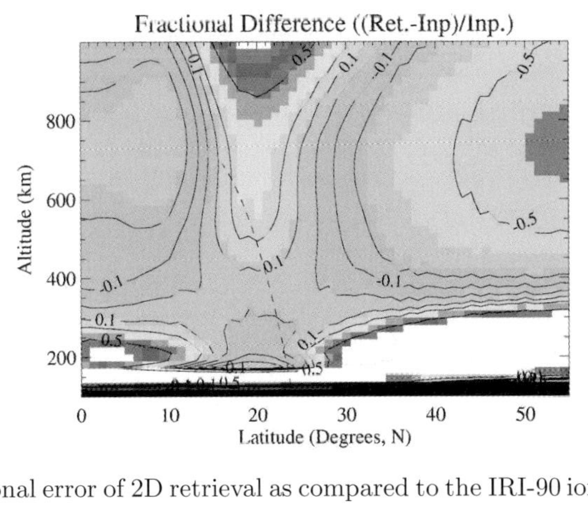

Fig. 3. Fractional error of 2D retrieval as compared to the IRI-90 ionosphere. Dashed line rising from approximately 24°N represents the tangent point trajectory.

Figure 3 shows the fractional reconstruction error as compared to the "truth" (the IRI-90 ionosphere) in the simulation experiment. Although large fractional errors occur far from the occultation tangent points, the result near the tangent points indicates the value of the TIP measurements in conjunction with the occultation data. Results from simulation experiments combining space-based UV radiance measurements with ionospheric occultation data have also been reported in [3, 8, 16].

5 Ionospheric Scintillations

One of the objectives of the TBB is global monitoring of ionospheric scintillations [1]. Ionospheric scintillations on satellite to ground links are often associated with plasma bubbles or sharp electron density gradients. Measurements of phase and amplitude scintillations at 150 MHz, 400 MHz, and 1067 MHz, will provide valuable data for scintillation studies and for generation of global scintillation maps.

Another kind of scintillation will be measured with the GPS occultation receiver at tangent altitudes around 100 km. It is hypothesized that this kind of scintillation arises as a result of sporadic E-layers [6, 7]. Figure 4 shows an example from the proof-of-concept GPS/MET radio occultation experiment (launched in April 1995) where the phases and amplitudes at the beginning of a setting occultation (50 Hz sampling rate starting at about 120 km) exhibit large oscillations, characteristic of multipath propagation, presumably caused by a sporadic E-layer. Simulations of radio occultation data affected by

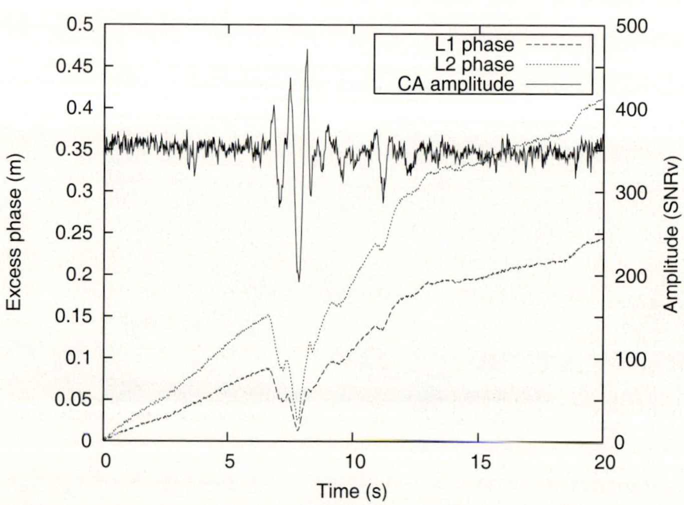

Fig. 4. Excess phase and amplitude over the first 20 s of a GPS/MET occultation (50 Hz) which occurred near 30°N, 105°W at 8:09 UTC on February 4, 1997.

multipath propagation in the lower troposphere [2, 5] show similar characteristic excess phase depletions as the ones seen around 8 s (at tangent altitudes around 100 km) in Fig. 4. Thus, it might be possible to detect sporadic E-layers globally using the occultation data. Additionally, it may be possible to localize ionospheric irregularities along the occultation path [15] and investigate the vertical structure associated with sporadic E-layers by inversion based on thin screen model wave propagation.

6 Ionospheric Tomography and Assimilation

The ionospheric RO data from COSMIC will contain valuable high-resolution information about the vertical electron density gradients, but also entangled information about the horizontal structure in the occultation plane. One way of separating the vertical and horizontal information is to combine the RO data with a priori information from an ionospheric model [8]. This can be done within the framework of ionospheric tomography using the RO TEC data. Figure 5 shows the result of combining the data from a GPS/MET occultation with the NeUoG climatological ionospheric model [13]. In the tomographic reconstruction algorithm, the ionosphere was divided into 1000 layers and 45 horizontal bins over a 60° span. The inversion took into account very large a priori uncertainties and error correlations in the NeUoG model, such that the occultation data were heavily weighted, while the NeUoG model mostly contributed with important information about large-scale horizontal gradients (see [9] for more details).

An alternative approach which will be considered for the COSMIC data is 2D variational analysis (or assimilation) of the retrieved electron density profiles using a refractive index mapping operator [17]. Within this framework, it

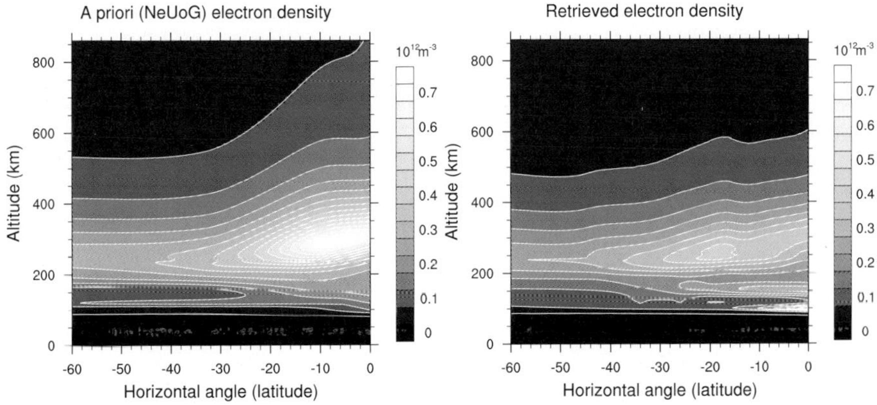

Fig. 5. A priori (left) and tomographically reconstructed (right) electron density in the occultation plane for an ionospheric GPS/MET occultation which occurred near 28°S at ~9:30 LT on February 20, 1997.

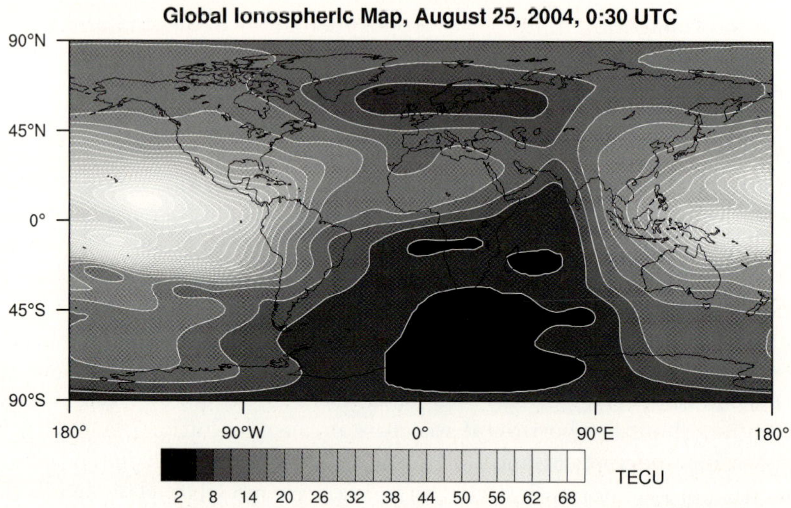

Fig. 6. Example of Global Ionospheric Map; data by courtesy of JPL.

will also be considered to include the 50 Hz data collected by the limb antennas at tangent altitudes below ~ 120 km to produce electron density profiles in the lower part of the ionosphere with very high vertical resolution. Using the mapping operator, it should be possible to include correction for multipath propagation generated by sharp E-layer gradients (cf. Fig. 4), something which tomographic reconstruction does not allow for.

It will also be considered to combine the RO data with data from Global Ionospheric Maps (GIMs) (Fig. 6), as well as – when applicable – the data of similar nature from the TIP and the TBB transmitter. GIMs of vertical TEC are generated on a regular basis from a global network of ground-based GPS receivers. For general near real-time processing, CDAAC will most likely implement a simple approach [10] using the vertical TEC from GIMs to mitigate the effect of horizontal gradients in the retrieval of electron density profiles. At the same time, this approach gives a rough estimate of the three-dimensional electron density distribution in the vicinity of the occultation tangent points. The GIMs currently available from the Jet Propulsion Laboratory (JPL) has a temporal resolution of one hour and a spatial resolution of 2° by 2°.

7 Summary

The six satellite COSMIC mission, scheduled for launch in early 2006, is expected to provide a large amount of data useful for atmospheric sciences, numerical weather prediction, climate research, and space weather studies. In this paper we have given an overview of the COSMIC mission with a focus on the ionospheric data products that will be used for ionospheric monitoring and space weather research. Three instruments on board each COSMIC satellite will provide ionospheric data. The GPS receiver payloads will probe the ionosphere up to about 800 km using the RO technique, and beyond that they will measure the TEC to all GPS satellites in view. The Tiny Ionospheric Photometers will measure the nadir intensity from radiative recombination emission along the sub-satellite tracks, providing valuable information about horizontal gradients on the night-side ionosphere. Finally, the Tri-Band Beacons will provide TEC and measure scintillations on satellite-to-ground links. It is expected that the information on horizontal electron density gradients from ionospheric models, GIMs, TIP, and/or TBB observations, in combination with the occultation data, will improve electron density profiling for COSMIC, and perhaps even allow high-resolution estimates of the two- and three-dimensional electron density distributions in the vicinity of the occultations. In combination, it is anticipated that the ionospheric data from the COSMIC constellation will provide researchers with unprecedented high-resolution, global coverage information about the ionosphere and its spatial and temporal variations.

Acknowledgements. Development of the COSMIC Data Analysis and Archival Center (CDAAC) is primarily supported by NSF-ATM (ATM #0410018), and by NOAA (NAO1AANEG0362). ONR is supporting ionospheric research at the CDAAC under contract #N000014-00-C-0291.

References

[1] Bernhardt PA, Selcher CA, Basu S, Bust G, Reising SC (2000) Atmospheric studies with the Tri-Band Beacon instrument on the COSMIC constellation. Terrestrial, Atmospheric and Oceanic Sciences 11:291–312

[2] Beyerle G, Gorbunov ME, Ao CO (2003) Simulation studies of GPS radio occultation measurements. Radio Sci 38:1084, doi:10.1029/2002RS002800

[3] Dymond KF, Thomas RJ (2001) A technique for using measured ionospheric density gradients and GPS occultations for inferring the nighttime ionospheric electron density. Radio Sci 36:1141–1148

[4] Dymond KF, Nee JB, Thomas RJ (2000) The Tiny Ionospheric Photometer: An instrument for measuring ionospheric gradients for the COSMIC constellation. Terrestrial, Atmospheric and Oceanic Sciences 11:273–290

[5] Gorbunov ME, Gurvich AS (1998) Algorithms of inversion of Microlab-1 satellite data including effects of multipath propagation. Int J Remote Sens 19:2283–2300

[6] Gorbunov ME, Gurvich AS, Shmakov AV (2002) Back-propagation and radioholographic methods for investigation of sporadic ionospheric E-layers from Microlab-1 data. Int J Remote Sens 23:675–685

[7] Hajj GA, Romans LJ (1998) Ionospheric electron density profiles obtained with the Global Positioning System: Results from the GPS/MET experiment. Radio Sci 33:175–190

[8] Hajj GA, Lee LC, Pi X, Romans LJ, Schreiner WS, Straus PR, Wang C (2000) COSMIC GPS ionospheric sensing and space weather. Terrestrial, Atmospheric and Oceanic Sciences 11:235–272

[9] Herman BM, Kursinski ER, Feng D, Flittner D, Ward D, Syndergaard S, Lane E (2003) Active tropospheric ozone and moisture sounder (ATOMS). Science Report, NASA contract no. NAS1-99055, Institute of Atmospheric Physics, The University of Arizona, Tucson, Arizona

[10] Hernández-Pajares M, Juan JM, Sanz J (2000) Improving the Abel inversion by adding ground GPS data to LEO radio occultations in ionospheric sounding. Geophys Res Lett 27:2473–2476

[11] Jakowski N, Wehrenpfennig A, Heise S, Reigber C, Lühr H, Grunwaldt L, Meehan TK (2002) GPS radio occultation measurements of the ionosphere from CHAMP: Early results. Geophys Res Lett 29:1457, doi:10.1029/2001GL014364

[12] Lee LC, Rocken C, Kursinski R (eds) (2001) Applications of Constellation Observing System for Meteorology, Ionosphere & Climate. Springer, Hong Kong

[13] Leitinger R, Titheridge JE, Kirchengast G, Rothleitner W (1996) Ein "einfaches" globales empirisches Modell für die F-Schicht der Ionosphäre. Kleinheubacher Berichte 39:697–704, English version available from reinhart.leitinger@uni-graz.at

[14] Schreiner WS, Sokolovskiy SV, Rocken C, Hunt DC (1999) Analysis and validation of GPS/MET radio occultation data in the ionosphere. Radio Sci 34:949–966

[15] Sokolovskiy S, Schreiner W, Rocken C, Hunt D (2002) Detection of high-altitude ionospheric irregularities with GPS/MET. Geophys Res Lett 29:1033, doi:10.1029/2001GL013398

[16] Straus PR (1999) Correcting GPS occultation measurements for ionospheric horizontal gradients. In: Goodman JM (ed) Proceedings of the Ionospheric Effects Symposium. Natl Telecommun and Inf Serv, Alexandria, Virginia, 700–709

[17] Syndergaard S, Kursinski ER, Herman BM, Lane EM, Flittner DE (2005) A refractive index mapping operator for assimilation of occultation data. Mon Weather Rev (in press)

The Operational EPS GRAS Measurement System

J.-P. Luntama

EUMETSAT, Am Kavalleriesand 31, D-64295 Darmstadt, Germany
luntama@eumetsat.de

Abstract. GRAS (**G**lobal Navigation Satellite System **R**eceiver for **A**tmospheric **S**ounding) is an advanced GPS receiver that has been developed by the European Space Agency (ESA) and EUMETSAT for Radio Occultation (RO) measurements in the framework of the EUMETSAT Polar System (EPS). The space segment of the EPS program consists of a series of three Metop satellites, to be flown successively in the years from 2006. The GRAS data products retrieved by the EPS Ground Segment can be assimilated into NWP (Numerical Weather Prediction) models. The GRAS products will potentially have a positive impact on the forecast if they can be delivered within 3 hrs from each observation made by the GRAS instrument with the required accuracy. A positive impact on the forecast by using RO data from the CHAMP mission has already been demonstrated (Healy et al. 2005; Healy and Thepaut 2005). The operational Near Real Time (NRT) processing and dissemination of RO products has not yet been demonstrated and is a challenge, expected to be met for the first time by the EPS/Metop mission. GeoForschungsZentrum Potsdam (GFZ) has shown good product timeliness results for the CHAMP mission by being able to disseminate data within 4 hrs from the measurements (Wickert et al 2005).

1 GRAS Product Requirements

GRAS data products generated and disseminated by the EPS Ground Segment are split into two categories based on the processing level: GRAS level 1b products containing bending angle profiles, and GRAS level 2 products containing refractivity, temperature and humidity profiles.

The accuracy and timeliness requirements for the GRAS level 1b and level 2 products are shown in Table 1. The accuracy requirement for the level 1b total bending angle profile is 1 μrad or 0.4 %, which ever is higher. This enables retrieval of the temperature and humidity profiles in the level 2 products with the specified accuracies. It should be noted that the rising occultations will typically not start from the surface level but from the height of a few kilometers because the receiver requires some time to acquire the GPS signal.

Table 1. GRAS product accuracy requirements.

		Level 1b products	Level 2 products	
		Total bending angle	Specific humidity	Temperature
Coverage		Global	Global	Global
Horizontal sampling		The average distance between individual soundings over a period of 12 hrs is less than 1000 km		
Vertical range		Surface – 80 km	Surface – 100 hPa	Surface – 1 hPa
Vertical sampling rate[1]	0 – 5 km	2 – 5 Hz	0.4 – 2 km	0.3 – 3 km
	5 – 15 km	2 – 5 Hz	1 – 3 km	1 – 3 km
	15 – 35 km	2 – 5 Hz	-	1 – 3 km
	35 – 50 km	2 – 5 Hz	-	1 – 3 km
RMS accuracy[2]	0 – 5 km	1 μrad or 0.4 %[3]	0.25 – 1 g/kg[4]	0.5 – 3 K
	5 – 15 km	1 μrad or 0.4 %[3]	0.05 – 0.2 g/kg[4]	0.5 – 3 K
	15 – 35 km	1 μrad or 0.4 %[3]	-	0.5 – 3 K
	35 – 50 km	1 μrad or 0.4 %[3]	-	0.5 – 5 K
Timeliness		2 h 15 min	3 h	3 h

[1] After noise filtering.
[2] With no systematic biases.
[3] Whichever is greater.
[4] Equivalent to a requirement of 5 % in Relative Humidity.

Table 1 contains also the timeliness constrains for the GRAS level 1b and level 2 product dissemination, and the required effective vertical sampling rates for the product profiles.

2 GRAS Measurement System

The elements of the GRAS measurement system are presented in Fig. 1. The system design drivers have been the product accuracy and the timeliness of dissemination (see Table 1).

The main elements of the GRAS system developed for the EPS mission are the GRAS receiver onboard the Metop spacecraft, the satellite receiving antennas at the EPS Polar Site, the EPS Core Ground Segment (CGS), the GRAS Ground Support Network (GSN) Service, and the GRAS Meteorology Satellite Application Facility (SAF). More detailed descriptions of these elements are provided below.

The GRAS Receiver

The GRAS instrument performs occultation measurements with two high gain antennas pointing to the flight and to the anti-flight directions of the Metop

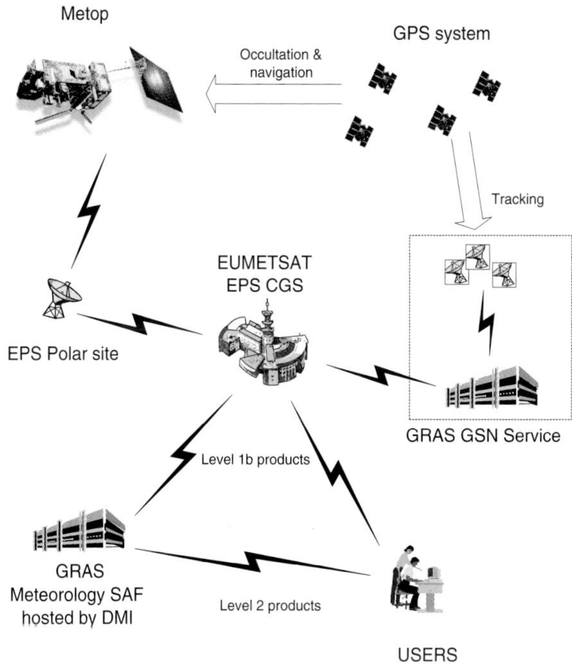

Fig. 1. Elements of the EPS GRAS measurement system.

spacecraft. These antennas will enable GRAS to measure about 500 occultation measurements (rising and setting) per day. A third, hemispherical antenna pointing to the zenith direction is used for navigation measurements. With the navigation data and the GPS navigation messages, GRAS can automatically determine the positions of the Metop and the GPS satellites and predict the times and azimuth and elevation angles of the occultations. GRAS has been designed to be able to operate in codeless-mode ensuring that successful measurements are also obtained when the GPS AS (Anti-Spoofing) is activated. The GRAS receiver has also a special "raw sampling" measurement mode for conditions where the signal-to-noise ratio (SNR) of the GPS transmission is very small. This mode will typically be entered when the signal propagation path passes through the lower troposphere. A more detailed description of the GRAS receiver can be found in (Loiselet et al. 2000).

The EPS Polar Site

The downlink of the measurement and telemetry data from the Metop spacecraft takes place via a ground station called EPS Polar Site, which is located in Svalbard, Norway. From the polar site the data is transferred to the EUMETSAT Head Quarters (HQ) in Darmstadt, Germany. The data transfer from the ground station to EUMETSAT HQ takes place approximately at the same rate as the measurement data collection onboard the satellite is done. As a result any data will suffer an initial delay of at least one full Metop orbit period, i.e., 102 minutes from the observation before it is available for processing. With some overhead due to the data buffering and transfer, the delay can be several minutes longer. This makes the product dissemination timeliness requirement significantly more difficult to fulfill.

The slow data transfer from the ground station to EUMETSAT HQ impacts especially the Metop NRT Precise Orbit Determination (POD). In the GRAS instrument data format the navigation data required for the NRT POD and the occultation data are interleaved. The slow data rate together with the GRAS product dissemination timeliness requirements mean that the NRT POD processing must solve the orbit in short arcs of 3 to 12 minutes (Fadrique and Herrero 2001). In order to make this process more efficient and to allow a more accurate solution for the GRAS clock offset, a Square Root Information Filter (SRIF) based POD algorithm has been selected for the NRT POD implementation (EUMETSAT 2004).

The GRAS Ground Support Network Service

The GRAS GSN (Ground Support Network) Service has been developed to ensure that the positions and velocities of the GPS satellites, and the ground-based measurements to support the GPS clock correction are provided in time to allow NRT occultation data processing (see Sect. 3). The GSN Service collects ground based GPS measurement data from about 25 globally distributed fiducial GPS stations and performs GPS orbit determination. The NRT products provided by the GSN service are:

- GPS position and velocity vectors;
- GPS and fiducial ground station clock offset estimates;
- GPS tracking data from the fiducial stations;
- Earth Orientation Parameters (EOP);
- Troposphere Zenith Delay (TZD) estimates for each fiducial station;
- 1 Hz Sounding Support Data (SSD) of the occulting GPS satellite carrier phase, pseudorange, and SNR for L1 and L2 channels.

The GPS position and velocity vectors provided in NRT will be based on orbit predictions because the processing of the large amount of ground based tracking data is not possible in NRT. The GRAS GSN will also provide more accurate GPS orbits from an off-line orbit determination process. These off-line GSN products can potentially be used to provide more accurate inputs for non-NRT applications like climate monitoring. The off-line products will also be used for the NRT product quality monitoring.

The EPS Core Ground Segment

The EPS Core Ground Segment (CGS) is located at EUMETSAT Headquarters in Darmstadt, Germany. The CGS performs the level 1 processing of the

GRAS data by using the raw measurements from the Metop spacecraft and the NRT POD products provided by the GRAS GSN (see Sect. 3).

The GRAS level 1b products are disseminated directly to the NWP users and to the GRAS Meteorology SAF for level 2 processing.

The GRAS Meteorology SAF

The GRAS Meteorology SAF is the final element in the GRAS data processing chain. The GRAS SAF is a consortium lead by the Danish Meteorological Institute (DMI). The GRAS SAF consortium includes also the Met Office, and the Institut d'Estudis Espacials de Catalunya (IEEC). The mission of the GRAS SAF is to retrieve the geophysical data products (refractivity, temperature and humidity profiles) using the GRAS level 1b products from the EPS CGS. GRAS SAF will also perform off-line re-processing of the GRAS measurement data to generate products for climate applications.

More information about the GRAS Meteorology SAF can be found from the SAF web page: http://dmiweb.dmi.dk/pub/GRAS_SAF/.

The Users

EPS level 1b and level 2 products are disseminated to the EUMETSAT user community in NRT. All EPS products are also made available to the users via the U-MARF (Unified Meteorological Archive and Retrieval Facility) facility. The U-MARF will be integrated with the EUMETSAT User Services. U-MARF provides users access to raw instrument output data (GRAS level 0 products), level 1b products, and level 2 products. This allows users to access GRAS data for scientific research and e.g., for climate monitoring applications.

3 GRAS Level 1 Data Processing

The GRAS level 1 data processing retrieves total bending angle profiles and impact parameter estimates from the raw measurement data of the instrument. The level 1 processing also includes the NRT Precise Orbit Determination (POD) of the Metop spacecraft and reassembly of the GRAS instrument telemetry data.

The first part of the GRAS level 1 processing is called level 1a processing (Fig. 2). This part of the processing includes collecting complete time series of the occultation and navigation measurements from the raw data stream, quality checking of the input data, reassembling the raw instrument output into phase, amplitude, and code phase measurements, and assigning an identification code to each measurement. Level 1a processing includes also an instrument correction, where the GRAS phase, amplitude, and code measurements are corrected for the impacts of the receiver hardware. This correction

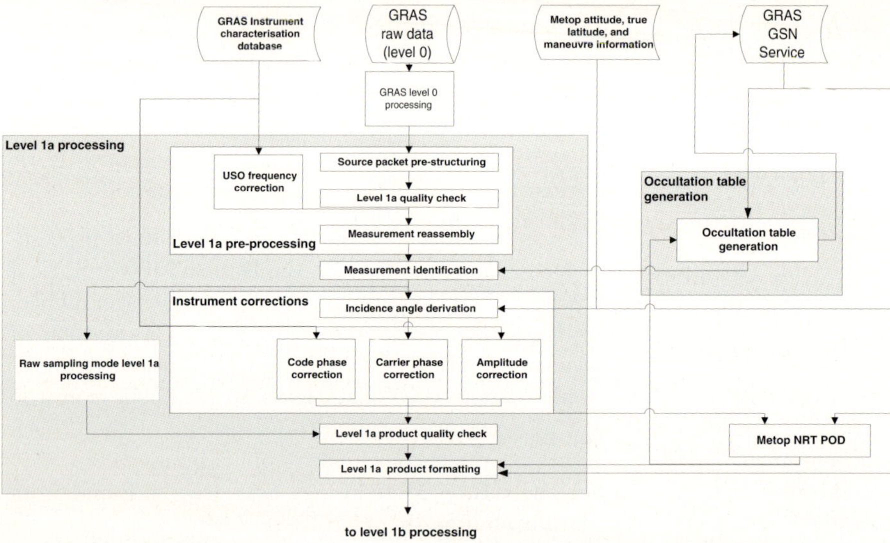

Fig. 2. Block diagram of the GRAS level 1a data processing.

is based on the receiver characterization measurements performed by the instrument manufacturer. The last steps of the level 1a processing are quality check and formatting of the level 1a products.

Metop NRT POD is not nominally a part of the level 1a processing of the occultation data from the GRAS instrument. However, the precise Metop orbit is required in order to retrieve bending angle profiles. This is why the Metop NRT POD has to be performed in parallel with the level 1a processing. The NRT POD processing is using the navigation data measured by the GRAS Zenith Antenna (GZA). The navigation data is passed through the measurement re-assembly and instrument correction functions before it can be ingested into the POD.

A SRIF (Square Root Information Filter) based processing has been selected for the GRAS NRT POD in order to make the data processing computationally efficient. This allows inclusion of a clock model into the POD formulation. The task of the clock model is to ensure that the expected stability of the GRAS Ultra Stable Oscillator (USO) is not corrupted by the measurement noise in the determination of the GRAS clock offset. The clock offset estimates can be used to correct the clock errors in the occultation data if No Differencing (ND) or ground based Single Differencing (SD1) modes are used in the level 1b processing (see Fig. 3). The clock offset estimates are also used to synchronize time stamps of the GRAS data samples with the reference time of the measurement system.

The occultation table generation function produces a table, which contains predictions of all occultation and navigation measurements to be performed by the GRAS instrument during the next 24 to 36 hours. It is based on the predicted GPS and Metop satellite orbits, and on the measurement geometry. The occultation table is used by the GSN Service to select the Sounding Support Data (SSD) to be sent to the EPS CGS for the GRAS level 1 processing.

The second part of the level 1 processing is called level 1b processing (Fig. 3). This part contains the actual retrieval of the bending angle profiles. The beginning of the level 1b processing contains the selection of the best reference GPS satellite (pivot satellite) and fiducial stations for clock correction by single or double differencing. The GRAS level 1b processing is designed to allow five different clock correction options ranging from No Differencing (ND) to Double Differencing using two ground stations (DD2). The level 1b processing contains also an isolation of the auxiliary measurement data, and a number of corrections to remove impacts of the measurement system. The data corrections remove the geometrical path and clock errors from the measured phase values, and filter out noise outside the bandwidth containing the atmospheric information. The impact of the Earth oblateness is removed by transforming the measurement geometry into a coordinate system defined by the local radius of curvature of the ellipsoid shape of the Earth (Syndergaard 1998).

The retrieval of the bending angle profile in level 1b processing is performed in parallel by three methods: Geometrical Optics (GO), Wave Optics (WO), and processing of the Raw Sampling (RS) mode data.

GO retrieval is the baseline processing method for the full measured profile from the height of about 80 km down as near to the surface as the instrument can track the GPS signal. Normally the tracking of the L1 signal should be possible down to a few kilometers above the surface. The GRAS instrument may loose the tracking of the L2 signal significantly higher due to a lower SNR. When this happens, the ionosphere correction of the measurement data is performed by extrapolating the ionosphere correction from the heights when both L1 and L2 signals were tracked downwards (EUMETSAT 2004). The Total Electron Content (TEC) along the occultation ray path is calculated as part of the level 1b processing for all occultation measurements.

The purpose of the WO processing is to increase the quality of the retrieved total bending angle in the case of atmospheric multipath propagation. The WO processing is based on the Phase Transform (PT) technique (Jensen et al. 2003). In practice it may be necessary to approximate the PT method by the Full Spectrum Inversion (FSI) technique in order to make the processing faster.

The third bending angle retrieval method in the GRAS level 1b processing is based on the use of the GRAS Raw Sampling (RS) mode data. This data can potentially produce bending angle profiles from the lowest parts of the troposphere. However, the data processing algorithm for the RS mode data has not yet been defined.

A full specification of the GRAS level 1 data processing algorithm is provided in (EUMETSAT 2004).

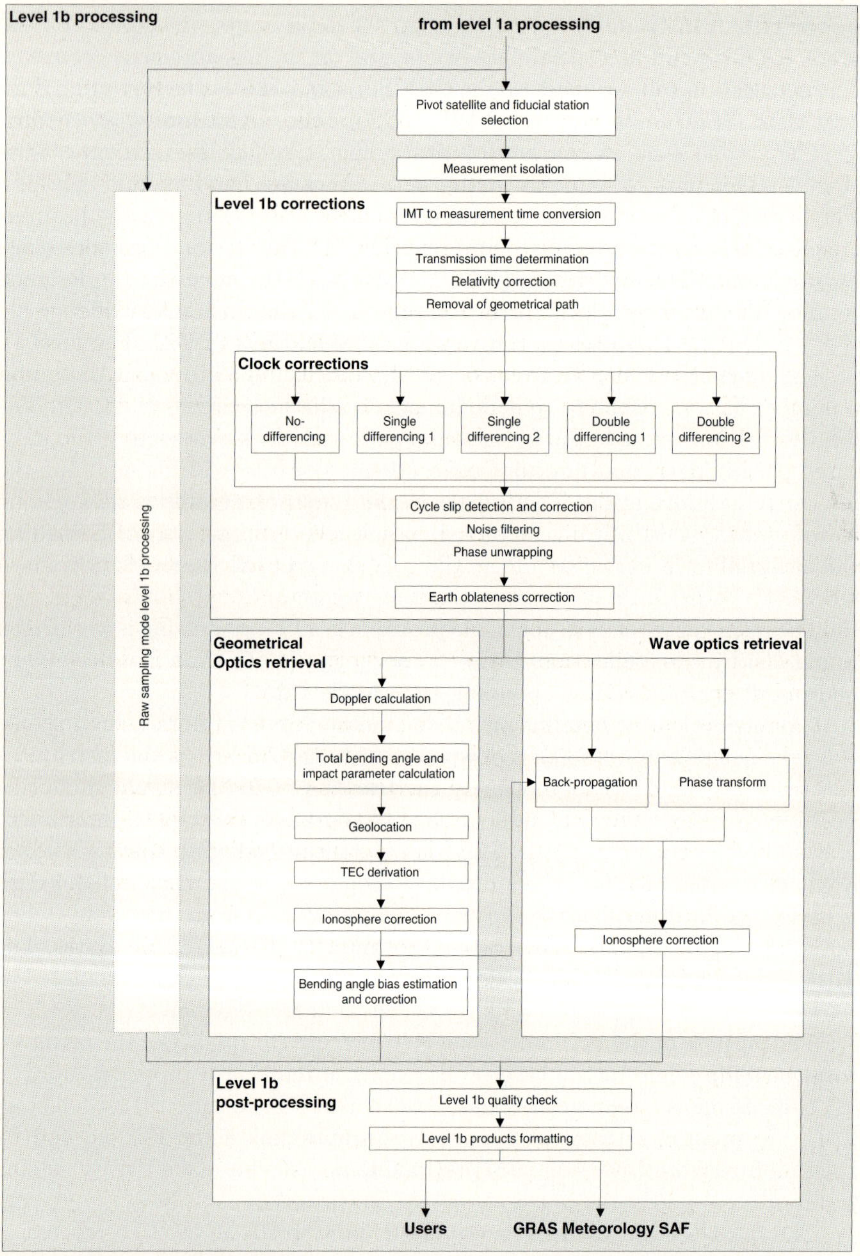

Fig. 3. Block diagram of the GRAS Level 1b data processing.

4 GRAS Level 1 Product Validation

GRAS level 1b product validation is based on using 1D-Var to forward model Numerical Weather Prediction (NWP) background into bending angle profiles. The 1D-Var approach automatically compares the measured bending angle profile to the model background to find an assimilation solution (Healy and Eyre 2000). In the solution the measurement and the background should match within the expected error margins. The 1D-Var retrieval provides diagnostics about the convergence of the iterative adjustment of the NWP background, difference between the measurement and the final solution, correctness of the expected error levels, and a retrieved geophysical data profile. With a global NWP model, every occultation measurement by GRAS can be validated. This enables fast generation of validation statistics. A detailed description about the use of 1D-Var technique for RO bending angle validation is provided in Marquardt et al. (2005).

The noise levels in the GRAS measurement data can be validated by comparing them to theoretically predicted noise levels. A robust way of doing this is Generalised Cross Validation (GCV). GCV is a well established method for the objective estimation of parameters for smoothing splines when the standard deviation of the noise in the data is not known. The use of the Generalised Cross Validation (GCV) in noise level estimation is described in Marquardt et al. (2005).

Direct intercomparisons between GRAS measurements and co-located[1] independent measurements will also be used in the GRAS level 1 validation. RO measurements by other missions (e.g., CHAMP, COSMIC), radiosonde measurements, and geophysical data profiles retrieved from satellite or ground based remote sensing measurements are potential reference measurements. The correct approach for the direct intercomparison of independent data are discussed in Marquardt et al. (2005).

Validation of the NRT GRAS POD solutions is planned to be carried out by intercomparison with independent POD solutions.

5 Summary

EPS GRAS mission will provide radio occultation data for operational NWP users. The GRAS level 1b products will be disseminated to the users within 2 hours 15 minutes and the level 2 products within 3 hours from the observation. GRAS GSN Service will provide GPS POD solutions and ground based measurement data for the GRAS data processing.

The processing of the GRAS measurement data to bending angle profiles and to geophysical products will be performed by the EPS CGS and GRAS meteorology SAF, respectively. The EPS CGS is located at the EUMETSAT

[1] Measurements within 3 hrs and 300 km.

HQ in Darmstadt, Germany. The GRAS Meteorology SAF is hosted by DMI in Copenhagen, Denmark.

1D-Var retrieval is planned to be used for the validation of the GRAS level 1b data products. This approach allows a direct validation of the bending angle profiles against NWP background. 1D-Var retrieval also allows verification of the GRAS measurement error characterisation for NWP applications. Intercomparisons of GRAS data products with independent observations are planned to complete the validation activities.

References

EUMETSAT (2004) Gras level 1 product generation function specification. Tech. Rep. EPS/SYS/SPE/990010, EUMETSAT, Am Kavalleriesand 31, D-64295 Darmstadt, Germany, (Available from EUMETSAT)

Fadrique FMM, Herrero OL (2001) Gras/metop precise orbit determination: Final report. EUMETSAT Study EUM/CO/00/818/DK, GMV-GRASPOD-FR-01, EUMETSAT, Am Kavalleriesand 31, D-64295 Darmstadt, Germany, (Available from EUMETSAT)

Healy SB, Eyre JR (2000) Retrieving temperature, water vapor and surface pressure information from refractive-index profiles derived by radio occultation: A simulation study. Quart J Roy Meteorol Soc 126:1661–1683

Healy SB, Thepaut JN (2005) Assimilation experiments with CHAMP GPS radio occultation measurement. Quart J Roy Meteorol Soc, in press

Healy SB, Jupp AM, Marquardt C (2005) Forecast impact experiment with GPS radio occultation measurements. Geophys Res Lett 32(3), doi:10.1029/2004GL020806

Jensen AS, Lohmann MS, Benzon HH, Nielsen AS (2003) Full spectrum inversion of radio occultation signals. Radio Sci 38(3), doi:10.1029/2002RS002763

Loiselet M, Stricker N, Menard Y, Luntama JP (2000) GRAS – Metop's GPS-Based Atmospheric Sounder. ESA Bulletin (102):38–44

Marquardt C, Healy SB, Luntama JP, McKernan E (2005) Gras level 1b product validation with 1d-var retrieval. Technical Memorandum No.12, EUMETSAT, Am Kavalleriesand 31, D-64295 Darmstadt, Germany, (Available from EUMETSAT)

Syndergaard S (1998) Modeling the impact of the Earth's oblateness on the retrieval of temperature and pressure profiles from limb sounding. J Atmos Sol-Terr Phys 60:171–180

Wickert J, Beyerle G, Schmidt T, Grunwaldt L, Heise S, König R, Michalak G, Reigber C (2005) GPS radio occultation with CHAMP and GRACE: A first look at a new and promising satellite configuration for global atmospheric sounding. Ann Geophys 23:653–658

ROSA: The Italian Space Agency GPS Radio Occultation Receiver. Signal Tracking Characteristics and Terrestrial Measurement Campaign

R. Notarpietro[1], A. Zin[2], G. Perona[1], L. Corgnati[1], and M. Gabella[1]

[1] Politecnico di Torino, Dipartimento di Elettronica, Torino, Italy
riccardo.notarpietro@polito.it
[2] Alenia Spazio – Laben, Milano, Italy

Abstract. In the framework of an Italian Space Agency (ASI) contract, Alenia Spazio – Laben developed a new GPS radio occultation receiver: the ROSA receiver. It will manage both "rise" and "set" occultation events, tracking signals coming from the stratosphere in closed-loop and switching to the open-loop when they emerge from the troposphere. It is well known that the open-loop signal tracking needs a good Excess-Doppler prediction, and that it has to be evaluated in advance (and in quasi real time) by the occultation on-board software. The adopted prediction strategy is a simplified and optimized version of the technique suggested by Sokolovskiy (2001). This work provides the results of the validation of this technique, carried on through comparisons with about 200 Excess-Doppler observations collected by the CHAMP Radio Occultation GPS receiver during the winter season 2002/2003 above Europe. Some insights about the prediction of the end occultation time for setting events (or start occultation time for rising events) are also given. The strategy implemented by the on-board ROSA occultation manager for the open-loop tracking is quickly introduced. Also a mountain top measurement campaign with the ROSA breadboard receiver is discussed. It is planned for 2006. This experiment is encouraged by ASI in order to test and to validate the ROSA tracking capabilities and the algorithms necessary for the extraction of the GPS occultation observables from raw data.

1 Introduction

The importance of the radio occultation (RO) technique applied to GPS measurements for the terrestrial atmospheric remote sensing has been increased during the last decade, since the first related mission (the GPS/MET concept mission, see Melbourne et al. 1994; Ware et al. 1996) highlighted its possibilities. Given the good results obtained, several other satellites carrying on-board a GPS receiver for the "radio occultation" sounding of the atmosphere were planned and launched. Currently, at least two missions operatively supply atmospheric profiles retrieved with such a technique: the German CHAMP mission, and the Argentine SAC-C. But, due to the presence of water vapor in the troposphere, defocusing, multipath and other anomalous propagation phenomena are quite common. When the GPS receiver tracking devices are PLL-based, the main consequence in reception can be the early loss of the phase lock and the following tracking failure. Hence it is not possible to extract the carrier phase information needed for the measurement. As a consequence, most of the retrieved atmospheric profiles are characterized by a lack of data in the lower atmospheric layers. Moreover a negative refractivity bias is often observed, in particular for the tropical retrieved atmospheres (see Ao et al. 2003; Beyerle et al. 2004). In the framework of the ASI contract for the development of ROSA, a new space-based GPS Radio Occultation receiver, the Alenia Spazio – Laben will implement an alternative phase tracking approach based both on the closed loop (CL) and on the open loop (OL) technique. When the receiver works in OL, the tracking circuitries will quickly search and follow the received frequency only if they are aided by a model. This model should be able to predict in advance the Excess-Doppler experienced by the GPS carrier. It can be used to down-convert the signal which, in turn, is low-pass filtered. Its in-phase (I) and in quadra-phase (Q) components are sampled at 100 Hz as suggested by Sokolovskiy (2001, 2004), in order to allow for the full spectral reconstruction of the "residual" excess-phase, without aliasing of harmonics. The real excess-phase will then be reconstructed (in post-processing) from the sampled I and Q components, using also the information given by the model adopted for the real down-conversion. In this case the signal could be analyzed for a longer time and a better characterization of the lower atmospheric layers could become possible.

In the following section the Excess-Doppler prediction technique implemented in the ROSA on-board occultation software is described. In Sect. 3 the results of its validation are shown. The comparisons are carried against real CHAMP Excess-Doppler observations. Sect. 4 deals with the optimizations applied to the prediction technique in order to make it suitable for a quasi real-time applicability. In the last section, a ground measurement campaign, which is planned for the next 2006 year in order to test a bread-board of the ROSA receiver, will be introduced and discussed.

2 Excess-Doppler Prediction Technique and ROSA OL Management Approach

Assuming that GPS and LEO orbits can be accurately known in advance with respect to the occultation event and given an atmospheric model approximating the propagation media, it is always possible to predict the Excess-Doppler evolution by means of ray-tracing techniques. But this approach is not efficient, since the prediction has to be computed in a quasi real-time and just prior each event by the on-board software. As described by Sokolovskiy (2001), it is more efficient to "model" in advance the trajectories bendings α characterizing the overall event in function of the impact parameters a and of the position in the atmosphere (i.e., the latitude) of the geometrical tangent point (TP_{geom} in Fig. 1). This task may be accomplished using the following formulation describing the dependence between α, a and the refraction index profile $n(h)$. Assuming a locally spherical distributed

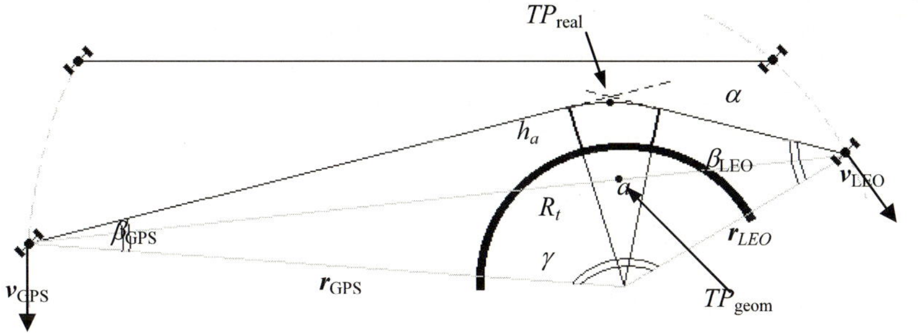

Fig. 1. Radio Occultation geometry.

atmosphere and a single ray Geometric Optics (GO) propagation, it is quite easy to show that:

$$\alpha(a) = -2(R_t + h_a) \int_{R_t+h_a}^{\infty} \frac{1}{\sqrt{(R_t + h)^2 n^2(h) - (R_t + h_a)^2}} \frac{1}{n(h)} \frac{\mathrm{d}n(h)}{\mathrm{d}r} \mathrm{d}r \qquad (1)$$

$$\text{with } h_a + R_t = a \text{ and } r = R_t + h$$

(R_t is the Earth's mean radius and h_a is called impact height). The $\alpha(a)$ function computed using (1) does not depend on the orbital parameters, but it only depends on the atmospheric refraction index profile. This means that it can be evaluated in advance on the basis of some atmospheric models taken as reference. The results of this calculation can then be stored in a database from which the required bending angle profiles can be loaded by the occultation managing software. The size and the complexity of this database depend on the atmospheric model adopted to represent the atmosphere. For this purpose, we chose the CIRA86aQ_UoG model developed at the University of Graz (Kirchengast et al. 1999). This model is defined by 12 monthly zonal-mean vertical profiles of temperature $T(h)$, water vapor partial pressure $e(h)$ and dry pressure $p_D(h)$, distributed on a grid of 10° step in latitudes from 90°N to 90°S. These atmospheric profiles are defined from the surface up to 120 km altitude, with a vertical step of 1 km until about the tropopause (15 km altitude) and of 5 km above. The conversion into refractivity profiles $N(h)$ through the well known Smith and Weintraub formulation (1953) allows the definition of the refractivity model from which the $\alpha(a)$ database can be derived.

For each month and for each latitude the entire $\alpha(a)$ profile is evaluated through (1) and it is stored on the database as a column vector. The first four elements are the impact parameters computed (using the Bouguer's rule) for rays whose real tangent points (TP_{real} on Fig. 1) are placed on the surface and at 2 km, 4 km and 6 km above the surface. The first one is used to state the end (start) of a set (rise) "ideal" Geometric Optics occultation prediction. One among the other three impact parameters stored before the minimum one, is used by the occultation manager to define the start (end) point of the OL tracking mode. The other variables stored on the database vector are the bending values α evaluated for that particular model atmosphere, associated to impact parameters varying from the minimum one up to 120 km, with a constant vertical step of 1 km.

The prior knowledge of the instantaneous value of the TP_{geom} latitude addresses to the right column of the database and the overall $\alpha(a)$ evolution can be extracted from the receiver mass memory and can be used to predict the instantaneous Excess-Doppler value. This prediction can be done in two steps. The first allows the calculation of the angles β_{LEO} and β_{GPS} between the real trajectories and the satellites vector positions, solving the following system derived from the geometry:

$$\begin{cases} r_{\text{LEO}}(t)\sin[\beta_{\text{LEO}}(t)] = R_t + h_a(t) \\ r_{\text{GPS}}(t)\sin[\beta_{\text{GPS}}(t)] = R_t + h_a(t) \\ \beta_{\text{LEO}}(t) + \beta_{\text{GPS}}(t) + \gamma(t) - \alpha(h_a, t) - \pi = 0. \end{cases} \tag{2}$$

Knowing the instantaneous positions of the satellites ($\boldsymbol{r}_{\text{LEO}}(t)$ and $\boldsymbol{r}_{\text{GPS}}(t)$), at each instant time $t = \tau$ the overall $\alpha(a)$ evolution (extracted from the database) is used to iteratively solve the system (2). The outputs are the values of the $\beta_{\text{LEO}}(\tau)$ and $\beta_{\text{GPS}}(\tau)$ angles and the values of the bending $\alpha(\tau)$ and of the impact parameter $a(\tau)$ related to that particular GO trajectory. Now, knowing the instantaneous velocity vectors of both the satellites, the Excess-Doppler value $\Delta f_d(\tau)$ can be easily computed. The end of the occultation event is stated when the $a(\tau)$ parameter extracted from the database is less then the minimum one (for which the tangent height of the correspondent ray perigee is equal to 0).

Details of the ROSA occultation management (i.e., issues about the event selection, its prediction and channel allocation) are given by Zin et al. (2004). In the same reference, considerations regarding the sensitivity of the prediction model to numerical inaccuracies, to LEO orbital errors and to receiver clock instability are also reported. The ideal open-loop tracking approach in ROSA is shown in Fig. 2.

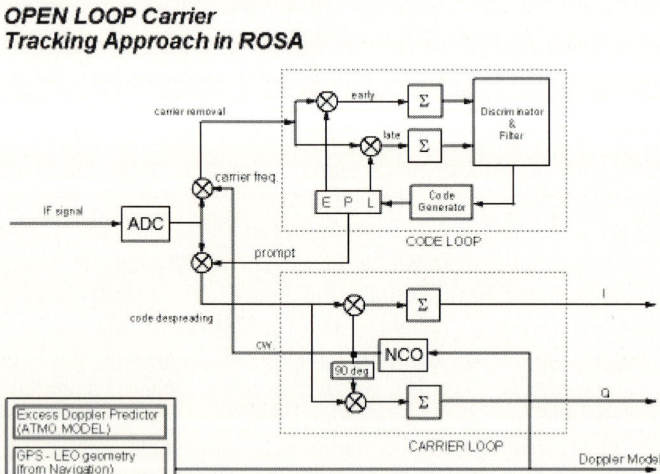

Fig. 2. Ideal open-loop tracking channel.

In the real case, the open-loop (OL) channel runs in parallel to the close-loop (CL) one, with the starting time of raw (I, Q) sampling being selectable by the user input switch linked to the guessed tangent point height. The code settings from the CL channel are used to control the code Numerically Controlled Oscillator (NCO) settings of the OL channel. The "prompt" branch of the code delay line is then used for de-spreading. The OL carrier NCO is driven by the ROSA navigation processing plus the Excess-Doppler computation from the on-board atmospheric model. Then the I and Q samples are given in output at a relatively high frequency, selected according to the suggestions of Sokolovskiy (2001), in order to allow the full signal spectrum reconstruction in post-processing. It is worth noting that the code tracking process is still carried out in close-loop.

3 Validation Analysis

The prediction technique described in the previous section has been validated using real CHAMP observations. This analysis wants to be complementary and wants to complete the results already obtained by Sokolovskiy (2001, 2004) through simulations. The "BlackJack" GPS receiver on-board the CHAMP satellite is based on the PLL tracking of the signal phase. Even if the implementation of "fly-wheeling" may allow the signal tracking beyond the PLL capabilities, CHAMP observations often suffer from tracking errors, in particular when signals emerge from atmospheric layers which introduce severe multipath. Since we want to compute the extent of the frequency mismodeling between predicted and observed Excess-Doppler evolutions (and since real OL data are still not available), we have preferred to carry on our analysis on "quite good" real data, collected from relatively dry atmospheres where multipath conditions were rare. Moreover, it is known that the spread of the Doppler induced by atmospheric effects on the signal propagation is not so different when tropics are taken into consideration instead of higher latitudes (always around ±10 Hz or ±15 Hz, see Sokolovskiy (2001); Notarpietro et al. (2003)). However, a test of ROSA OL capabilities based on real measurements will be lead in the next future from the top of a mountain (see Sect. 5).

Therefore, we carried on our analysis using the about 200 settings occultation events observed by CHAMP above Europe (in the area between –10° and 26° of longitude and between 30° and 58° of latitude) from December 2002 until March 2003. For each occultation event the real Excess-Doppler evolution is computed considering the derivative of the second order "moving" polynomial fit of the 50 Hz L_1-excess phase data. In particular, the fitting has been computed considering at each time an appropriate number of phase samples, so that the polynomial used to fit the data could be able to filter out high frequency features introduced on samples by small scale irregularities inside each first Fresnel zone (see Hajj et al. 2002). Moreover, using the same satellites positions and velocities of the real event, the corresponding Excess-Doppler variation is predicted with a 50 Hz sampling rate. Fig. 3a qualitatively shows both "real" and predicted Excess-Doppler evolutions related to the overall dataset taken into account. Two characteristics are quite evident. The first one is the spread of the evolutions, due to the different atmospheric sounding directions which, in turn, are due to the different orbital geometries characterizing each occultation event. The second concerns the evolutions of the "real" observations, in particular when the "fly-wheeling" is activated. In these cases the phase delay is not directly measured by the receiver but it is extrapolated. However, the "fly-wheeling" helps the PLL circuitries of the CHAMP GPS receiver after a tracking failure. And, in most cases, a useful signal has been tracked also after the end of the "GO" prediction (Fig. 3b – case B, OCC#0071_2002_336). These contributions are very important if wave optics techniques are used for the inversion (the Canonical Transform (Gorbunov 2002) or the Full Spectrum Inversion (Jensen et al. 2003)). Therefore, two different situations arise when we want to evaluate the distribution of the maximum displacements between predicted and observed Excess-Doppler data, as depicted in Fig. 3b. For each event this maximum displacement has been evaluated until the last "fly-wheeling" activation inside each time interval when predicted values are available (Fig. 3b – case A, OCC#0191_2002_335). Alternatively, when the "BlackJack" receiver has been able to track some useful signals beyond the GO limits, the maximum displacement has been evaluated at the end of the prediction interval (Fig. 3b – case B). The end of the comparison intervals are shown with arrows in Fig. 3b for both these cases. As shown in Fig. 4, the maximum displacements so computed are distributed between ±15 Hz. And this confirms and completes the results observed by Sokolovskiy (2001). We can therefore conclude that also using this CIRA86aQ_UoG model-based technique, the center frequency of the incoming signal can be supplied to the open-loop circuitries with an accuracy of ±15Hz.

Another feature that this prediction technique has allowed us to put in evidence is the identification of the end occultation time for a given setting event, under a "GO" point-of-view. Since a rising event can be viewed as a "reversed" setting event, the following results can also be applied to detect the start of an occultation event. Thanks to the fairly good CHAMP occultation observations analyzed, we have noted that useful data have been observed after the time limit evaluated using the prediction technique previously described. And this situation happened for about the 27 % of the events considered. In Fig. 5 we have displayed the distribution of the differences between the time of the last "fly-wheeling" activation and the end time of the predicted "GO" occultation. We can in fact see that for about the 27 % of the events, this difference is positive. Therefore, in order to take into consideration the important contributions related to multipath phenomena in the lower and most turbulent part of the troposphere, we can suggest to enable the OL data to be stored for an "extra-time" of about 4 s (the mean value computed considering the positive part of the distribution shown in Fig. 5) after (before) the last (first) setting (rising) trajectory has been evaluated by the prediction algorithm. The Excess-Doppler model can be deduced extrapolating the evolution computed using the prediction technique. In the future a similar analysis will be carried on using CHAMP tropical observations. In this case we can give a more accurate estimate of the Excess-Doppler prediction extra-time. Probably it will be more than the figure of 4 s found considering mid-latitude winter observations.

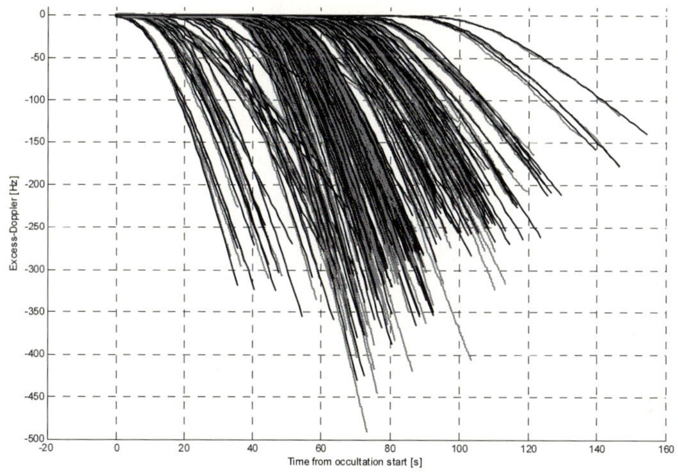

Fig. 3a. Qualitative comparisons between IInd order polynomial fittings of CHAMP Excess-Doppler observations (gray lines) and Excess-Doppler predictions (black dotted lines).

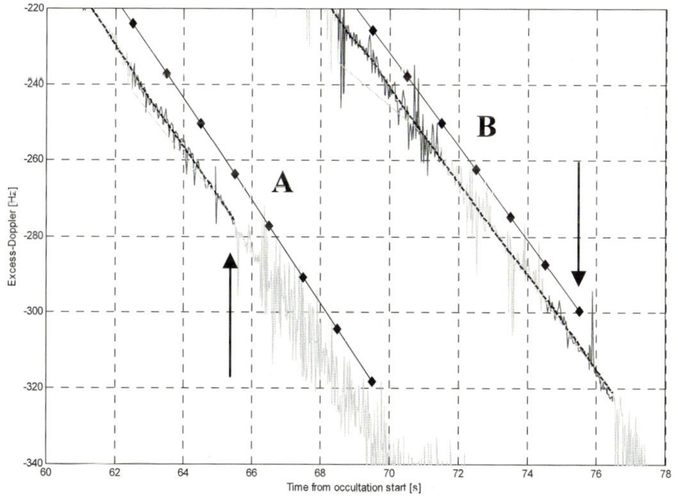

Fig. 3b. Details of two different observations (gray and light "noisy" gray continuous lines: real data without and with "fly-wheeling", respectively; black dotted line: polynomial fitting) and predictions (diamonds).

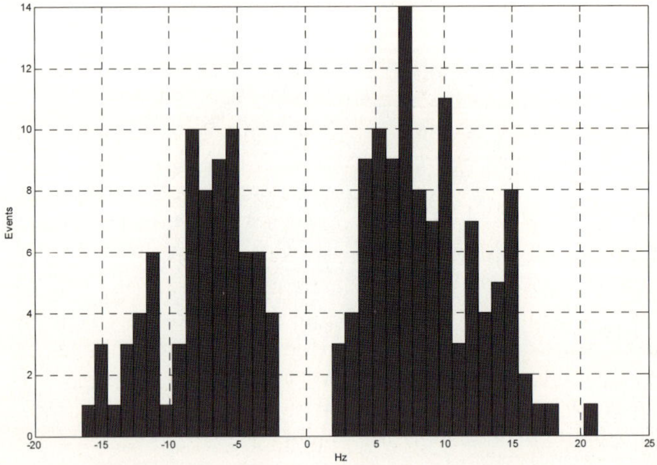

Fig. 4. Maximum displacements distribution between predicted and "real" Excess-Doppler evolution.

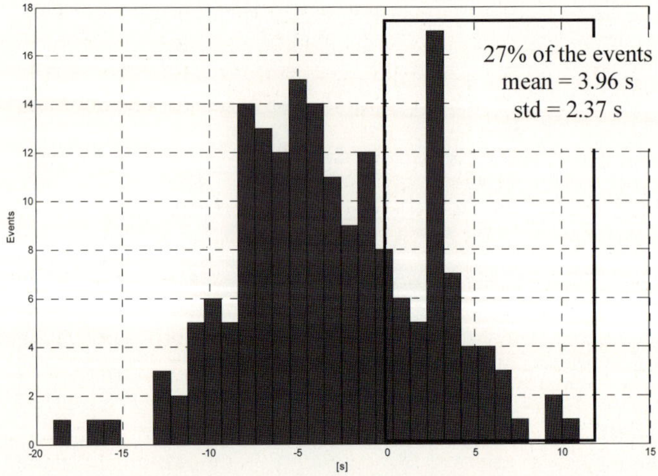

Fig. 5. Distribution of the differences in the "end" occultation times evaluated considering real and predicted Excess-Doppler evolutions (the highlighted area shows the positive distribution part).

4 Optimization of the Prediction Technique

For the prediction technique described above, the mean computational time is 610 s (with a 150 s of standard deviation) for a single CHAMP occultation event when the simulations are carried out in a Matlab environment, using a Pentium III – 1.1 GHz clock frequency processor.

In order to make this technique suitable for quasi real time applications, some optimizations have been implemented. Since the atmosphere mainly influences the signal propagation at the lower altitudes, we have chosen to extract only one $\alpha(a)$ evolution for each occultation event. In particular we have imposed to extract the $\alpha(a)$ profile relative to the latitude of the geometrical tangent point on the Earth surface. Considering only this profile for the overall occultation event, it means only one interrogation and extraction from the database. A mean error of about 0.5 Hz has been evaluated considering the maximum displacements between observed and predicted Excess-Doppler, computed with and without this optimization. Adopting this approach, the mean computational time decreases to about 50 s (15 s of standard deviation).

Another important improvement is obtained by decimating the Excess-Doppler prediction "sampling rate". Computing one Excess-Doppler estimate for each second of the occultation event, the observed mean computational time decreases to about 6.3 s, with an 88 % of reduction with respect to the previous case. Adopting a compiled instead of an interpreted programming code (like Matlab) more than two orders of magnitude of reduction are expected (for more details about the CPU load observed using a ROSA receiver DSP simulator see Zin et al. 2004).

5 Measurement Campaign from a Mountain Site

In order to test the receiver performances (in particular as far as the OL tracking capabilities are concerned) and to validate the data processing chain (at least at the level of extraction of SNRs and carrier phases from raw GPS data), a ground measurement campaign (called ROSA VIRTUAL SATELLITE) with a breadboard of the ROSA receiver is planned for the next 2006 year. A quite similar approach for the OL tracking and for the GPS observables generation will be adopted analyzing data observed from the top of a mountain. The main difference with the prediction scheme described in Sect. 2 is related to the pre-computation of the $\alpha(a)$ database, since a slightly modified version of (1) has to be used.

As far as the measurement site is concerned, one of the best choice (under a costs/benefits point of view) could be the top of the M.te Lema (1620 m a.s.l.), inside the MeteoSwiss radarmeteorological facility. The RO antenna can be mounted on a porthole of the building looking at the south direction, toward the Pianura Padana (a map is given in Fig. 6). Despite the lower altitude of the site

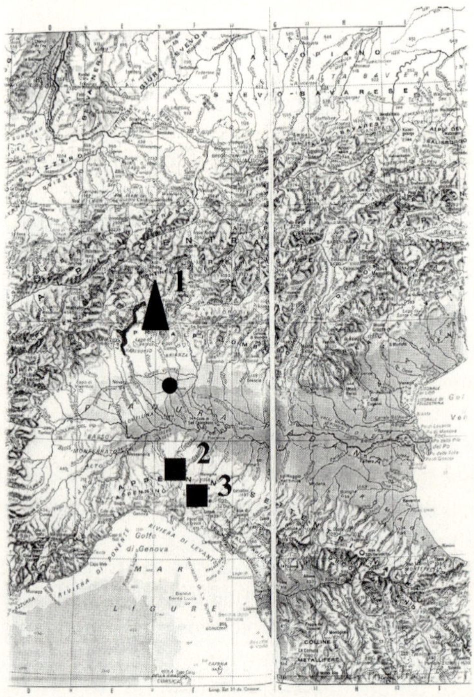

Fig. 6. Map of the proposed measurement site location. Triangle (1) shows the position of M.te Lema (1620 m a.s.l.); squares show the positions of the main obstacles lying in the field of view: (2) M.te Lesima (1724 m a.s.l.) and (3) M.te Maggiorasco (1799 m a.s.l.). Circle shows Milano's position.

and the presence of some "important" orographic obstacles which will limit the signal propagation in some directions (M.te Lesima and M.te Maggiorasco are the higher mountains in the antenna's field-of-view), we think we can however properly test the ROSA OL performances, since we can expect a quite noisy signal coming from the local horizon. Given the smaller size of the first Fresnel zone, a single measurement sample is representative of small-scale (and therefore, stronger) turbulences; moreover, small scale refractivity structures are less filtered out by the smaller Fresnel size. Probably, given the experimental geometry, many difficulties in applying the overall retrieval scheme will appear; however the inversion algorithms validation is not so important in this context (wave optics algorithms are not applicable for fixed receivers and, furthermore, they are still well tested using real satellite occultation observations). Indeed, it is more important to develop the software for the extraction of the GPS observables (carrier phases and SNRs) from raw data and for the ground processing chain. From this point of view, M.te Lema can be a very good choice, given its logistical "appealing". It is nearby and easily accessible by the people who will be operating the receiver. Furthermore, the entire ROSA VIRTUAL SATELLITE (safely left inside the radar-meteorological building) will be remotely controlled.

6 Conclusion

In this paper the Excess-Doppler prediction technique implemented on the on-board software for the open loop operations of the ROSA receiver has been described. It has been validated through the comparisons with about 200 fairly "good" occultation events observed by CHAMP, in order to complete the results obtained by Sokolovskiy (2001) using simulated data. A bandwidth allocation of about ±15 Hz is confirmed for the OL tracking of the signal. An analysis of the prediction of the end (start) time for setting (rising) occultation events has been carried out. The choice of at least an "extra-time" of 4 s for the Doppler prediction seems to be necessary, in order to take into account for the important contributions given by radio waves diffracted by laminated structures in the moist troposphere. The 4 s "extra-time" has been obtained considering the winter mid-latitude CHAMP soundings only. In the future we will analyze tropical occultations, in order to verify the basis of this figure. The optimizations described have significantly reduced the computation time, making this technique suitable for quasi real-time applications. Concluding, a mountain top measurement campaign with the ROSA bread-board receiver is planned for 2006, in order to test the ROSA OL capabilities and to validate the algorithms for the generation of GPS observables.

Acknowledgements. This work has been funded with the contract ASI-Laben I/026/01/0 (in particular with the sub-contract 238/2002 drawn up between Alenia Spazio – Laben and the Electronics Department of the Politecnico di Torino). We kindly acknowledge Alenia Spazio – Laben team for their support. Thanks to CHAMP analysts team in making available the data used in our analysis.

References

Ao CO, Meehan TK, Hajj GA, Mannucci AJ, Beyerle G (2003) Lower-troposphere refractivity bias in GPS occultation retrievals. J Geophys Res 108(D18), doi:10.1029/2002JD003216

Beyerle G, Wickert J, Schmidt T, Reigber C (2004) Atmospheric sounding by global navigation satellite system radio occultation: An analysis of the negative refractivity bias using CHAMP observations. J Geophys Res 109(D1), D01106, doi:10.1029/2003JD003922

Gorbunov ME (2002) Canonical transform methods for processing of Radio Occultation data in the lower troposphere. Radio Sci 37(5) 1076, doi:10.1029/2000RS002592

Hajj GA, Kursinski ER, Romans LJ, Bertiger WI, Leroy SS (2002) A technical description of atmospheric sounding by GPS occultations. J of Atmosph and Solar-Terr Phys 64(4):451–469

Jensen AS, Lohmann MS, Benzon HH, Nielsen AS (2003) Full Spectrum Inversion of Radio Occultation Signals. Radio Sci 38(3):1040, doi:10.1029/2002RS002763

Kirchengast G, Hafner J, Poetzi W (1999) The CIRA86aQ_UoG model: an extension of the CIRA-86 monthly tables including humidity tables and a Fortran95 global moist air climatology model. Tech Rep for ESA/ESTEC Nr 8

Melbourne WG et al. (1994) The application of spaceborne GPS to atmospheric limb sounding and global change monitoring. JPL Publ 94-18

Notarpietro R, Gabella M, Perona G, Zin A (2003) On the prediction of the Excess-Doppler evolution for the bandwidth optimization of the LAGRANGE RO receiver. Symposium on Atmospheric Remote Sensing using Satellite Navigation Systems, Matera

Smith EK, Weintraub S (1953) The constants in the equation for atmospheric refractive index at radio frequencies. Proc IRE 41:1035–1037

Sokolovskiy SV (2001) Tracking tropospheric radio occultation signals from low Earth orbit. Radio Sci 36(3):483–498

Sokolovskiy SV (2004) Open loop tracking and inverting GPS radio occultations signals: simulation study. In: Kirchengast G, Foelsche U, Steiner AK (eds) Occultations for probing atmosphere and climate. Springer, Berlin, pp 39–52

Ware R et al. (1996) GPS sounding of the atmosphere from Low Earth Orbit: preliminary results. Bull Am Meteor Soc 77(1):19–40

Zin A, Crosta P, Scaciga L, Notarpietro R (2004) GNSS Radio Occultation From Leo: Occultation Management And Open-Loop Tracking Aspects. 2nd ESA Workshop on Satellite Navigation User Equipment Technologies (NAVITEC '2004), The Netherlands

Tropospheric Water Vapor from LEO-LEO Occultation: Estimation by Differential Attenuation Measurements near 20 GHz

F. Argenti[1], F. Cuccoli[1], L. Facheris[1], E. Martini[2]

[1] CNIT research unit at the Department of Electronics and Telecommunications of the University of Firenze, Via di Santa Marta 3, Firenze, Italy
cuccoli@achille.det.unifi.it
[2] CNIT research unit at the Department of Information Engineering, University of Siena, Via Roma 56, Siena, Italy

Abstract. A differential measurement concept is presented for retrieving the total content of water vapor (*Iwv*, Integrated water vapor) along the propagation path between two counter-rotating Low Earth Orbiting (LEO) satellites. This approach, referred to as DSA (Differential Spectral Attenuation) method, is based on the simultaneous measurement of the total attenuation at two relatively close frequencies in the K band, and on the estimate of a "spectral sensitivity parameter" that is highly correlated to the *Iwv* content of the LEO-LEO link in the low troposphere. The DSA approach has the potential to overcome all spectrally "flat" and spectrally correlated phenomena, including atmospheric scintillation.

1 Introduction

In this paper, we first describe how a differential measurement concept recently introduced by the authors for Earth-satellite link (Cuccoli et al. 2001; Cuccoli and Facheris 2002), here referred to as DSA (Differential Spectral Attenuation) can be applied to the radio occultation geometry to estimate the water vapor content along a LEO-LEO satellite link, and to retrieve water vapor profiles from received power measurements during a LEO-LEO radio occultation. It is shown that the DSA concept has the potential to overcome all spectrally "flat" and spectrally correlated phenomena, including atmospheric scintillation. Such potential comes from the tight relationship between the spectral sensitivity functions near 20 GHz and the *Iwv* along the LEO-LEO link, at any tangent altitude up to 10 km. In this paper we show that such a relationship is straightforward and invariant in its form at any latitude and season (which affects only the coefficients of the form itself): Therefore, it allows a direct measurement of the integrated water vapor through simple power measurements at the receiver site, without requiring any a priori information about any other atmospheric or climatic parameter.

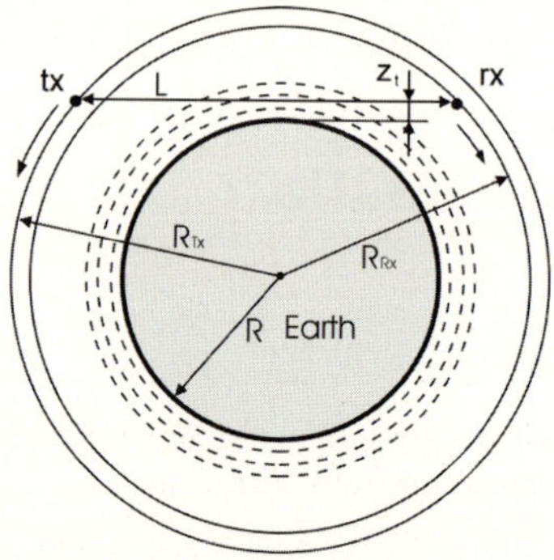

Fig. 1. Geometry and parameters of a LEO-LEO link (with simplified rectilinear radio-propagation path).

2 The Spectral Sensitivity Function and the DSA Approach

For a microwave transmitter-receiver link, the radiative transfer equation providing the power spectral density $P_{rx}(f)$ at the receiver can be arranged in the following form (Cuccoli et al. 2001):

$$P_{rx}(f) = \alpha(f) P_{tx}(f) \exp(-\tau_a(f)), \tag{1}$$

where f is the frequency, $P_{tx}(f)$ is the transmitted spectral power, $\tau_a(f)$ is the optical depth related to the propagation link and $\alpha(f)$ is the contribution due to all effects different from gaseous absorption (i.e., scattering, antenna gains, defocusing, etc.). In the frequency range of interest (around 20 GHz), the optical depth depends on the absorption due to water vapor, O_2 and N_2. The total spectral attenuation $A(f)$ for the given link is consequently defined as:

$$A(f) = \frac{P_{tx}(f)}{P_{rx}(f)} = \frac{\exp(\tau_a(f))}{\alpha(f)}. \tag{2}$$

Throughout the paper, the 2D geometry of Fig. 1 is considered, with two counter rotating LEO satellites on two different orbits (circular for the sake of simplicity) and the propagation path assumed rectilinear. An immediate justification of the exploitability of a differential spectral attenuation measurement approach is given by Fig. 2, where the basic MLS (Mid Latitude Summer) atmospheric model profiles have been assumed (McClatchey et al. 1972). Under the hypothesis of a spherical symmetry of the vertical profiles of temperature, pressure and water vapor, Fig. 2 shows the total spectral attenuation together with the separate contribu-

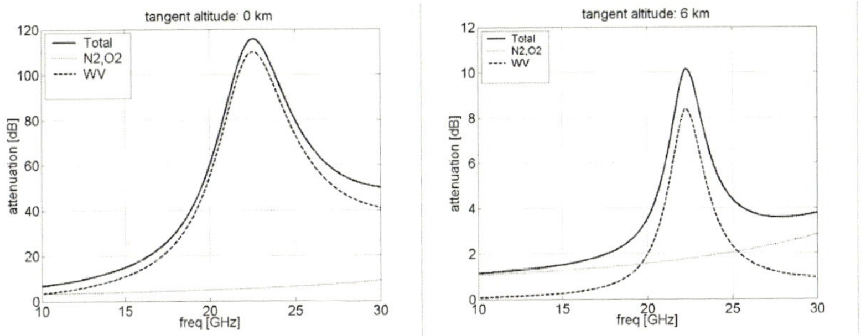

Fig. 2. Total spectral attenuation $A(f)$ in dB and its contributions due to water vapor only and due to N_2/O_2 for 0 km and 6 km of z_t. Hypotheses: MLS atmospheric model profiles (McClatchey et al. 1972), spherical symmetry for the atmospheric structure, and MPM93 propagation model (Liebe et al. 1993).

tions by water vapor and O_2/N_2, for 0 km and 6 km of the tangent altitude z_t. Evidently, though the shape of the curves changes with z_t, measuring the derivative of the spectral attenuation allows (in principle) to remove all absorption contributions that are frequency independent and to obtain estimates that are correlated only to the water vapor content. However, the derivative of the spectral attenuation still brings the contribution of $\alpha(f)$, which is typically unknown.

Using a normalized differential approach can help to solve this problem. Let us therefore define the spectral sensitivity function $S(f)$ as:

$$S(f) = \frac{1}{A(f)} \frac{\mathrm{d}}{\mathrm{d}f}\left[A(f)\right]. \tag{3}$$

If $\mathrm{d}\alpha(f)/\mathrm{d}f \cong 0$ in the frequency region of interest, $S(f)$ becomes:

$$S(f) = \frac{\mathrm{d}}{\mathrm{d}f}\left[\tau_a(f)\right] \cong \frac{\mathrm{d}}{\mathrm{d}f}\left[\tau_{a\mathrm{H_2O}}(f)\right], \tag{4}$$

which shows that the spectral sensitivity function $S(f)$ is independent of any frequency-flat contribution, while directly depending on the absorption optical depth contributed by water vapor only. In practical measurements, the following approximate version of $S(f)$ (as from Eq. (3)) can be used:

$$\hat{S}(f_0) = \frac{A(f_+) - A(f_-)}{2\Delta f\ A(f_+)}, \tag{5}$$

where f_o is a central frequency (the carrier) and $f_+ = f_o + \Delta f$, $f_- = f_o - \Delta f$ are two opportunely spaced frequencies. Assuming that the transmitted signals are two sinusoidal tones at f_+ and f_- with powers P_{tx+} and P_{tx-}, then $A(f_+) = P_{tx+}/P_{rx+}$ and $A(f_-) = P_{tx-}/P_{rx-}$ are determined by separately bandpassing the corresponding received signals (with powers P_{rx+} and P_{rx-}). Assuming for the sake of simplicity that the two spectrally separated transmitted signals have the same power, the

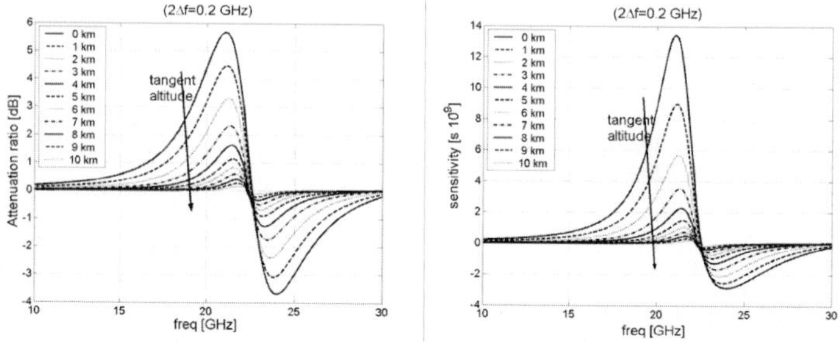

Fig. 3. Left plots: Differential attenuation (in dB) versus center frequency f_o for $2\Delta f = 200$ MHz. Right plots: The corresponding spectral sensitivity (as from (Cuccoli et al. 2001). Hypothesis: MLS atmospheric profiles with spherical symmetry and MPM93 propagation model (Liebe et al. 1993).

sensitivity estimated according to Eq. (5) can be expressed in terms of the received powers only (and of the frequency difference):

$$\hat{S}(f_o) = \frac{P_{rx\_} - P_{rx_+}}{2\Delta f \, P_{rx\_}}. \tag{6}$$

The frequency difference $2\Delta f$ should be selected according to two opposite practical requirements: the first is to have Δf small enough so that any "frequency-flat" contribution can be realistically considered as such; on the other hand, Δf must be sufficiently great to guarantee the practical detectability of the differential attenuation in the presence of signal disturbances.

With regard to this last issue, f_o plays a major role since it would be evidently highly desirable to select f_o in a frequency range where the derivative of the total spectral absorption curve is a maximum, which is however not possible in general since such a point depends on the tangent altitude z_t. This is well evidenced by Fig. 3, where the differential attenuation (in dB) and the corresponding spectral sensitivity function are reported versus f_o, for tangent altitudes from 0 km to 10 km and for $\Delta f = 100$ MHz. All plots are obtained adopting the MLS vertical profiles and assuming that they do not change with their position above the Earth's surface (spherical symmetry hypothesis).

It can be noticed that: for any value of z_t; frequencies lower than 15 GHz are hardly exploitable for relative differential spectral attenuation measurements; the maxima of the differential attenuation and of the sensitivity curves fall between 21 GHz and 22 GHz, depending on z_t; increasing altitudes increase the practical problem of measuring the differential attenuation/sensitivity, due to increased requirements in terms of receiver's sensitivity and signal to noise ratio at the receiver. However, while the qualitative behaviors described above hold in general, a profile change – even keeping the spherical symmetry hypothesis – has a quantitative impact on all the aforementioned quantities. For this reason, the feasibility analysis of the differential spectral attenuation approach cannot neglect the issue of natural variations of temperature, pressure and water vapor profiles.

3 The ITU Constraints

ITU (International Telecommunication Union) radio-regulations allow: 17.2 GHz to 17.3 GHz frequency band for Earth exploration satellite and space research and 20.1 GHz to 21.2 GHz for fixed/mobile satellite-Earth (ITU-R, 2001). For DSA measurements we selected the carrier $f_o = 17.25$ GHz with $\Delta f = 0.05$ GHz in order to use the whole 17.2 GHz to 17.3 GHz band. Instead, the 20 GHz band can be exploited without being subject to particular restrictions. Therefore, for the simulations shown here we chose the carrier at 20.25 GHz and a frequency separation of 0.2 GHz ($\Delta f = 0.1$ GHz) since that seemed the most appropriate compromise between the two aforementioned opposite requirements of allowing a sufficiently great spectral separation (to get a sufficiently high differential attenuation) and of keeping a sufficiently high spectral channel correlation (for limiting scintillation effects). The notation for the spectral sensitivities at these two frequencies will be the following:

$$S_{17.25} = \hat{S}(f_o)\Big|_{\substack{f_0=17.25\\ \Delta f=0.05}}, \qquad S_{20.25} = \hat{S}(f_o)\Big|_{\substack{f_0=20.25\\ \Delta f=0.1}}. \tag{7}$$

4 The Relationship Between *Iwv* and Spectral Sensitivity

4.1 The Radiosonde Data Set

Natural variations of the atmospheric conditions were simulated by means of real radiosonde data, extended to the whole Earth's atmosphere through the spherical symmetry hypothesis. Specifically, vertical profiles of temperature $T(z)$, pressure $p(z)$ and water vapor concentration $N_w(z)$ were used from a data set of 565 radiosonde observations gathered through the whole year 2000 at the meteorological station of S. Pietro Capofiume (Bologna, Italy, lat = 44.65 N, lon = 11.62 E, WMO code: 16144), with a variable vertical resolution, always better than 125 m. The Millimeter wave Propagation Model-version 1993 (MPM93) (Liebe et al. 1993) was used to simulate the total attenuation at the two frequencies (f_+ and f_-) along the LEO-LEO propagation link. Then the spectral sensitivity was simulated according to Eq. (5). The radiosonde profiles are quite uniformly distributed over the four seasons and almost all of them were collected at 11.00 am and 11.00 pm local time. In order to limit the effects of variable vertical resolution in each radiosonde profile and from one profile to the other, the profiles were interpolated with a constant step of 125 m. These additional interpolated data were then added to the original ones so that every radiosonde profile could exhibit at least one data point every 125 m.

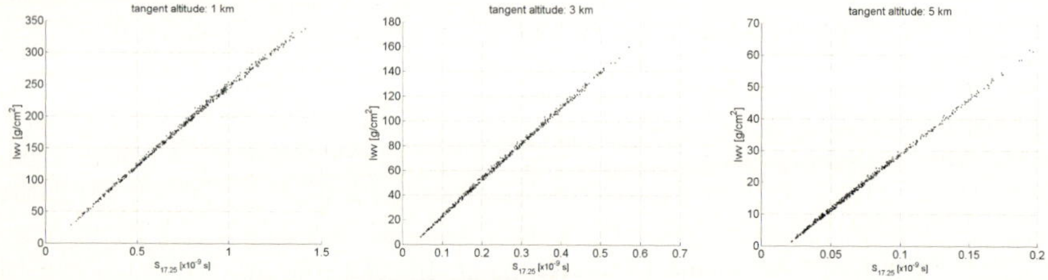

Fig. 4. 565 Iwv-$S_{17.25}$ data pairs at tangent altitudes 1,3 and 5. In abscissa $S_{17.25}$ in ns, in ordinate the Iwv in g cm^{-2}.

4.2 The *Iwv-S* Relationships Coefficients

The measurement of $S_{17.25}$ and $S_{20.25}$ in ideal conditions (no receiver noise and no signal fluctuations nor impairments) was simulated in correspondence of 6 tangent altitudes, 1 km to 6 km, step 1 km, for $S_{17.25}$ and of 10 tangent altitudes, 1 km to 10 km, step 1 km, for $S_{20.25}$. At the same tangent altitudes the "true" Iwv was computed, achieving in this manner 565 Iwv-$S_{17.25}$ pairs and the same number of Iwv-$S_{20.25}$ pairs for each of the mentioned tangent altitudes.

Some Iwv-$S_{17.25}$ plots are reported in Fig. 4, whereas some of Iwv-$S_{20.25}$ are reported in Fig. 5. By simply observing the trend of the scatter plots it appears that: $S_{17.25}$ and Iwv are linearly related; $S_{20.25}$ and Iwv are not linearly related up to 5 km; $S_{20.25}$ and Iwv are linearly related from 5 km to 10 km, even at 10 km the relationship begins to be not so tight. A linear relationship was therefore utilized to convert $S_{17.25}$ to Iwv:

$$\hat{Iwv}(z) = a_1(z) \cdot S_{17.25}(z) + a_0(z) \tag{8}$$

and a quadratic one to convert $S_{20.25}$ to Iwv:

$$\hat{Iwv}(z) = a_2(z) \cdot [S_{20.25}(z)]^2 + a_1(z) \cdot S_{20.25}(z) + a_0(z). \tag{9}$$

The conversion coefficients (a_0, a_1 and a_2) were computed through standard regression methods applied on Iwv-S pairs. The hypotheses under which they were computed are: rectilinear propagation between the two LEO-LEO satellites and spherical symmetry hypothesis for the atmospheric structure.

In order to evaluate the error caused by the estimation the Iwv through Eqs. (8) and (9), three different error parameters (functions of the altitude z) were used. The first two are the mean and the standard deviation of the per cent relative error:

$$\varepsilon(z) = \frac{\hat{Iwv}(z) - Iwv(z)}{Iwv(z)} \cdot 100 \tag{10}$$

where $Iwv(z)$ and $I\hat{w}v(z)$ are respectively the true and the estimated integrated water vapor at a given tangent altitude.

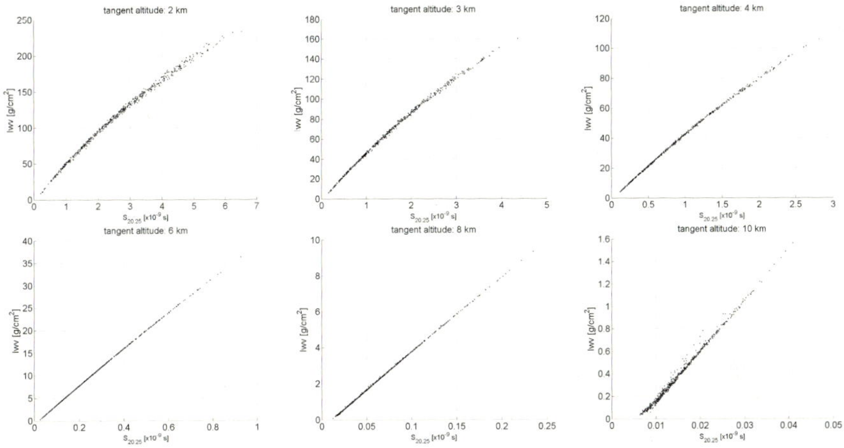

Fig. 5. 565 *Iwv*-$S_{20.25}$ data pairs at tangent altitudes 2, 3, 4, 6, 8, and 10 km. In abscissa $S_{20.25}$ in ns, in ordinate the *Iwv* in g cm^{-2}

Table 1. a_1 and a_0 coefficients (linear fit) for the yearly data plots of Fig. 4. *Mean* and *std* stand for mean and standard deviation of ε while *rms* is the error parameter defined in Eq. (11).

Tangent altitude z [km]	$a_1(z)$ [g cm^{-2} ns^{-1}]	$a_0(z)$ [g cm^{-2}]	$mean(\varepsilon(z))$ [%]	$std(\varepsilon(z))$ [%]	$rms(z)$ [%]
1	266	–7.2	0.2	1.9	1.6
2	283	–7.6	0.3	2.9	1.8
3	302	–7.2	0.4	3.2	1.8
4	323	–6.5	0.8	4.8	2.0
5	351	–6.0	1.1	7.7	2.4
6	383	–5.4	1.1	6.8	2.6

Table 2. Coefficients based on quadratic fit for the yearly data plots of Fig. 5. Symbols as in Table 1.

Tangent altitude z [km]	$a_2(z)$ [g cm^{-2} ns^{-2}]	$a_1(z)$ [g cm^{-2} ns^{-1}]	$a_0(z)$ [g cm^{-2}]	$mean(\varepsilon(z))$ [%]	$std(\varepsilon(z))$ [%]	$rms(z)$ [%]
2	–2.35	51	1.1	0.3	3.5	2.1
3	–2.92	50	–1.2	0.2	2.1	1.6
4	–3.18	47	–1.3	0.1	1.3	1.2
5	–3.45	45	–1.0	0.1	0.8	0.7
6	–3.83	44	–0.8	0.1	0.6	0.3
7	–4.40	43	–0.6	0.0	1.5	0.3
8	–5.14	43	–0.5	–0.1	3.2	0.8
9	–1.95	43	–0.4	–0.3	6.8	2.1
10	75	42	–0.3	–0.4	12.0	5.3

In alternative to the standard deviation of $\varepsilon(z)$, a third error parameter $rms(z)$ (root mean square) is used:

$$rms(z) = \frac{std\left(I\hat{w}v(z) - Iwv(z)\right)}{mean\left(Iwv(z)\right)} \cdot 100 \qquad (11)$$

where *mean* and *std* stand for mean and standard deviation, respectively. The results of the linear fit for the $S_{17.25}$ case are summarized in Table 1. In the case of $S_{20.25}$, the coefficients $a_2(z)$, $a_1(z)$ and $a_o(z)$ were computed by quadratic fit, thus providing the results reported in Table 2 together with the pertinent errors.

5 End-to-End Simulation Scenario

5.1 Receiver Scheme

As mentioned before, we assumed that two continuous wave (CW) signals with the same amplitude and with constant frequencies f_+ and f_- are transmitted. The duration of the simulation is that of a radio occultation of two counter-rotating satellites orbiting at 650 km and 850 km, namely the time needed for the propagation path to immerge in the atmosphere and for the signal to vanish after path blocking due to the Earth. The receiver is assumed to be coherent. Two branches are needed for separating the contribution of each of the two CW signals from the total received signal and for estimating the received powers. However, the frequency of the received CW signals is not constant due to the relative motion of the satellites. Therefore, additional circuitry is required to recover frequency and phase of each of them, so that their amplitude can be estimated through mixing and integration, and consequently their powers P_{rx+} or P_{rx-}. In particular, we need of a PLL (Phase Locked Loop, that starts tracking the signals before the link immerges in the atmosphere) that provides a reference signal with perfect recovery of frequency and phase, including frequency deviation from the nominal value due to Doppler shifts and slow phase variations due to channel phase distortion. The PLL shares the estimate of the Doppler frequency shift (that can be deduced from geometrical calculations) with an automatic frequency control (AFC) block that performs the signal frequency tracking. The PLL tracks the carrier phase rotation, thus recovering also a residual carrier frequency offset coming from the AFC. The estimated phase is then fed to a coherent amplitude detector.

5.2 Modeling Signal Disturbances

In addition to Doppler phase shift and thermal noise at the receiver (modeled as AWGN, Additive White Gaussian Noise), the CW signals on each of the two receiving channels are degraded by two additional phenomena related to the random nature of the propagation channel: defocusing and scintillation. Defocusing is a modulation of the signal amplitude received during an occultation, due to "large scale" variations of the atmospheric index of refraction (Sokolovskiy 2000) and can be therefore modeled in general as a power loss depending on the tangent altitude of the LEO-LEO link. The power loss is maximum at the lowest altitudes, and typically the power loss function is obtained experimentally as an average of different atmospheric conditions. The expression used in our simulations is:

$$A_{\mathrm{DEF}}(z) = 10\exp(-0.089z) \quad [\mathrm{dB}] \tag{12}$$

where z is expressed in km. While defocusing is simply an additional attenuation factor to be accounted for on both receiving channels, related to variations of the mean value of the refraction index in a turbulent medium, diffraction effects related to fluctuations of the refraction index cause signal fluctuations referred to as tropospheric scintillation. This is the most important reason for introducing the DSA concept to estimate water vapor content in the atmosphere through a LEO-LEO link. Scintillation introduces a random variation on the power received in each channel of the link: while the entity of the variations depends on the tangent altitude and is a relatively slow process with a maximum bandwidth of a few Hertz, a high correlation is expected to exist between the variations introduced on two channels that are relatively close in frequency. In practice, the combined effect on DSA measurements of scintillation disturbance and thermal noise at the receiver was simulated assuming that the two received signals $r_+(t)$ and $r_-(t)$ are:

$$\begin{aligned} r_+(t) &= \chi_+(t)s_+(t) + n_+(t) \\ r_-(t) &= \chi_-(t)s_-(t) + n_-(t) \end{aligned} \tag{13}$$

where $s_+(t)$ and $s_-(t)$ are the two band-pass filtered signals in absence of any disturbance while $n_+(t)$ and $n_-(t)$ are thermal noise disturbances produced by band-passing the AWGN disturbance generated at the receiver. $\chi_+(t)$ and $\chi_-(t)$ are multiplicative disturbances that model the effect of scintillation at the receiver. It is assumed that both $\chi_+(t)$ and $\chi_-(t)$ are wide sense stationary processes exhibiting the same mean and log-amplitude variance σ_χ^2, this latter defined as:

$$\sigma_\chi^2 = Var\left[20 \cdot \log_{10}\left(\frac{\chi(t)}{mean(\chi(t))}\right)\right]. \tag{14}$$

The fluctuations of the received signal power in the two channels used for DSA measurement were modeled as random processes with a unit mean log-normal distribution and characterized by a 5 Hz bandwidth power spectral density. To generate the time series corresponding to the multiplicative disturbances of the two channels, $\chi_+(n)$ and $\chi_-(n)$ were generated starting from two time series – $x_-(n)$ and $x_+(n)$ – of zero-mean independent Gaussian variates. Such time series were filtered by two identical low pass filters with a 5 Hz cutoff frequency (a Butterworth filter was used for convenience); the following linear transformation was then applied to the two filtered time series in order introduce a zero-lag correlation ρ between the outputs:

$$\begin{cases} y_+(n) = x_+(n) \\ y_-(n) = \rho\, x_+(n) + \sqrt{1-\rho^2}\, x_-(n). \end{cases} \qquad (15)$$

$y_-(n)$ and $y_+(n)$ are two linearly correlated, zero-mean independent Gaussian, time series. Finally, $\chi_+(n)$ and $\chi_-(n)$ are: $\chi_+(n) = 10^{\,0.1\, y_+(n)}$ and $\chi_-(n) = 10^{0.1\, y_-(n)}$.
The e.m. models needed for estimating the scintillation effects in terms of ρ and σ_χ suffer from a basic lack of information that prevents the researchers from estimating those parameters with the accuracy that would be needed to provide reliable error estimates (see Facheris et al. 2004 for a detailed e.m. analysis of the scintillation in LEO-LEO radio occultations). For this reason, simulations were carried out for a range of values of the two scintillation parameters, within which the performance may vary significantly. In a very conservative way, values of σ_χ up to 3 dB were considered, though it is envisaged that they are quite pessimistic, if not unrealistic. The standard deviation (in dB) of the fluctuations of the received signal powers in the two channels used for DSA measurement was assumed to decrease linearly with altitude.

5.3 Reference Link Budget and Simulation Parameters

It is assumed that a pair of power estimates is provided every T_s seconds, and a spectral sensitivity estimate sample is provided as from Eq. (5. To reduce fluctuations on the final estimate, N_A samples are averaged to provide a sensitivity estimate every $T_A = N_A T_s$ seconds.

Table 3. Reference link budget parameters used for 17.25 GHz and 20.25 GHz LEO-LEO link.

Parameter	17.25 GHz	20.25 GHz
Tx Power (on each channel)	33 dBm	33 dBm
Tx antenna gain	28 dB	29 dB
Rx antenna gain	28 dB	29 dB
Implementation margin	–1 dB	–1 dB
Rx equivalent noise temperature	347 K	398 K
C/N_0 (without atmospheric absorption)	68 dBHz	68.1 dBHz

The value of T_S must be large enough to allow sufficiently stable power measurements. The value of T_A must be chosen small enough to allow stationarity of power measurements, but large enough to have a high number of independent sensitivity samples. Also based on the spectral width and correlation time of the scintillation phenomenon, all the experimental results reported in the following were obtained by using $T_S = 0.01$ s, $N_A = 10$, $T_A = 0.1$ s.

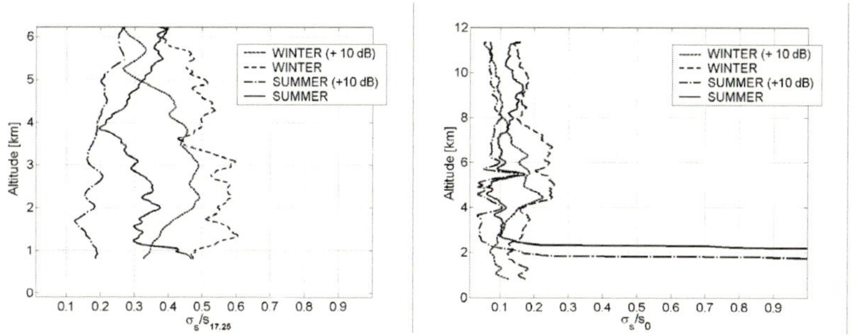

Fig. 6. Relative estimation error σ/S vs. tangent altitudes obtained through simulations with 50 runs considering 2 reference radiosonde observations (winter and summer) at 17.25 GHz (left) and 20.25 GHz (right). Scintillation parameters: $\sigma\chi$ max = 1 dB and ρ = 0.99.

6 Simulation Results

Monte Carlo simulations with 50 runs were performed for each case. Let $S_k(z)$ be the value of an average DSA measurement, computed at tangent altitude z during the k-th radio occultation realization. This value is compared with a reference value $S(z)$ obtained by considering the received signal with no distortion introduced by the channel, apart from atmospheric attenuation. Let $\overline{S}(z)$ and $\sigma_s(z)$ be the mean value and the standard deviation computed over all the realizations $\{S_k(z)\}_{k=1}^{50}$. $\overline{S}(z)$ vs. $S(z)$ yields a measure of the bias of the estimator, whereas $\sigma_s(z)$ vs. $S(z)$ yields an evaluation of the estimation error.

No significant bias was evidenced. Therefore, only $\sigma_s(z)$ vs. $S(z)$ is considered in this paper. Figure 6 shows the results of simulations run with $\rho = 0.99$, 1 dB for σ_χ at ground and a surplus power gain of 0 dB and +10 dB with respect to the link budget parameters in Table 1. Notice that the performance at 20.25 GHz is better than that at 17.25 GHz for both atmospheric conditions between 2 km and 6 km tangent altitude. Summer conditions give better results since in winter the amount of water content in the atmosphere is smaller, which causes lower differential attenuation. This leads to a greater relative fluctuation of the spectral sensitivity at the parity of scintillation disturbance and thermal noise at the receiver. Moreover, notice that at 20.25 GHz the higher level of atmospheric attenuation in summer condition gives bad performance below 2 km tangent altitude. In any case, a surplus power gain of +10 dB allows a reduction of the estimation error of about 30 %.

7 Conclusions

The DSA approach seems to be well applicable for estimating the water vapor content along a LEO-LEO satellite link, and for retrieving water vapor profiles from received power measurements during a LEO-LEO radio occultation. Such potential comes from the tight relationship between the spectral sensitivity functions at 17.25 GHz and 20.25 GHz and the *Iwv* along the LEO-LEO link, at any tangent altitude up to 10 km. Such relationship, though depending on latitude and season, allows a direct measurement of integrated water vapor through simple power measurements at the receiver site, without requiring any a priori information about any other atmospheric or climatic parameter. The optimal candidate signals for performing DSA measurements are deemed to be two tones symmetrically located with respect to a "virtual" carrier frequency: this solution is at the same time the closest to the ideal DSA concept and the simplest in terms of implementation (both in transmission and reception), while also allowing a good rejection of possible signal interference due to terrain multipath at the lowest altitudes through Doppler filtering.

Further investigation should be made in order to quantify the true level of scintillation. Moreover, the impact of other diffraction effects on the DSA approach, should be accounted for, especially those that can not be included inside the scintillation model proposed here.

Acknowledgements. This work has been supported by ESA under ESTEC contract No. 17831/03/NL/FF.

References

Cuccoli F, Facheris L (2002) Estimate of the tropospherical water vapor through microwave attenuation measurements in atmosphere. IEEE Transactions on Geoscience and Remote Sensing 40(4):735–741

Cuccoli F, Facheris L, Tanelli S, Giuli D (2001) Microwave attenuation measurements in satellite-ground links: the potential of spectral analysis for water vapor profiles retrieval. IEEE Transactions on Geoscience and Remote Sensing 39(3):645–654

Facheris L et al. (2004) Alternative Measurements Techniques for LEO-LEO Radio Occultation (AlMeTLEO). ESA-ESTEC contract No. 17831/03/NL/FF Mid-term Project report

International Telecommunication Union Volume 3 (2001) Resolutions - Recommendations Volume 4 - ITU-R Recommendations, Edition of 2001 Radio Regulations

Liebe HJ, Hufford GA, Cotton MG (1993) Propagation Modeling of Moist Air and Suspended Water/Ice Particles at Frequencies below 1000 GHz. In: AGARD, 52nd Specialists Meeting of the Electromagnetic Wave Propagation Panel on "Atmospheric Propagation Effects through Natural and Man-Made Obscurants for Visible to MM-Wave Radiation", Palma de Mallorca, Spain, May 1993

McClatchey RA, Fenn RW, Selby JEA, Volz FE, Garing JS (1972) Optical properties of the atmosphere. Environmental Research Papers 411, AFCRL-72-0497, Air Force Cambridge Research Laboratories
Sokolovskiy SV (2000) Inversion of radio occultation amplitude data. Radio Science 35(1): 97–105

Processing X/K Band Radio Occultation Data in Presence of Turbulence: An Overview

M. E. Gorbunov[1,2] and G. Kirchengast[2]

[1] Institute of Atmospheric Physics, Moscow, Russia
gorbunov@dkrz.de
[2] Wegener Center for Climate and Global Change (WegCenter) and Institute for Geophysics, Astrophysics, and Meteorology (IGAM), University of Graz, Austria

Abstract. Canonical Transform (CT) and Full-Spectrum Inversion (FSI) methods can be used in processing radio occultation (RO) data to retrieve transmission profiles, which are necessary for the retrieval of atmospheric humidity. LEO-LEO occultations with X/K band frequencies can provide transmission data along the wing of the 22 GHz water vapor absorption line. The retrieval of transmission requires spherical symmetry of the atmospheric refractivity. This condition is broken in presence of turbulence, which can result in significant errors in transmission. We suggest the computation of the differential transmission from the differential CT/FSI-amplitude to correct for the effect of turbulence and horizontal gradients. The efficiency of the method is tested by realistic numerical end-to-end simulations. We modeled turbulence based on power form of the spectrum. The simulations demonstrate that the error in the retrieved transmission can be significant in a single frequency channel, while the differential transmission can be retrieved with high accuracy.

1 Introduction

Radio occultations using GPS signals proved to be a very powerful technique of sounding the Earth's atmosphere [14]. However, atmospheric refractivity indicates very weak absorption and dispersion at GPS frequencies. This makes it impossible to separate the dry and wet terms of the retrieved refractivity without employing additional a priori information. Use of an observation system of Low Earth Orbiters (LEOs) implemented with transmitters and receivers of radio signals in 9 GHz to 30 GHz band can solve this problem [12, 13, 15]. Water vapor has an absorption line centered near 22 GHz. Therefore, from measurements of phase and amplitude, complex refractive index can be retrieved. Then, pressure, temperature, and water vapor profiles can be solved for, using a spectroscopic model of the water vapor line and the hydrostatic equation. [12] describe this retrieval processing in detail, but restricted to the geometric-optics approach for the transmission and bending angle retrieval.

Here the focus is on transmission and bending angle retrieval by wave-optical methods, which will be the approach generally required with real data, since the retrieval scheme will encounter a significant challenge in case of turbulence. The amplitude of a radio occultation signal is significantly more sensitive to small scale turbulence than the phase [16, 17]. In presence of turbulence, the amplitude of the wave field undergoes strong scintillations, which can overwhelm the effect of absorption. In order to reduce the effect of scintillations, it was suggested to use twin frequencies [1, 15].

Given the measurements of the wave field $u_1(t)$ and $u_2(t)$ for two frequencies f_1 and f_2, the difference $\Delta f = f_1 - f_2$ being small enough, we consider the ratio $|u_1(t)| \,/\, |u_2(t)|$. Because for neighbor frequencies the effects of diffraction and interference will not differ significantly, it is expected that they will be reduced in the ratio. The amplitude is proportional to the absorption factor $\exp(-\tau_{1,2})$, where $\tau_{1,2}$ is the integral absorption along a ray for frequency f_1 or f_2. In this case $\ln(u_1(t)/u_2(t))$ equals differential absorption, $\tau_2 - \tau_1$. However, this is only valid for Δf being small enough. On the other hand, choice of too small Δf will result in too low values of the differential absorption, which will increase the noise sensitivity.

The wave optics processing methods such as Canonical Transform (CT) [2, 5, 6, 7] or Full-Spectrum Inversion (FSI) [10, 11] transform the wave field into the representation of impact parameter. In this representation, the amplitude describes the distribution of the energy with respect to impact parameters. For a spherically-symmetric atmosphere, the amplitude of the transformed wave field is proportional to the exponential function of the integral absorption along the ray. Therefore, CT and FSI techniques can be used for the retrieval of atmospheric absorption [3, 11]. These techniques will significantly reduce retrieval errors due to multipath and diffraction. However, the problem of turbulence still persists. Since turbulence is a 3D inhomogeneous structure, the amplitude of the transformed wave field also indicates scintillations [16].

In this study, we suggest the differential method for retrieval of absorption in combination with the CT or FSI techniques. These methods define the transformed wave field $\hat{\Phi} u_{1,2}(p)$, where p is the impact parameter. The transformed field at each single frequency is then to a very significant extent free from the effects of diffraction and multipath. These effects may only be significant for atmospheric inhomogeneities with scales below 50 m [7]. The logarithmic ratio of the transformed amplitudes $\ln\left(\left|\hat{\Phi} u_1(p)\right| \,/\, \left|\hat{\Phi} u_2(p)\right|\right)$ will then further suppress the scintillations due to small scale turbulence and will be equal to the differential absorption $\tau_2 - \tau_1$ with a much higher accuracy then the direct amplitude ratio $\ln(|u_1(t)| \,/\, |u_2(t)|)$.

We tested the performance of the new type of the differential method in numerical simulations. In the simulations we used 3D global fields from an ECMWF analysis with superposition of random 2D turbulence. The simulations include modeling realistic receiver noise using the ACE+ baseline values of 67 dBHz.

2 Description of Method

Given measurements of the complex wave field during a radio occultation experiment, $u(t)$, its transform to the representation of the impact parameter is given by the following Fourier Integral Operator (FIO):

$$\hat{\Phi} u = \sqrt{\frac{-ik}{2\pi}} \int a(p,Y) \exp(ikS(p,Y)) u(Y(t)) \mathrm{d}Y, \tag{1}$$

where $k = 2\pi/\lambda$ is the wavenumber, $a(p,Y)$ and $S(p,Y)$ are the amplitude and phase function of the FIO, respectively, Y is a coordinate along the trajectory, which generalizes the use of coordinate θ in the FSI method [5, 6, 10, 11]. The amplitude function has the following form [6]:

$$a(p,Y) = \left(\sqrt{r_\mathrm{R}^2 - p^2} \sqrt{r_\mathrm{T}^2 - p^2} \frac{r_\mathrm{R} r_\mathrm{T}}{p} \sin\theta \right)^{1/2}. \tag{2}$$

The asymptotic solution for the wave field can be expressed in terms of the inverse FIO with the following amplitude function [6]:

$$a^*(Y,p) = \left(\frac{1}{\sqrt{r_\mathrm{R}^2 - p_\mathrm{R}^2} \sqrt{r_\mathrm{T}^2 - p_\mathrm{T}^2}} \frac{p_\mathrm{T}}{r_\mathrm{R} r_\mathrm{T} \sin\theta} \frac{\mathrm{d}p_\mathrm{T}}{\mathrm{d}p} \right)^{1/2}, \tag{3}$$

where r_T and r_R are (LEO-)Transmitter and (LEO-)Receiver satellite radii, and θ is the angular distance between the satellites, p_T and p_R are the impact parameters at transmitter and receiver satellite, and p is the effective impact parameter.

For a spherically symmetric atmosphere, $p_\mathrm{R} = p_\mathrm{T} = p$. Generally, due to horizontal gradients, these three impact parameters are different, and the following equation can be established [4]:

$$p_\mathrm{R} = p_\mathrm{T} + \int \frac{\partial n}{\partial \theta} \mathrm{d}s, \tag{4}$$

where the integral is taken along the ray. p_R and p_T can be expressed as functions of effective impact parameter p. The relation between p, p_R, and p_T includes the horizontal gradient of refractive index $\partial n / \partial \theta$, which is unknown a priori.

The amplitude of the transformed wave field retrieved by the CT or FSI method equals the following expression:

$$A_{1,2}(p) = \bar{A}_{1,2} \frac{\sqrt{r_\mathrm{R}^2 - p^2} \sqrt{r_\mathrm{T}^2 - p^2}}{\sqrt{r_\mathrm{R}^2 - p_\mathrm{R}^2} \sqrt{r_\mathrm{T}^2 - p_\mathrm{T}^2}} \frac{p_\mathrm{T}}{p} \frac{\mathrm{d}p_\mathrm{T}}{\mathrm{d}p} \exp(-\tau_{1,2}(p)) \equiv$$
$$\equiv \bar{A}_{1,2} K(p) \exp(-\tau_{1,2}(p)). \tag{5}$$

This expression is represented as a composition of the normalizing constant $\bar{A}_{1,2}$, integral absorption along the ray $\exp(-\tau_{1,2}(p))$, and the factor $K(p)$ that depends on the horizontal gradients. For a spherically layered medium, factor $K(p)$ equals unity, and absorption can be retrieved from the CT amplitude. For a 3D medium with horizontal gradients, factor $K(p)$ approximately equals $\mathrm{d}p_{\mathrm{T}}/\mathrm{d}p$, differing from unity. This shows that in presence of 3D inhomogeneities, in particular, atmospheric turbulence, the amplitude of the transformed wave field, $A_{1,2}(p)$ will undergo scintillations. Since factor $K(p)$ is unknown a priori, the error of the retrieved absorption in each channel will equal $\ln K(p)$.

It is important that $K(p)$ does not depend on the frequency of the wave field. Therefore, the term $\ln K(p)$ will cancel in the differential transmission $\tau_2(p) - \tau_1(p)$ and only some residual diffractive effects will be left. These residual errors are best assessed by numerical simulations as discussed below. The normalizing constants $\bar{A}_{1,2}$ are estimated from the wave field at heights from 25 km to 30 km. The logarithmic ratio of the normalized amplitudes can be expressed as follows:

$$\ln \frac{A_1(p)/\bar{A}_1}{A_2(p)/\bar{A}_2} = \tau_2(p) - \tau_1(p). \tag{6}$$

Here transmissions $\tau_{1,2}(p)$ are measured in Neper. Multiplication with a factor of $20/\ln 10$ will convert them to dB.

The described method is assessed and verified below by rigorous wave optical forward-inverse simulations with fields containing severe small-scale random turbulence and thermal receiver noise.

3 Numerical Simulations

A realistic model of turbulent atmosphere includes regular part from ECMWF analyses and anisotropic turbulence with a magnitude chosen according to radiosonde measurements. Turbulence was modeled as a random anisotropic relative perturbation of the refractivity field with a power form of the spectrum:

$$\tilde{B}(\kappa) = \begin{cases} \tilde{c}\kappa_{\mathrm{ext}}^{-\mu} & \kappa < \kappa_{\mathrm{ext}} \\ \tilde{c}\kappa^{-\mu} & \kappa_{\mathrm{ext}} \le \kappa \le \kappa_{\mathrm{int}} \\ \tilde{c}\kappa^{-\mu} \exp\left[-\left(\dfrac{\kappa - \kappa_{\mathrm{int}}}{\kappa_{\mathrm{int}}/4}\right)^2\right] & \kappa > \kappa_{\mathrm{int}} \end{cases} \tag{7}$$

where $\kappa = \left(\kappa_z^2 + q^2 \dfrac{\kappa_\theta^2}{r_E^2}\right)^{1/2}$, κ_z and κ_θ are the spatial frequencies (wave numbers) conjugated to the polar coordinates z and θ in the occultation plane, q is the anisotropy coefficient [8, 9]. The factor of $\tilde{c}$ normalizes the rms turbulent fluctuations to unity. In the coordinate space we use an additional factor of

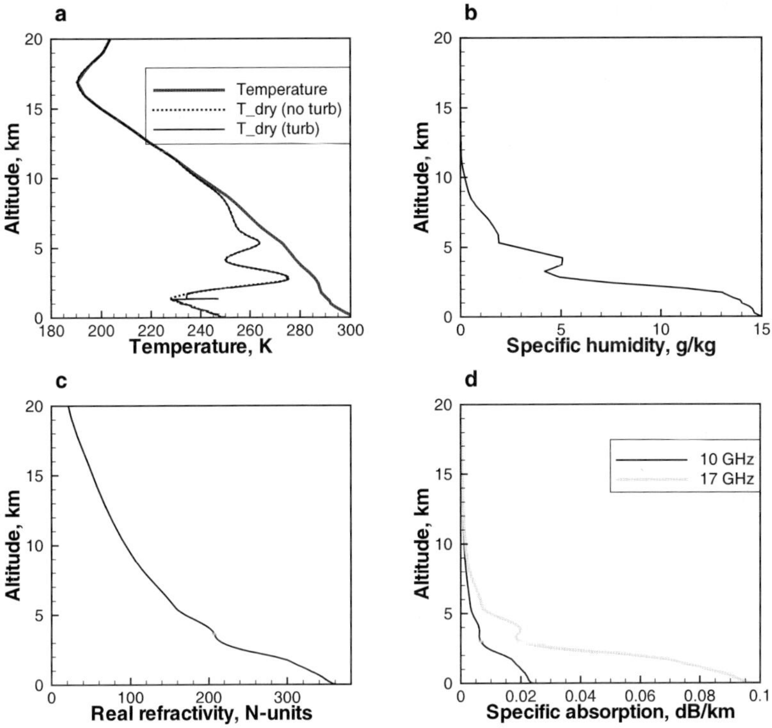

Fig. 1. Simulated occultation event 0118, May 29, 2001, UTC 13:12, 10.4°S 140.7°E, ECMWF field with superimposed power turbulence, frequency channels 10 GHz and 17 GHz: (a) local temperature of the modeled atmosphere T and retrieved dry temperatures, T_{dry} for the regular medium and dry temperature $T_{\mathrm{dry}}^{(\mathrm{turb})}$ for the turbulent medium, (b) specific humidity, q, (c) real refractivity, N, and specific absorptions, $(20/\ln 10)kN_I$, for the two frequencies.

$c(z)$, which describes the relative magnitude of turbulent perturbations as a function of altitude. We performed single-run simulations, because each simulation is a time-consuming procedure, and a montecarlo simulation would require too large a computation time.

Figures 1, 2, and 3 show the results of simulations for a tropical occultation. The turbulence is characterized by external scale $2\pi/\kappa_{\mathrm{ext}} = 0.3$ km, internal scale $2\pi/\kappa_{\mathrm{int}} = 0.03$ km, exponent $\mu = -4$ (-5 for 3D spectrum), and anisotropy $q = 20$. The magnitude $c(z)$ was set according to a low-latitude radiosonde observation on St. Helena island (Stephan Bühler, private communication, 2004): $c(z)$ equals 0.006 at a height of 2.0 km, and logarithmic-linearly decreases to 0.0005 at a height of 7.5 km. Beyond this interval, $c(z)$ is constantly extrapolated. The refraction angle profile indicates strong multipath propagation. The random error of CT differential transmission is about 0.2 dB (4 % to 5 % std.deviation).

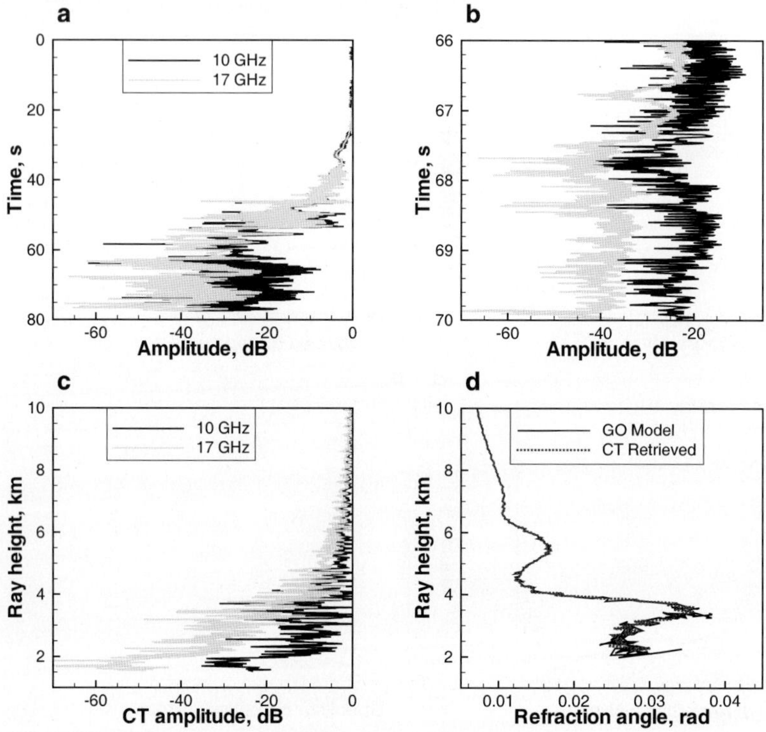

Fig. 2. The same simulated event with superimposed power turbulence, frequency channels 10 GHz and 17 GHz: (a) amplitudes in the two channels, (b) enlarged fragment of amplitude records in multipath area, (c) CT amplitudes for the two channels, and (d) refraction angles, computed by the GO model and retrieved by the CT method.

Figure 4 shows the results of processing the same simulated event with superimposed receiver noise with a magnitude of 67 dBHz. The increase of transmission retrieval errors due to the noise is most visible below a ray height of about 3.5 km, in the area of strong multipath propagation, where the amplitude is low.

4 Conclusions and Outlook

Processing X/K band radio occultation data in presence of turbulence poses a significant challenge, due to the scintillations imposed by the turbulence in the measured amplitude profiles. In earlier transmission retrieval approaches, a possibility was discussed of retrieving differential absorption from the direct ratio of measured amplitudes for two different (closely spaced) frequency

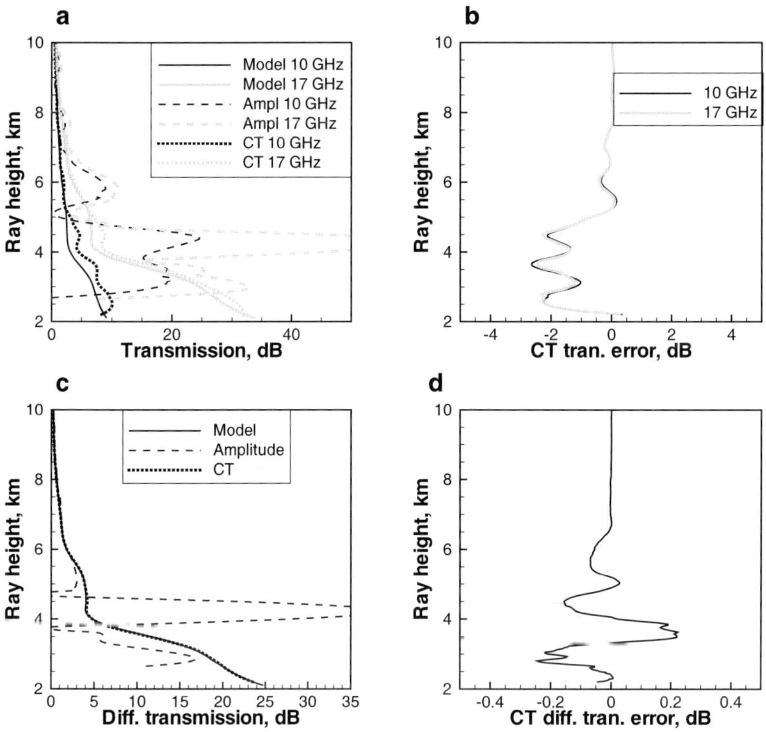

Fig. 3. The same simulated event with superimposed power turbulence, frequency channels 10 GHz and 17 GHz: (a) transmissions for the two channels, true model, computed from the amplitudes, and computed from the CT amplitudes, (b) errors of the CT transmission, (c) differential transmission, true model, computed from the measured amplitudes $|u_{1,2}(t)|$, and computed from the CT amplitudes, and (d) errors of the CT differential transmission.

channels. Here we discussed an advanced differential method of retrieval of atmospheric transmissions based on the ratio of the CT amplitudes.

The new method results in much more accurate correction for turbulence scintillations, as compared to taking the direct ratio of the measured wave fields. This is due to the following reasons: 1) The CT mapping corrects for diffraction and multipath propagation effects, 2) The resulting transformed field is independent from diffraction except for small scales below about 50 m. 3) The ratio of the transformed amplitudes then further corrects for small-scale scintillations and effects of the non-sphericity of the atmosphere. 4) The new method does not impose any significant restriction for the frequency difference Δf between the channels, and there is no requirement that Δf is small (e.g., clearly smaller than 1 GHz). This has important technical advantages and provides very good differential transmission sensitivity if spacings of a few GHz are chosen (e.g., 10 GHz and 17 GHz or 17 GHz and 20 GHz, or similar).

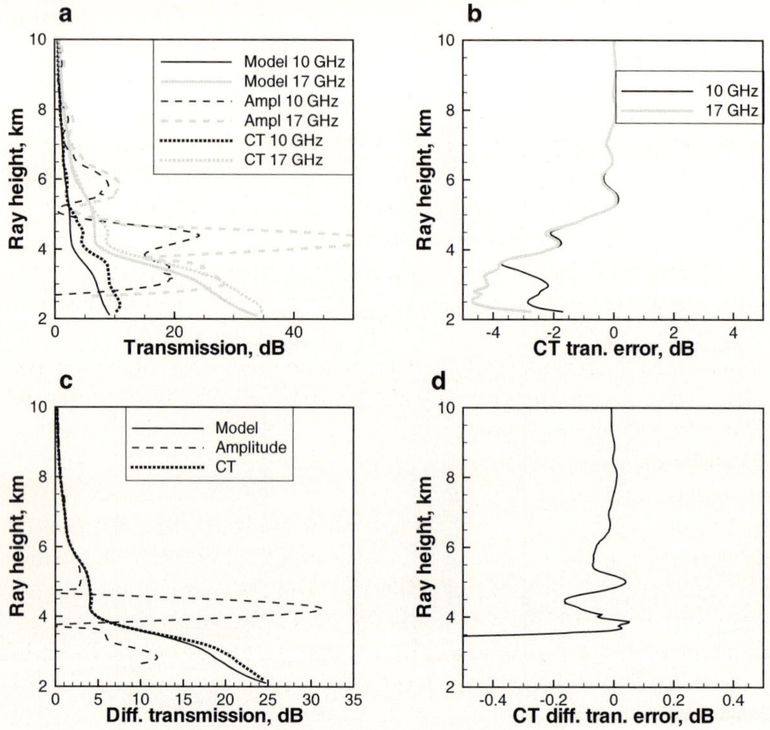

Fig. 4. The same simulated event with superimposed power turbulence, frequency channels 10 GHz and 17 GHz, with a model of receiver noise 67 dBHz: (a) transmissions for the two channels, true model, computed from the amplitudes, and computed from the CT amplitudes, (b) errors of the CT transmission, (c) differential transmission, true model, computed from the measured amplitudes $|u_{1,2}(t)|$, and computed from the CT amplitudes, and (d) errors of the CT differential transmission.

In practice the frequency separation is limited by a finite SNR and the large attenuation for the 22 GHz line, which broadens out in the lower troposphere. Using more frequencies may then be required to limit the dynamic range.

We performed numerical simulations with a realistic model of the turbulent atmospheric refractivity field in a rigorous forward-inverse modeling framework. The model also included receiver noise at a realistic level (carrier-to-noise 67 dBHz, ACE+ baseline). These numerical simulations showed the high capabilities of the CT differential method. The influence of the noise is only significant below a ray impact height of 4 km (below 2 km to 3 km altitude), where the carrier-to-noise ratio is becoming very low due to strong absorption and defocusing in the lower troposphere. In this context 1 kHz sampling rate is the minimum required rate for adequate wave optics processing of X/K band occultation data. Effects of small amplitude drifts of 0.5 % over

20 s (ACE+ specification for maximum drift) were also assessed and found of minor relevance compared to thermal noise and of no further concern.

Processing differential transmissions further to imaginary refractivity and, in turn, together with real refractivity derived from bending angles, to atmospheric profiles is a procedure identical to using single-channel transmissions [12, 13, 15]. Due to the differencing, there is one differential transmission profile less, however, than single-channel transmission profiles. In the case of ACE+ with 3 frequencies this implies availability of two differential transmission profiles, which are still sufficient in combination with the real refractivity profile to separate water vapor and liquid water from temperature, down into the lower troposphere.

Further efforts to perform the explicit turbulence modeling at higher resolution (e.g., down to inner scale of turbulence of 10 m instead of 30 m, and down to an-isotropy coefficients as small as 5) will be worthwhile, for assessment of the theoretical expectation that residual errors will become smaller when approaching isotropic turbulence.

Independent of such further work, there is clear evidence from the present study already that in those turbulent cases, where single-channel transmissions might be too noisy to be processed directly, the use of differential transmissions is an adequate alternative. It can be expected, based on the experience from single-channel transmission processing [13], that also differential transmissions will allow to meet X/K band occultation observation requirements such as laid out for ACE+.

Acknowledgements. The authors gratefully acknowledge valuable discussions related to this work with A.S. Gurvich (IAP Moscow, Russia), S.V. Sokolovskiy (UCAR, Boulder, CO, USA), F. Cuccoli (Univ. of Florence, Italy), S. Buehler (Univ. of Bremen, Germany), P. Silvestrin (ESA/ESTEC, Noordwijk, Netherlands), and further colleagues of the ESA-ACEPASS study team. The work was funded by ESA under ESTEC Contract No. 16743/02/NL/FF.

References

[1] Facheris L, Cuccoli F (2003) Analysis of differential spectral attenuation measurements applied to a LEO-LEO link. ESA-ACEPASS Report (contract 16743/02/NL/FF), Inst of Elect and Telecommunications, Univ of Florence

[2] Gorbunov ME (2002) Canonical transform method for processing radio occultation data in lower troposphere. Radio Sci 37(5), doi:10.1029/2000RS002592

[3] Gorbunov ME (2002) Radio-holographic analysis of Microlab-1 radio occultation data in the lower troposphere. Journal of Geophysical Research 107, doi:10.1029/2001JD000889

[4] Gorbunov ME, Kornblueh L (2001) Analysis and validation of GPS/MET radio occultation data. Journal of Geophysical Research 106:17 161–17 169

[5] Gorbunov ME, Lauritsen KB (2002) Canonical transform methods for radio occultation data. Scientific Report 02-10, Danish Meteorological Institute, Copenhagen, Denmark, http://www.dmi.dk/dmi/Sr02-10.pdf

[6] Gorbunov ME, Lauritsen KB (2004) Analysis of wave fields by Fourier Integral Operators and its application for radio occultations. Radio Sci 39(4), RS4010, doi:10.1029/2003RS002971

[7] Gorbunov ME, Benzon H-H, Jensen AS, Lohmann MS, Nielsen AS (2004) Comparative analysis of radio occultation processing approaches based on Fourier integral operators. Radio Sci 39(6), RS6004, doi:10.1029/2003RS002916

[8] Gurvich AS, Brekhovskikh VL (2001) Study of the turbulence and inner waves in the stratosphere based on the observations of stellar scintillations from space: A model of scintillation spectra. Waves in Random Media 11, 163–181, doi:10.1088/0959-7174/11/3/302

[9] Gurvich AS, Chunchuzov IP (2005) Estimates of characteristic scales in the spectrum of internal waves in the stratosphere obtained from space observations of stellar scintillations. J Geophys Res 110(D3), D03114, doi:10.1029/2004JD005199

[10] Jensen AS, Lohmann MS, Benzon H-H, Nielsen AS (2003) Full spectrum inversion of radio occultation signals. Radio Sci 38, doi:10.1029/2002RS002763

[11] Jensen AS, Lohmann MS, Nielsen AS, Benzon H-H (2004) Geometrical optics phase matching of radio occultation signals. Radio Sci 39(3), RS3009, doi:10.1029/2003RS002899

[12] Kirchengast G, Fritzer J, Schwaerz M, Schweitzer S, Kornblueh L (2004) The atmosphere and climate explorer mission ACE+: Scientific algorithms and performance overview. Tech Report for ESA/ESTEC No 2/2004, Inst for Geophys, Astrophys, and Meteorol, Univ of Graz, Austria

[13] Kirchengast G, Schweitzer S, Ramsauer J, Fritzer J, Schwaerz M (2004) Atmospheric profiles retrieved from ACE+ LEO-LEO occultation data: Statistical performance analysis using geometric optics processing. Tech Report for ESA/ESTEC No 1/2004, Inst for Geophys, Astrophys, and Meteorol, Univ of Graz, Austria

[14] Kursinski ER, Hajj GA, Leroy SS, Herman B (2000) The GPS radio occultation technique. Terrestr Atmosph Ocean Sci 11(1):53–114

[15] Kursinski ER, Syndergaard S, Flittner D, Feng D, Hajj G, Herman B, Ward D, Yunck T (2002) A microwave occultation observing system optimized to characterize atmospheric water, temperature and geopotential via absorption. Journal of Atmospheric and Oceanic Technology 19(12):1897–1914

[16] Sokolovskiy SV (2001), Tracking tropospheric radio occultation signals from low earth orbi. Radio Sci 36(3):483–498, doi:10.1029/1999RS002305

[17] Yakovlev OI, Matyugov SS, Vilkov IA (1995) Attenuation and scintillation of radio waves in the Earth's atmosphere from radio occultation experiment on satellite-to-satellite links. Radio Science 30(3):591–602

Use of GNSS Occultation Data in Numerical Weather Prediction and in Atmospheric Studies

Assimilation of GNSS Radio Occultation Data into Numerical Weather Prediction

P. Poli

Centre National de Recherches Météorologiques (CNRM),
Groupe de Modélisation, d'Assimilation, et de Prévision (GMAP),
Météo France, 42 av. G. Coriolis, 31057 Toulouse, France
paul.poli@meteo.fr

Abstract. Today's Numerical Weather Prediction (NWP) systems assimilate meteorological observations from a variety of sources. This information is used operationally to update the analysis of global or regional three-dimensional meteorological fields. The resulting analysis enables then to issue weather forecasts of prime importance for many users. Occultation data collected by refraction of signals from Global Navigation Satellite Systems (GNSS) in the atmospheric limb contain information in temperature and water vapor content. This paper reviews data assimilation techniques, the information contained in GNSS occultation data and how it pertains to the observational needs of today's NWP systems. Recent forecast impact experiments of GNSS occultation data are reviewed, and specific areas for further impact are discussed.

1 Introduction

The year 2004 marks the centennial of Bjerknes' 1904 [3] paper on the presentation of weather forecasting as a problem in mechanics and physics. In a hundred years' time, what appeared originally as a combination of equations impossible to initialize and solve in the vast domain of our atmosphere has been turned into routine operations in forecasting centers around the world. These centers update their weather forecasts on a regular basis, based on meteorological observations dispatched by the Global Telecommunication System (GTS): in situ measurements as well as observations from meteorological satellites. Today, the data from those satellites have an impact on the accuracy of Numerical Weather Prediction (NWP) comparable to the impact of in situ observations [25].

However, it was not always that way. Twenty years after the launch of the world's first sounding instrument in 1969 (followed by many passive infrared and microwave sounders), it was still unclear whether the atmospheric soundings from satellites had fulfilled their promise of improving NWP accuracy [20]. It is only in the last fifteen years that a firm, positive answer has been formulated as we gained a better understanding of information theory and the sounders' observation characteristics (errors).

The observations collected by means of Global Navigation Satellite System (GNSS) Radio Occultation (RO) represent an exotic new source of data when compared to the traditional in situ measurements or even current sounders' data [11]. GNSS RO data may soon be available via the GTS as the European Meteorological Polar orbiting satellite Metop and the planned 6-satellite Constellation Observing System for Meteorology, Ionosphere, and Climate (COSMIC) are scheduled to deliver about 3 000 daily occultation profiles in near real-time. Past and current GNSS RO missions such as the GPS for meteorology experiment (GPS/MET), the CHAllenging Mini-satellite Payload experiment (CHAMP), and the Satelite de Applicaciones Cientificas (SAC-C) have provided opportunities to evaluate the impact of GNSS RO on NWP. We review here the impact studies done so far and give some prospects for possibly more impact in the future.

2 Modeling the Weather

2.1 General Circulation Models

Bjerknes [3] identified seven variables and seven equations as necessary to characterize the weather at any time t and at any location (x, y, z) in the Earth's atmosphere. In order to obtain a realistic weather prediction, there are three crucial aspects to consider when solving these equations: discretization, boundary conditions (BCs), and initial conditions (ICs). Discretization refers to the reduction of the seven 4D-variables to arrays defined for discrete times and locations; it coined the term Numerical Weather Prediction (NWP), using General Circulation Models (GCMs) to solve for these equations. The BCs constrain the numerical GCM solutions at the surface or at the top of the atmosphere, while the ICs provide the necessary starting point of integration. Creating such ICs requires to know the atmospheric state and project that information onto the GCM grid (for a geographical grid), or onto the set of possible GCM solutions (for spectral models).

Several sources of information are available to prescribe the ICs: observation, climatology, and first-guess. The "few" (on the order of 10^5) observations available today within a few hours of ICs' given date/time are unfortunately insufficient to constrain alone all the degrees of freedom of today's GCM (on the order of 10^7), leading to an under-determined problem. A climatology is obtained from a combination of observations averaged over a long period of time and originating from a variety of observing systems. A first-guess is typically a numerical forecast obtained by running a GCM for a few hours from an old set of ICs. As such, a first-guess may be designed to reproduce a prescribed climatology in usually observation-void areas. Data assimilation is used to combine all these sources of information and create ICs.

2.2 Data Assimilation: A Random-Error Reduction Process

In 1963, Lorenz [12] showed the high sensitivity of weather forecasts to the ICs. His work suggested reduced IC errors should result in increased forecast accuracy. At the same time, steady increases in available computing power enabled improvements by orders of magnitude in the resolution of the GCMs used in NWP. This scaled into an increase in the number of grid-points to be initialized before each forecast run. Although the number of observations has picked up since the 1960s, the GCM degrees of freedom still outnumber the observations by at least two orders of magnitude (sometimes more depending on the GCM resolution). The acute problem of creating IC fields from the available observations has thus become more and more difficult. Techniques such as krieging were attempted but with limited success.

Data assimilation (DA) designates a process by which the random errors in the first-guess are reduced by bringing in information from the observations. DA can be summarized as a random-error reduction process. A consequence of this is that DA cannot force a GCM to reproduce phenomena that the GCM cannot simulate, such as gravity waves. Also, DA cannot either fix problems intrinsic to a NWP system (e.g., stratosphere too warm), or resolve problems associated with a particular observation data-type (e.g., instrument biases).

DA will work best if the observations to assimilate already resemble the first guess they are supposed to help improve, and if some strong assumptions about the errors are respected. Namely, DA typically assumes that the observation and first-guess errors are uncorrelated with one another, present constant biases, and the statistics of those errors are Gaussian and perfectly known.

2.3 Operational Data Assimilation

In practice, first-guess and observations are merged in a statistically optimal way by minimizing a cost (or penalty) function constructed as the sum of three terms [24]. The first term (denoted J_0 hereafter) represents the distance between the newly created ICs and the assimilated observations. The second term is the distance between the first-guess and the new ICs, thus constraining the ICs to ensure that they remain close to the first-guess. The third (optional) term may include additional constraints to prevent the creation of unrealistic or unstable modes in the ICs.

If $\mathbf{x}$ designates the new set of ICs, the observation cost can be written

$$J_0[\mathbf{x}] = (\mathbf{y^0} - h(\mathbf{x}))^{\mathrm{T}} \mathsf{R}^{-1} (\mathbf{y^0} - h(\mathbf{x})) \tag{1}$$

[24], where $\mathbf{y^0}$ is the set of observations, h is an observation operator which maps $\mathbf{x}$ into the observations' space, and R is the observation error covariance matrix (combines measurement and representativeness errors). Assimilating a new data-type with success requires an accurate knowledge of R and an accurate, yet computationally efficient operator h whose tangent linear model is also required in order to minimize the total cost function.

Today's operational NWP centers rely on a so-called Three or Four Dimensional Variational analysis procedure (3DVAR or 4DVAR). The 3DVAR uses all the observations within a time window as if they all originated from the center of the time window, while the 4DVAR uses the observations at the time they are valid. The resulting set of new ICs is called the analysis.

The time sequence for Météo France operational NWP "Production" system is the following, for issuing an analysis and forecasts based on 0000 GMT. The cut-off for accepting observations via the GTS is 1 hr 50 min. All the observations received by 0150 GMT and corresponding to events between then and 2100 GMT the day before are assimilated in the 4DVAR which produces an analysis within 30 minutes. Forecasts valid between 0000 GMT+3 hrs and 0000 GMT+102 hrs are then issued. All the products must be available no later than 0340 GMT.

A "Memory" system runs in parallel to the "Production" system, but with a cut-off time of 3 hrs instead of 1 hr 50 min after the synoptic hour. This allows more observations to be accepted and results in analyses and 6-hr forecasts of better quality. That 6-hr forecast serves as the first-guess for the "Production" run of the next synoptic hour.

Observations assimilated at Météo France include in situ observations (surface, radiosondes, aircraft), passive sounder data from geostationary and polar-orbiting satellites, and satellite winds.

Assimilation of GNSS RO data builds on the knowledge learned by implementing the assimilation of in situ and sounder data. Significant differences between these data and GNSS RO data are presented next.

3 Assimilation of GNSS RO Data for NWP

3.1 A New Breed of Data

GNSS occultation data present major differences with respect to other satellite data currently assimilated in NWP systems.

First, GNSS data are mostly all-weather, although intense precipitation systems are usually associated with sharp water vapor gradients where GNSS occultation processing may encounter difficulties (e.g., multipath, signal tracking, super-refraction). Second, the GNSS occultation technique monitors the Earth's atmosphere from the side, yielding a higher vertical resolution (sub-km) than passive nadir-viewing sounders but with a horizontal resolution elongated in the direction of the ray (about 300 km). Accounting for such horizontally smoothed observations represents a challenge for the European Centre for Medium-Range Weather Forecasts (ECMWF) and Météo France 4DVAR systems which assume at the most vertically averaged observations, making the use of horizontally averaged observations a difficult task. Third, the GNSS occultation data yields atmospheric observations independently of the surface type, while the vertical sounders' lower-peaking channels usually carry a contribution from the surface through the emissivity term. Fourth, the GNSS occultation data can be provided as a function of altitude, unlike other measurement techniques which yield measurements as a function of pressure.

3.2 Options for Assimilation

The processing of GNSS occultation measurements yields several levels of data, each of which is a potential candidate for data assimilation. We review here the important points for each data-type [10].

(1) Amplitude and Phase

The raw amplitude and phase data collected as time series can present a very high vertical sampling for a receiver tracking frequency of 50 Hz. The raw measurement errors are mainly Gaussian and vertically uncorrelated, which matches the DA hypotheses. However, noise reduction in the raw data is required via smoothing/filtering, which could introduce vertical correlations in the measurement errors, depending on the smoothing parameters [23]. The observation operator calculations require exact orbit information from the GNSS transmitter and receiver as well as a ray-shooting procedure or wave optics propagation. To the best of our knowledge, this level of data has not yet been used in DA experiments.

(2) Bending Angle

The bending angle data can be used as a function of impact parameter. Derivation of the impact parameter assumes local spherical symmetry. Bending angles are obtained by differentiation from the phase measurements. Consequently, bending angle measurements present negative error correlations with neighboring measurements in the vertical [23]. Original assimilation attempts of bending angle data required a computationally expensive ray-tracing observation operator to simulate the bending angles from the first-guess [10]. This proved computationally expensive and Palmer et al. [14] developed a pure vertical observation operator (neglecting horizontal gradients of refractivity) as a computationally more affordable alternative, but which required to consider also bending angle representativeness errors. Recently, a fast observation operator for bending angle was developed [15]; this operator showed improved fit with the CHAMP and SAC-C bending angle observations as compared to the pure vertical observation operator, while reducing the computation time as compared to a ray-tracer.

(3) Refractivity

The derivation of refractivity as a function of altitude assumes a bending angle profile for the mesospheric altitudes and above, and local spherical symmetry. The latter assumption poses a problem because (a) during a GNSS occultation, the tangent points drift slowly in the horizontal, and (b) the bending of the rays includes a contribution from the horizontal refractivity gradients. Rieder and Kirchengast [19] showed that refractivities presented measurement error vertical correlations very similar to those of bending angles. As for the refractivity representativeness errors induced by a local refractivity observation operator, they were shown to be more vertically correlated than bending angle representativeness errors [22]. A clean assimilation of refractivity under DA hypotheses using a local refractivity operator would require to specify these refractivity representativeness errors. Previous work with CHAMP and SAC-C refractivity observations [15] has shown that the refractivity representativeness errors might be somehow reduced by using an observation operator which takes into account the horizontal gradients from the first-guess.

(4) Temperature and/or Humidity

Finally, temperature or humidity can be obtained as a function of pressure by assuming hydrostatic equilibrium and ancillary data. Optimal estimation methods can also be used to derive the two simultaneously with the help of a priori information and error estimates to constrain the retrievals. The errors in the retrieved atmospheric parameters suffer from all the approximations made in the processing steps above, and are convolved vertically. These errors incorporate a priori errors and are hence difficult to estimate in an assimilation system which would use the same short-term forecast in both (a) the a priori for the temperature/humidity retrieval and (b) the first-guess for the assimilation of that retrieval.

4 Review of GNSS RO Impact Studies on NWP

We now review some forecast impact experiments which assimilated GNSS RO data at the levels (4), (3), and (2).

4.1 Assimilation of Temperature

Poli and Joiner [16] applied a One-Dimensional Variational analysis (1DVAR) approach to GPS/MET refractivity data in order to retrieve simultaneously temperature and humidity. They used 6-hour forecasts issued by the Data Assimilation Office's (DAO) Physical State statistical Analysis System (PSAS) as the a priori. The retrieved temperatures were then assimilated in the DAO PSAS system (close to a 3DVAR) assuming the same error characteristics as for radiosondes. Two experiments were run for a two-week period (June–July 1995). The control experiment assimilated all the operational data (conventional and TOVS observations), while the GNSS experiment assimilated the same observations plus the GPS/MET 1DVAR-retrieved temperatures. Because the GPS/MET refractivity data presented a bias with respect to the first-guess, no 1DVAR-retrieved humidity was assimilated. Also, no 1DVAR-retrieved temperature was assimilated below the 500 hPa level in the tropics.

The GNSS experiment showed slightly improved (degraded) tropospheric forecasts in the Northern (Southern) Hemisphere, as compared with the control. The overall differences were small, suggesting a mostly marginal impact of the GPS/MET data in the troposphere.

The following limitations were noted in this work:

(i). A small dataset of GNSS occultations (less than 200 per day) was assimilated when compared to other assimilated data-types (tens of thousands per TOVS satellite per day).

(ii). The GNSS occultation data presented a bias in the lower troposphere as compared to the first-guess.

(iii). The GNSS occultation data had been processed using the Geometrical Optics (GO) technique, while recent results have shown reduced errors with advanced processing techniques [2].

(iv). The observation operator for simulating GNSS occultation data did not account for the refractivity horizontal gradients. Consequently, the refractivity error covariance matrix should have included adequate representativeness errors to reflect the larger errors induced in the case of intense horizontal gradients.

(v). The correlation of errors between the 1DVAR retrievals and the 6-hr forecast used both as the a priori in the 1DVAR and as the first-guess in the PSAS assimilation was not taken into account, thus leading to a suboptimal weight of the GNSS occultation data in the global analyses [9].

(vi). No Quality Control (QC) was applied to the GNSS data.

4.2 Assimilation of Refractivity

Healy et al. [5] conducted an assimilation experiment of CHAMP refractivity data into the Met Office's 3DVAR system, for a period of two weeks (May–June 2001). An observation operator embedded in the 3DVAR system allowed for a direct assimilation by updating both temperature and humidity fields in the 3DVAR analyses. This direct assimilation approach addressed the issue (v) above. A 1DVAR procedure was used as a pre-processor to perform QC, addressing point (vi). The limitations (i)–(iv) remained however. The refractivity bias [13] prompted the authors to discard the CHAMP refractivity data below 4 km altitude.

The results of Healy et al. indicated improved forecast skill for temperature in the stratosphere for the GNSS experiment. When compared with the control, the GNSS experiment demonstrated in the stratosphere a reduction up to 0.2 K in the RMS of temperature differences forecast minus radiosonde. However, no significant impact was found in the lower to mid troposphere.

4.3 Assimilation of Bending Angle

Zou et al. [26] conducted an assimilation experiment of CHAMP bending angle data into the National Centers for Environmental Prediction's (NCEP) 3DVAR system for a period of 2 weeks (July 2002). The limitation (vi) was addressed by a QC performed by the data producer; (v) was avoided by the direct assimilation; (iv) was addressed by the use of a ray-tracing observation operator within the 3DVAR. Limitations (i)–(iii) remained however. Namely, the bias in the CHAMP occultation data has been found to be related with receiver software on-board [1] as well as super-refraction effects in the atmosphere [21].

The results of Zou et al. indicated slightly improved forecast skill for tropospheric temperatures in both hemispheres in the GNSS experiment as compared to the control. However, the control run did not assimilate all the other satellite observations assimilated operationally at NCEP for the same time period, and hence showed degraded accuracy as compared to the NCEP operational model. It is hence not clear whether using the GNSS data in that context would have helped improve the NCEP operational scores.

At ECMWF, Healy and Thépaut [6] conducted a 4DVAR assimilation experiment of CHAMP bending angle data, using a one-dimensional bending angle observation operator and two months of CHAMP observations processed using the advanced Full Spectrum Inversion method [8]. For the same reason as [5], they removed all the CHAMP observations below 5 km impact height. Healy and Thépaut obtained a clear positive impact on upper tropospheric and lower stratospheric temperature forecasts, validating their results by comparison with radiosondes.

5 Prospects for Impact

This section lists specific areas where GNSS occultation may help improve NWP forecast accuracy.

5.1 Focusing on Forecast-Sensitive Regions

With the advent of GCM adjoint models, methods have been developed to search for forecast-sensitive regions or structures in the atmosphere [17], given a specific forecast term, area, and parameter. Experiments by Gelaro et al. [4] have shown that the positive impact gained from assimilating geostationary satellite wind data originated in the projection of the assimilation increments onto singular vectors located mostly in the lower troposphere. This work also showed that retaining only those increments below the 400 hPa level reduced wind analysis and forecast errors.

Reynolds and Gelaro [18] used adjoint-based techniques to map those regions where analysis errors had the largest impact on forecast errors. Their results stressed out the role of lower to mid-troposphere IC errors on forecast errors.

Noting that these results may be model-dependent, they seem to suggest that GNSS RO data may lead to a larger positive impact on forecasts when the lower to mid-tropospheric occultation data can be assimilated (i.e., after bias removal).

5.2 Importance of the Temperature Information

In the stratosphere, in the higher and mid-troposphere, in the lower tropospheric polar regions, and in the lower tropospheric winter mid-latitude regions, the refractive index is nearly uniquely related with total air density. This means that assimilating GNSS occultation observations will mostly impact the mass field in those regions.

Information about the mass field may as well translate into information about the circulation (wind), critical when it comes to NWP. Using the Rossby radius of deformation as a metrics to estimate the size perturbations need to reach in order to impact the wind circulation [7], one can find that (a) there is no influence of the mass field onto the wind field at the equator, whereas (b) the mass field mostly drives the wind field at the poles. In the mid-latitudes, only those structures with large horizontal extent (on the order of a few thousand kms) have a sensible impact on the wind circulation.

This would seem to suggest that the GNSS occultations could have an impact on extra-tropical cyclone prediction if they could help resolve large structures in the mid-latitudes missed by the other existing observing systems. In the tropics, a smaller impact of the mass information from GNSS occultation may be obtained.

5.3 Importance of the Humidity Information

In the lower tropospheric tropical and summer mid-latitude regions, GNSS occultation can report on information about the humidity. Although that information cannot be extracted directly without assumptions and introduction of significant errors, it may be indirectly assimilated via assimilation of bending angle or refractivity. Lower tropospheric humidity is important for tropical cyclone prediction.

Given the importance of parametrizations and the limited description of the water cycle in today's NWP models, it is unclear whether humidity information from GNSS occultation will be sufficient to improve the prediction of humidity fields for extended time ranges. However, an impact may be obtained on the prediction of intense weather events in short-range forecasts.

6 Conclusion

Owing to their original measurement principle, GNSS RO data represent an exotic data-type as compared to other satellite observations assimilated today in NWP, mostly vertical passive sounders. In the nearly ten year interval that spanned since GPS/MET, two successful missions have been launched and forecast experiments have been conducted with these limited datasets. The results have shown a positive impact in the stratosphere. In the troposphere however, the picture is still mixed. The GNSS occultation bias is a major impediment as it is situation-dependent but it is hoped to be partially removed with the new missions onboard Metop and COSMIC. Those two missions will carry a total of seven receivers, thus bringing many more GNSS occultation events to the NWP centers for them to assimilate.

Advances made to better simulate the GNSS RO data might help ensure that data assimilation hypotheses are verified by reducing representativeness errors (instead of neglecting them). These hypotheses may be regarded as sine qua non conditions for positive forecast impact in today's assimilation systems. GNSS RO occultation present a complex, non-trivial horizontal averaging that presents challenges for assimilation. As a reminder, it took a very long time before the passive sounders had a clear impact on forecast. GNSS occultation has already done better since a positive impact on stratospheric temperature forecasts has already been shown by Healy et al. [5] at the Met Office and by Healy and Thépaut [6] at ECMWF.

References

[1] Ao CO, Meehan TK, Hajj GA, Manucci AJ, Beyerle G (2003) Lower-troposphere refractivity bias in GPS occultation retrievals. J Geophys Res 108(D18):4577

[2] Beyerle G, Wickert J, Schmidt T, Reigber C (2004) Atmospheric sounding by GNSS radio occultation: An analysis of the negative refractivity bias using CHAMP observations. J Geophys Res 109(D1):D01106

[3] Bjerknes V (1904) Das Problem der Wettervorhersage, betrachtet vom Standpunkte der Mechanik und der Physik. Meteorol Zeitsch 21:1–7

[4] Gelaro R, Reynolds CA, Langland RH, Rohaldy GD (2000) A predictability study using geostationary satellite wind observations during NORPEX. Mon Wea Rev 128(11):3789–3807

[5] Healy SB, Jupp AM, Marquardt C (2005) Forecast impact experiment with GPS radio occultation measurements. Geophys Res Lett 32:L03804

[6] Healy SB, Thépaut JN (2005) Assimilation experiments with CHAMP GPS radio occultation measurements. Quart J Roy Meterol Soc, in press.

[7] Holton JR (1992) An introduction to dynamic meteorology. Academic Press, San Diego, CA, 3rd ed, 511 pp.

[8] Jensen AS, Lohmann MS, Benzon HH, Nielsen AS (2003) Full Spectrum Inversion of radio occultation signals. Radio Sci 38(3):1040

[9] Joiner J, Dee DP (2000) An error analysis of radiance and suboptimal retrieval assimilation. Quart J Roy Meteorol Soc 126(565):1495–1514

[10] Kuo, Y-H, Sokolovskiy SV, Anthes RA, Vandenberghe F (2000) Assimilation of GPS radio occultation data for numerical weather prediction. Terr Atm Ocean Sci 11(1):157–186

[11] Kursinski ER, Hajj GA, Schofield JT, Linfield RP, Hardy KR (1997) Observing Earth's atmosphere with radio occultation measurements using the global positioning system. J Geophys Res 102(D19):23429–23465
[12] Lorenz EN (1963) Deterministic nonperiodic flow. J Atmos Sci 20(2):130–141
[13] Marquardt C, Schöllhammer K, Beyerle G, Schmidt T, Wickert J, Reigber C (2003) Validation and data quality of CHAMP radio occultation data. In: Reigber C, Lühr H, Schwintzer P (eds) First CHAMP Mission Results. Springer, Berlin Heidelberg New York, pp 384–396
[14] Palmer PI, Barnett JJ, Eyre JR, Healy SB (2000) A nonlinear optimal estimation inverse method for radio occultation measurements of temperature, humidity and surface pressure. J Geophys Res 105(D13):17513–17526
[15] Poli P (2004) Effects of horizontal gradients on GPS radio occultation observation operators. II: a fast atmospheric refractivity gradient operator (FARGO). Quart J Roy Meteorol Soc 130(603):2807–2825
[16] Poli P, Joiner J (2003) Assimilation experiments of 1DVAR analyses with GPS/MET refractivity. In: Reigber C, Lühr H, Schwintzer P (eds) First CHAMP Mission Results. Springer, Berlin Heidelberg New York, pp 515–520
[17] Rabier F, Courtier P, Talagrand O (1992) An application of adjoint models to sensitivity analysis. Contrib Atmos Phys 65:177–192
[18] Reynolds CA, Gelaro R (2001) Remarks on Northern Hemisphere forecast error sensitivity from 1996 to 2000. Mon Wea Rev 129(8):2145–2153
[19] Rieder MJ, Kirchengast G (2001) Error analysis and characterization of atmospheric profiles retrieved from GNSS occultation data. J Geophys Res 106(D23):31755–31770
[20] Smith WL (1991) Atmospheric sounding from satellites – false expectation or the key to improved weather prediction? Quart J Roy Meteorol Soc 117(498):267–297
[21] Sokolovskiy S (2003) Effect of superrefraction and small-scale refractivity irregularities on inversion of radio occultation signals in the lower troposphere. Proc Intl Workshop GPS Meteorol, 14-17 Jan, Tsukuba, Japan: 2-07
[22] Steiner AK, Kirchengast G (2004) Ensemble-based analysis of errors in atmospheric profiles retrieved from GNSS occultation data. In: Kirchengast G, Foelsche U, Steiner AK (eds) Occultations for Probing Atmosphere and Climate. Springer, Berlin Heidelberg New York, pp 149–160
[23] Syndergaard S (1999) Retrieval analysis and methodologies in atmospheric limb sounding using the GNSS radio occultation technique. Danish Meteorol Inst Sci Rep 99-6
[24] Talagrand O (1997) Assimilation of observations, an introduction. J Meteorol Soc Japan 75(1B):191–209
[25] Thépaut JN (2003) Assimilation of satellite data at ECMWF: prospects for using GPS radio occultation measurements. Eumetsat Proc 2nd GRAS SAF User Workshop: 75–79
[26] Zou X, Liu H, Anthes RA, Shao H, Chang JC, Zhu YJ (2004) Impact of CHAMP radio occultation observations on global analysis and forecasts in the absence of AMSU radiance data. J Meteorol Soc Japan 82(1B):533–549

Observation Operators for the Assimilation of Occultation Data Into Atmospheric Models: A Review

S. Syndergaard[1], Y.-H. Kuo[1], and M. S. Lohmann[1,2]

[1] COSMIC Project Office, University Corporation for Atmospheric Research, Boulder, CO, U.S.A.
ssy@ucar.edu

[2] Joint Center for Satellite Data Assimilation, NOAA/NESDIS, Camp Springs, MD, U.S.A.

Abstract. During the past decade, researchers have been working to develop methods for the assimilation of Global Positioning System (GPS) radio occultation data into numerical weather prediction (NWP) models. Until recently, two strategies have received the most attention: 1) assimilation of the retrieved bending angle profiles, and 2) assimilation of the retrieved refractivity profiles. The assimilation of retrieved bending angle profiles, e.g., via a ray tracing model, has the advantage that horizontal refractivity gradients, affecting the observations, can be accounted for. However, accurate computation of ray paths through a NWP model is very time consuming in an operational system, and approximations to the ray tracing have to be made. In contrast, the assimilation of retrieved refractivity profiles, interpreted as being representative of the local vertical structure in the atmosphere (also known as assimilation of *local* refractivity), is simple and fast, but this approach does not account for the influence of the horizontal gradients. Recently, some new observation operators have been developed which are both fast and, to a large degree, capable of taking into account the influence of the horizontal gradients. These new observation operators rely on predefined ray trajectories, and could also become useful for future assimilation of other kinds of occultation data, e.g., ionospheric GPS occultation data, absorption data from low earth orbit constellations, or data from solar/stellar occultation experiments. In this review, a brief account of past and present efforts to develop efficient observation operators for the assimilation of GPS occultation data into atmospheric models will be given, ranging from early suggestions to the recently developed observation operators.

1 Introduction

The use of occultation data in operational weather prediction requires fast and accurate observation operators to be implemented in data assimilation systems [7, 36, 46]. Whereas occultation observations may present a very high vertical resolution, the nature of the observational geometry results in a horizontal resolution comparable to or coarser than the resolution of most numerical weather prediction (NWP) models. Also, because of the observational geometry, precise information in occultation data about vertical and horizontal atmospheric structure is entangled and not easy to separate. In addition, occultation measurements such as phase, amplitude, bending angle, optical depth, refractivity, etc., are non-traditional meteorological data. However, most of the information in occultation data is complementary to that of passive microwave/infrared nadir-viewing sounders [4], and it is anticipated that the assimilation of Global Positioning System (GPS) radio occultation data from upcoming satellite constellations like COSMIC [31] and EPS/MetOp [5], together with the data from new advanced infrared sounders, will have a positive impact on NWP in the near future.

Already more than a decade ago [6, 14] it was recognized that a major challenge regarding the assimilation of GPS radio occultation data would be to take into account the observational geometry and the inherent horizontal averaging in the direction of signal propagation, while keeping a relatively low computational cost. Since then, a large number of researchers have addressed the issue of horizontal gradients and their influence on GPS radio occultation data, and investigated how to best incorporate the data into atmospheric models. This paper is intended as an overview of these efforts, but will also at the end contribute with new insight and ideas related to some newly developed observation operators. Potentially, these operators can be used to assimilate all kinds of occultation data (e.g., neutral atmospheric and ionospheric GPS occultation data, cross-link absorption data from future low earth orbit (LEO) constellations, or data from solar/stellar occultation experiments) where the measurements basically are integrals of the atmospheric refractive index (real or imaginary part) along ray paths.

2 Assimilating Which Data Product?

The purpose of data assimilation is to extract the maximum information content of the data, and to use this information to improve the analysis of model state variables (wind, temperature, specific humidity, etc.), and thereby the next weather forecast. This can be attempted within many different frameworks of which three-dimensional (3D) and four-dimensional (4D) variational analysis probably are the most widely used in modern data assimilation systems. For the sake of simplicity, we shall here only describe the principle of 3D variational (3DVar) analysis. Briefly, 3DVar can be formulated as the minimization of the following cost function [24, 35]:

$$J(\mathbf{x}) = \frac{1}{2}(\mathbf{x} - \mathbf{x}_\mathrm{b})^\mathrm{T}\mathsf{B}^{-1}(\mathbf{x} - \mathbf{x}_\mathrm{b}) + \frac{1}{2}(H(\mathbf{x}) - \mathbf{y})^\mathrm{T}(\mathsf{O} + \mathsf{F})^{-1}(H(\mathbf{x}) - \mathbf{y}), \quad (1)$$

where $\mathbf{x}$ is a vector symbolizing the model state variables (on model grid points) to be solved for, $\mathbf{x}_\mathrm{b}$ is the corresponding background model field,

$\mathbf{y}$ is a vector of measurements, H is the observation operator which maps the model state into measurement space, and B, O, and F are matrices of the covariances of the expected errors in the background, measurements, and representativeness, respectively. For the assimilation of GPS radio occultation data, there are potentially many different choices of which data product to assimilate, and for each product at least a couple of different ways to construct an observation operator. Among the different factors to be considered are: 1) the introduction of a priori information in the data processing; 2) the ease to characterize the errors of the measurements; 3) the linearity of the observation operator; 4) the computational cost; 5) the accuracy (or representativeness) of the observation operator; 6) the ease to characterize the representativeness errors of the observation operator.

Figure 1 shows the retrieval chain for the processing of GPS radio occultation data into various retrieved products. The fundamental measurements are the phase paths and amplitudes at the two GPS frequencies, L1 and L2. These measurements can be processed into bending angles for each of the two frequencies, ionospheric-corrected bending angles, refractivity, and finally, temperature, pressure, and water vapor using auxiliary meteorological data. More details of the processing and inversion of GPS radio occultation data can be found in [28]. We shall use the term "measurements" to indicate both the fundamental measurements and the retrieved products. Measurement errors of the retrieved products are thus the fundamental measurement errors prop-

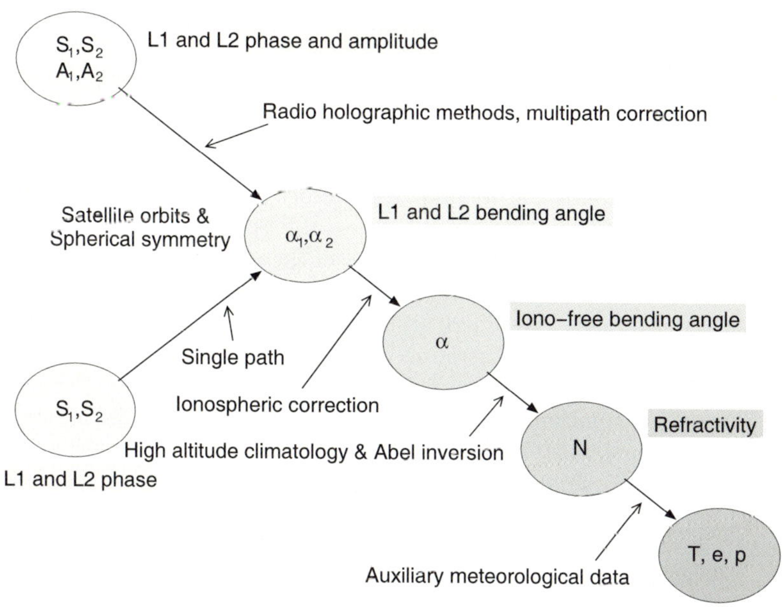

Fig. 1. Illustration of the GPS radio occultation retrieval chain.

agated through the retrieval chain, but measurement errors may also include the result of errors in a priori data used in the retrievals.

Reference [27] outlines the advantages and disadvantages of various assimilation strategies, ranging from the possible assimilation of raw phase and amplitude data to the assimilation of retrieved water vapor and temperature. However, two strategies have until recently been considered to be better choices than the others: one is the assimilation of the retrieved ionospheric-corrected bending angle profiles; the other is the assimilation of the retrieved refractivity profiles. Assimilation of higher level products such as geopotential height, temperature and/or humidity is generally considered a suboptimal choice, but may be necessary in data assimilation systems which are only able to assimilate conventional data.

It is almost meaningless to discuss observation operators without also discussing the corresponding errors of representativeness. Generally, errors of representativeness arise from two sources: 1) the limited resolution of the NWP model, and 2) the inability of the observation operator to derive a perfect measurement from a perfect model state [46]. In the following we will discuss the latter (also referred to as forward modeling errors by some authors), and when we use the term "representativeness errors", we do not consider the possible contribution arising from the limited resolution of the NWP model. In most of todays data assimilation systems there may also be a temporal misrepresentation of the data which we shall not address here.

3 Simple Observation Operators

3.1 Geopotential Height, Temperature, Humidity

Assimilation of GPS measurements in the form of retrieved profiles of geopotential height [3] or temperature and/or humidity [6, 40] may be done via very simple observation operators. In both cases, the operator just involves interpolation of model variables to the locations of the retrieved profiles. However, the processing of GPS occultation data to obtain these higher level products consists of many steps where assumptions and a priori data with their own error characteristics are introduced. Assumptions include local spherical symmetry of the refractive index field and atmospheric hydrostatic equilibrium. A priori information from climatology is usually introduced and merged with the data via statistical optimization in the upper part of the profiles [10, 12, 15, 19, 34], while a priori information from meteorological fields is necessary to obtain separate geopotential height, temperature, and humidity profiles in the troposphere, e.g., via 1DVar retrieval [20, 42]. Thus, there may be a significant correlation between the errors of retrieved geopotential height, temperature, and humidity which should be taken into account if more than one of these products are assimilated.

Because of the assumption of spherical symmetry, the retrieved profiles do not exactly represent the local vertical structure in the atmosphere. In the troposphere, interpretation of the retrieved profiles as vertical profiles introduces errors significantly larger than the measurement errors [29]. These interpretation-induced errors will be vertically correlated and they can be considered a result of the limited ability of the observation operator to represent how the actual observations are made and how the data are processed, and the errors should in principle be attributed to the observation operator as representativeness errors. Due to the inclusion of climatology, which can be assumed to have vertically correlated errors [34], and in particular due to error propagation through the hydrostatic integration, the measurement errors also become significantly vertically correlated [44, 52, 53].

3.2 Refractivity

A number of publications provide descriptions of *local* observation operators which can be used for the assimilation of refractivity profiles derived from GPS occultation data [6, 20, 23, 25, 26, 30, 41, 42, 45, 59, 61, 62]. There may be differences in the details, but essentially, a local refractivity observation operator consists of interpolation of model variables to the location of the retrieved profile as well as the calculation of the refractivity, N, via the following equation [48]:

$$N = k_1 \frac{p}{T} + k_2 \frac{e}{T^2}. \tag{2}$$

In (2), k_1 and k_2 are constants, p is pressure, e is water vapor pressure, and T is temperature. The observation operator may also include hydrostatic integration and conversion to geometrical height levels for models where geometrical or geopotential height is not already part of the model output. The assimilation of refractivity using a local observation operator is considered a better choice than the assimilation of the higher level products mentioned above, mainly because the measurement errors of the retrieved refractivity do not depend on a priori information otherwise introduced in the lower part of the profiles. However, a priori information from climatology is still introduced in the upper part of the profiles, and the retrieved refractivity profiles are still based on the assumption of spherical symmetry. Thus, errors of representativeness, using a local refractivity observation operator, dominate below 25 km to 30 km [28, 29].

To mitigate the representativeness errors, early works [6, 26, 61] suggested or introduced observation operators including simple horizontal averages, mimicking the observational geometry and to some extent taking into account that the fundamental measurement is an integrated quantity. Results from later simulation studies indicate a small reduction (about 30 %) in representativeness errors using either uniformly weighted [54] or Gaussian weighted [22] horizontal averaging. Another simulation study [9] of the sensitivity of retrieved profiles to the horizontal drift of the tangent points (the fact that the horizontal location of the tangent points vary with altitude for a general occultation geometry) indicates that the representativeness errors can be reduced by a factor of about two if the model refractivity is evaluated along the locus of the tangent points instead of along the vertical at a fixed location. The errors related to the tangent point drift have also been investigated in [29, 41].

3.3 Local Bending Angle

For the purpose of 1DVar retrievals and validation studies, many authors [17, 21, 37, 41, 57, 58, 62] have used what we shall refer to as a *local* bending angle observation operator (initially described in [6]). Such an operator has recently been tested in a 4DVar data assimilation system at the European Centre for Medium-Range Weather Forecasts (ECMWF) for the assimilation of bending angle profiles derived from GPS occultation data [21]. The first steps toward a local bending angle operator are identical to those of a local refractivity operator, but additionally the so-called *forward* Abel transform is applied to obtain the bending angle, α, as a function of impact parameter, a, neglecting horizontal gradients in the NWP model. The forward Abel transform is formally given by

$$\alpha(a) = -2a \int_{r_0}^{\infty} \frac{\mathrm{d}\ln n/\mathrm{d}r}{\sqrt{n^2r^2 - a^2}} \mathrm{d}r, \tag{3}$$

where $n = 1 + 10^{-6}N$ is the refractive index, r is the radius from the center of curvature, and r_0 is defined via the relation $a = r_0 n(r_0)$. This last relation introduces a significant amount of non-linearity into the observation operator because the "independent" variable, a, becomes a function of the model refractivity (or alternatively, for a given a, the lower limit of the integral, r_0, becomes a function of the model refractivity). In particular, the non-linearity may be appreciable in the lower troposphere, where vertical refractivity gradients and bending may be large.

Early arguments (as well as more recent ones) for assimilating bending angle instead of refractivity included the expectation that the measurement errors would be easy to characterize for the bending angle, perhaps with no or little vertical correlation between errors, whereas the *inverse* Abel transform [8], applied to obtain the refractivity, would increase the vertical correlation of the errors [6, 17, 22, 57]. Studies of the error covariance propagation [44, 53], however, suggest that the vertical correlations of the measurement errors for refractivity and for bending angle are very similar. In both cases the vertical correlation length is quite small (depending on how much smoothing is applied in the data processing), and in both cases there is a significant negative correlation between errors at adjacent heights. The negative correlations are primarily a result of a derivative operation necessary in the calculation of the bending angle. The Abel inversion (which can be considered approximately an operator of half integration [16]) basically just reduces the negative correlations.

As with the local refractivity observation operator, the representativeness errors of a local bending angle observation operator, due to the neglect of the horizontal gradients, are expected to be dominant throughout the troposphere and lower to mid stratosphere. The results from a simulation study assessing both measurement errors and representativeness errors [52] indicate that the representativeness errors of a local bending angle observation operator, down to about 5 km altitude, are almost uncorrelated in the vertical. For a local refractivity observation operator there is more vertical correlation between the errors. However, below $\sim$ 5 km, it may be difficult to characterize the representativeness errors of a local bending angle observation operator in a general way because of the non-linearity and probable multipath propagation.

Assimilation using a local bending angle observation operator has the advantage that it avoids the introduction of climatology in the retrievals. However, because of the limited vertical extent of a NWP model, and because the upper limit in (3) is infinite, it becomes necessary to extrapolate the model refractivity profile beyond the uppermost model level, e.g., assuming that the refractivity falls off exponentially with altitude [21]. The error introduced by such an assumption should probably also be attributed to the observation operator as a representativeness error. It then becomes a question whether this representativeness error is more significant than the error otherwise introduced by using climatology to retrieve a refractivity profile. The answer to this question will depend on the altitude of the uppermost level of the NWP model. The higher this level is, the smaller the error from the extrapolation is likely to be.

4 Two-Dimensional Ray Tracing Operators

4.1 Bending Angle

As pointed out in [6], neglecting the horizontal gradients in a bending angle observation operator can lead to errors which are comparable with the changes in the bending angle expected from typical errors in short-range forecast temperatures. For this reason, a lot of efforts over the years have been put into the development of bending angle observation operators using 2D ray tracing [11, 13, 14, 16, 18, 32, 33, 39, 41, 43, 47, 60, 62, 63, 64, 65]. Generally, ray tracing through a model atmosphere at radio frequencies involves numerical integration of the following coupled differential equations:

$$\frac{\mathrm{d}\mathbf{r}}{\mathrm{d}\tau} = \mathbf{n}, \qquad \frac{\mathrm{d}\mathbf{n}}{\mathrm{d}\tau} = n\nabla n, \tag{4}$$

where $\mathbf{r}$ denotes the coordinates of the ray and $\mathbf{n}$ is the refractive index *vector* (having the length of the refractive index and the direction of the wave normal). In principle, the most accurate forward modeling of bending angles (disregarding diffraction effects and noise due to turbulence [13]) requires 3D ray tracing, but this is prohibited in operational systems due to its very high computational cost. Two-dimensional ray tracing, where across-ray horizontal gradients are neglected, is usually accepted as a reasonable compromise [14, 64]. A 2D bending angle observation operator basically includes the following steps: 1) interpolation of NWP model variables into the 2D occultation plane; 2) calculation of the refractivity and its gradients in the occultation plane; 3) ray tracing in two opposite directions, starting at the tangent point (for each ray), toward the GPS satellite and toward the LEO satellite; 4) calculation of the bending angle as a function of impact parameter from the result of the ray tracing. The observation operator may also include hydrostatic integration in the occultation plane and transformation of geopotential height to geometrical height. These steps should be viewed as a very simplified outline and may not apply in detail to all 2D bending angle operators described in the above references.

A 2D bending angle observation operator, if properly implemented, takes into account most of the influence from horizontal gradients (depending on the horizontal resolution of the NWP model). The representativeness errors in the troposphere are therefore expected to be smaller than for the more simple observation operators described in the previous sections. Accurate and timely forward modeling of bending angles via 2D ray tracing is, however, not trivial. The bending angle as a function of impact parameter depends non-linearly on the refractivity field, and time-consuming ray tracing must be repeated several times in the minimization procedure of the cost function (1). For hydrostatic NWP models, this includes integration of the hydrostatic equation at a large number of locations along the ray paths [13, 18, 62]. A further complication is introduced because accurate calculation of the bending angles requires extrapolation of the NWP model above its highest level. Usually a climatological model is used for that purpose.

The assimilation, via 2D ray tracing, of bending angle measurements from the upcoming COSMIC mission, which is expected to provide data from about 2500 occultations per day, would require large computer resources. Depending on the particular implementation and the available computer power, the CPU time for the assimilation of the about 625 occultations anticipated within a six hour time window, may be a few days [60, 63], and many parallel processors would therefore be necessary. Nevertheless, a 2D ray tracing operator for bending angle assimilation, based on the work in [13], has been incorporated into a newly developed data assimilation system at the Deutscher Wetterdienst [43], and impact studies have been carried out using a 2D ray tracing operator in combination with the NCEP (National Center for Environmental Prediction) SSI (Spectral Statistical Interpolation) 3DVar system [32, 63, 65]. To reduce the computational cost, it has been suggested to assimilate bending angles using a 2D ray tracer in the troposphere only, while assimilating refractivity using a local refractivity observation operator at higher altitudes [27, 63].

A simulation study [18], using a high-resolution mesoscale model, investigated the errors caused by horizontal gradients in the troposphere in relation to the forward modeling of the bending angle as a function of impact parameter. Although the particular way the "true" impact parameter is defined and obtained in [18] may be questionable, an important finding of the study was that an uncertainty in the calculated impact parameter (or rather a misinterpretation of the retrieved impact parameter) of about 100 m can cause an effective bending angle error of up to 20 % near the surface (translated into refractivity this corresponded to a 2.4 % error). It is therefore essential that the impact parameter be modeled in a way consistent with how it is derived from real data [13, 50, 62], especially in the lowest few kilometers of the atmosphere and when large horizontal gradients are present.

4.2 Refractivity, Temperature, Humidity

Having the modeled bending angle as a function of impact parameter, it is possible to go a step further and calculate the corresponding refractivity profile via the inverse Abel transform. The 2D bending angle calculation plus the inverse Abel transform thus constitutes a refractivity observation operator which takes into account most of the influence from the horizontal gradients in refractivity [39, 62]. The retrieved refractivity profile can then be assimilated instead of the bending angle profile. This approach, however, is not considered an advantage over the assimilation of the bending angles [27]. Yet another observation operator can be obtained by adding yet another step and calculate profiles of temperature and humidity using the hydrostatic equation together with vertical profile information from the NWP model. As suggested in [15], the thus modeled temperature and humidity profiles can be used in data assimilation systems which are based on the *nudging* assimilation technique.

5 Advanced Alternatives to Ray Tracing

5.1 Bending Angle Integration

Reference [6] provides one of the first published description of a 2D observation operator for the assimilation of the bending angle as a function of impact parameter. The computation of the bending angle in this operator is, in principle, based on the integration of the following equation which can be derived from (4):

$$\mathrm{d}\alpha = -\frac{1}{n}\left(\frac{\partial n}{\partial r}\right)_{\theta} r\mathrm{d}\theta + \frac{1}{n}\left(\frac{\partial n}{\partial \theta}\right)_{r}\frac{\mathrm{d}r}{r}, \tag{5}$$

where θ is the angular coordinate in the occultation plane. The impact parameter is defined via the relation $a = r_0 n(r_0)$, where r_0 is the radius from the center of curvature at the tangent point. As noted in [17, 18], the operator can be simplified because the contribution to the bending angle from the second term on the right-hand-side of (5) is small.

More recently, a "fast" bending angle observation operator based on the model outlined in [6] was tested against a 3D ray tracer [22]. This enabled the assessment of the representativeness errors associated with various approximations used in the fast operator. It was found that the largest single component of the representativeness errors is due to the way the impact parameter is defined in the operator, which again stresses the importance of modeling the impact parameter in accordance with how it is derived from real data. As an alternative to ray tracing, this can be done via the equations given in [18].

5.2 FARGO-α and FARGO-N

Recent work [38, 39] describes a so-called Fast Atmospheric Refractivity Gradient Operator (FARGO). FARGO comes in two versions, FARGO-α for the assimilation of the retrieved bending angle, and FARGO-N for the assimilation of the retrieved refractivity. FARGO-α can be summarized by the following equations for the calculation of the bending angle:

$$\alpha(a) = \alpha_{\text{local}}(a) - \Delta\alpha(a), \quad \Delta\alpha(a) = \int_{-L}^{L} \cos\theta \left[\left(\frac{\partial n}{\partial r} \right)_{\theta} - \left(\frac{\partial n}{\partial r} \right)_{\theta=0} \right] \mathrm{d}s, \tag{6}$$

where $\alpha_{\text{local}}(a)$ is the result of the *forward* Abel transform (3) applied to the model refractivity profile along the locus of the retrieved tangent points. The integral in (6) is taken over a limited path ($L = 600$ km in the above references) determined from ray tracing in a spherically symmetrical atmosphere using a vertical profile of the model refractivity. FARGO-α is somewhat similar to the bending angle integration as described in [17], except that the bulk of the bending (given by α_{local}) has been separated from the perturbation due to the horizontal gradients (given by $\Delta\alpha$). The perturbation can then be estimated within a limited horizontal extent without compromising the accuracy of the calculation of the total bending angle too much.

FARGO-N is based on the calculation of $\Delta\alpha(a)$ and is approximately (we here use the approximation that $\ln n \approx 10^{-6}N$, and for convenience we shall disregard the factor of 10^{-6} here and in the following sections) given as

$$N(a) = N_{\text{local}}(a) - \frac{1}{\pi} \int_{a}^{\infty} \frac{\Delta\alpha(\rho)}{\sqrt{\rho^2 - a^2}} \mathrm{d}\rho, \tag{7}$$

where $N_{\text{local}}(a)$ is the model refractivity profile along the locus of the tangent points. FARGO-N is basically the result of applying an *inverse* Abel transform to FARGO-α, except that N_{local} is not obtained via (and is slightly different from) the *inverse* Abel transform of α_{local}.

Given a model refractivity field, the calculations of profiles of bending angle and refractivity, via FARGO-α and FARGO-N, as described in [38, 39], are non-linear operations. However, as shown in Sect. 5.5 below, the refractivity calculation via FARGO-N can be linearized by introducing a small modification, which also makes it very similar to *refractivity mapping* using finite straight line trajectories.

5.3 Refractivity Mapping

Refractivity mapping using finite straight lines [54, 55, 56] is based on the requirement that

$$S(y) = \int_{-L}^{L} N(x, y)\mathrm{d}x = \int_{-L}^{L} \bar{N}(r)\mathrm{d}x \tag{8}$$

for a given y. In (8), x and y are Cartesian coordinates in a 2D plane with the x-axis being parallel to the direction of propagation and the y-axis being parallel to the radius vector at the tangent point. Furthermore, $S(y)$ represents the so-called excess phase path, $N(x, y)$ is the model refractivity which is integrated along a line parallel to the x-axis ($x = 0$ corresponds to the location of the tangent point), and $\bar{N}(r)$ is the spherically symmetrical refractivity to be solved for. In [56], the two integrals in (8) are discretized to obtain the elements of matrices A and V such that $\bar{\mathbf{N}} = \mathsf{AV}\mathbf{N}$, where $\mathbf{N}$ symbolizes the model refractivity field, and $\bar{\mathbf{N}}$ the mapped 1D profile. The matrices A and V are the inverse and forward parts of the mapping operator, respectively. The approach thus roughly mimics how the observations are made and how the data are inverted into a refractivity profile assuming spherical symmetry, but using finite straight lines instead of curved ray trajectories. Alternatively, the ray path curvature is taken into account in [56] via a simple ad hoc modification of the straight line operator. The mapping operator, AV, can in principle be considered as a finite and discrete version of the 2D resolution kernel introduced in [1, 2].

The main advantage of refractivity mapping is that it allows for crude approximations as long as they are made in both the forward and the inverse parts. The errors from the approximations thus cancel to a large extent in the combined operator (e.g., the use of straight lines instead of ray paths). This simplifies many aspects of the implementation and consequently requires fewer computations. For example, because the mapping operator is based on finite integrations in both a forward and an inverse part, extrapolation of the NWP model above its highest level is avoided. The implementation in [56] also makes it possible to apply the mapping to a hydrostatic model without having to integrate the hydrostatic equation at a large number of locations.

In the special case where the integrals in (8) are limited only by the uppermost level of the NWP model (denoted here by $r_{\max}$), $\bar{N}(r)$ can be solved for via an Abel transform and written as

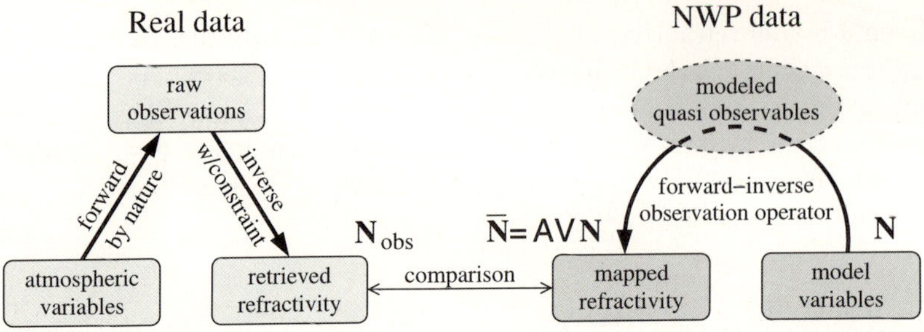

Fig. 2. Diagram showing the principle of assimilating refractivity via refractivity mapping.

$$\bar{N}(r) = -\frac{1}{\pi}\int_r^{r_{\max}} \frac{\mathrm{d}S/\mathrm{d}y}{\sqrt{y^2 - r^2}}\mathrm{d}y, \tag{9}$$

where $S(y)$ is obtained via the first integral in (8). Reference [50] also considers refractivity mapping and presents a generalized form of (9) for curved ray trajectories defined by a given refractive index profile that does not depend on the model refractive index. The results from two independent simulation studies [50, 56] suggest that refractivity mapping, using either straight lines or curved trajectories, results in a considerable reduction of the representativeness errors as compared to a local refractivity observation operator. Both studies indicate that refractivity mapping using curved trajectories may be able to account for about 90 % of the influence from horizontal gradients.

5.4 Quasi Excess Phase Modeling

As an alternative to assimilating the refractivity via a refractivity mapping operator, it has been suggested [49, 50] to assimilate what we shall refer to as the *quasi* excess phase (to distinguish it from the actual "raw" excess phase measurements). In [50], the retrieved quasi excess phase as a function of impact parameter, $S_{\mathrm{obs}}(a)$, is in principle obtained from the retrieved refractivity, N_{obs}, via the following equation:

$$S_{\mathrm{obs}}(a) = 2\int_a^{a_{\max}} \frac{N_{\mathrm{obs}}(\rho)[1 - \rho(\mathrm{d}\ln n_0/\mathrm{d}\rho)]\rho}{n_0(\rho)\sqrt{\rho^2 - a^2}}\mathrm{d}\rho, \tag{10}$$

where $\rho = rn_0$, and $n_0(\rho)$ is a predefined refractive index profile determining the trajectories along which the integration of the refractivity is made. The corresponding observation operator is obtained by integrating the model refractivity, $N(x, y)$, along these trajectories. For $n_0(\rho) = 1$ the observation operator uses straight line trajectories, and in principle reduces to the first integral in (8). Consequently, a quasi excess phase operator could also be written in matrix-vector form as $\mathbf{S} = \mathbf{V}\mathbf{N}$.

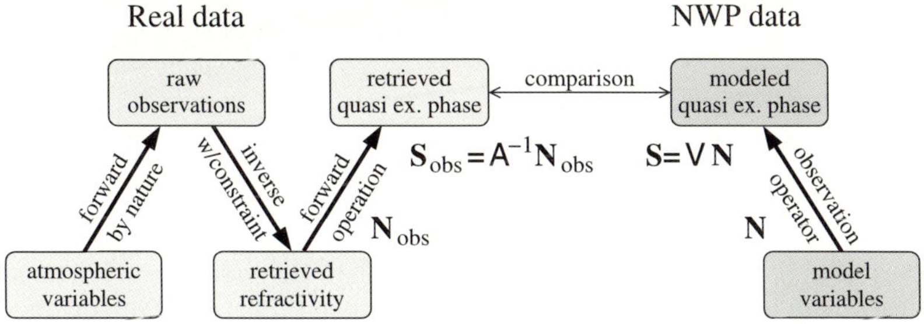

Fig. 3. Diagram showing the principle of assimilating the quasi excess phase.

Quasi excess phase modeling has the same advantages as refractivity mapping, since crude (but identical) approximations can be made in both the way the retrieved refractivity is integrated and the way the model refractivity is integrated. The implementation of a quasi excess phase operator in a data assimilation system, however, may be simpler because only the part corresponding to the V matrix is necessary. On the other hand, an extra step on the retrieval side, to obtain the retrieved quasi excess phase, is necessary instead, but this step does not have to be repeated in the minimization procedure of the cost function (1). The difference in computational cost between the two operators is probably small because the calculation of the model refractivity and interpolation to the points used in the forward model integrations (in both operators) may be the most time consuming part. The characteristics of the representativeness errors are different, e.g., it is shown in [49, 50] that the error standard deviation, in a fractional sense, of a quasi excess phase operator is about half of that of a refractivity mapping operator. However, as shown in the next section, the two operators, if properly implemented in a variational data assimilation system, should lead to the same assimilation result.

5.5 Relations Between Recently Developed Operators

In this section we show some relations between the recently developed observation operators described in Sects. 5.2–5.4. First, we consider the diagrams in Figs. 2 and 3. Figure 2 shows the principle of assimilating the refractivity via refractivity mapping, whereas Fig. 3 shows the principle of assimilating the quasi excess phase. Since the inverse part of the refractivity mapping operator, given by the A matrix, is invertible, and A^{-1} in principle corresponds to the operation to obtain the retrieved quasi excess phase via (10), we can write $\mathbf{S}_{\text{obs}} = \mathsf{A}^{-1}\mathbf{N}_{\text{obs}}$. In the following we shall disregard operations in the observation operators such as interpolation, the calculation of the refractivity via (2), and hydrostatic integration. Then, considering the observational part of the cost function in (1), we have for refractivity mapping:

$$J_{\rm obs}(\mathbf{N}) = \frac{1}{2}(\mathsf{AV}\mathbf{N} - \mathbf{N}_{\rm obs})^{\rm T}(\mathsf{O}+\mathsf{F})^{-1}(\mathsf{AV}\mathbf{N} - \mathbf{N}_{\rm obs}), \tag{11}$$

and for quasi excess phase modeling:

$$J_{\rm obs}(\mathbf{N}) = \frac{1}{2}(\mathsf{V}\mathbf{N} - \mathsf{A}^{-1}\mathbf{N}_{\rm obs})^{\rm T}(\widetilde{\mathsf{O}}+\widetilde{\mathsf{F}})^{-1}(\mathsf{V}\mathbf{N} - \mathsf{A}^{-1}\mathbf{N}_{\rm obs}). \tag{12}$$

In (12) $\widetilde{\mathsf{O}}$ and $\widetilde{\mathsf{F}}$ are introduced to indicate that the error covariances in the two cases will be different. However, since the two observation operators differ only via a linear, invertible operator A, the error covariances are related as

$$(\widetilde{\mathsf{O}}+\widetilde{\mathsf{F}})^{-1} = (\mathsf{A}^{-1}\mathsf{O}(\mathsf{A}^{-1})^{\rm T} + \mathsf{A}^{-1}\mathsf{F}(\mathsf{A}^{-1})^{\rm T})^{-1} = \mathsf{A}^{\rm T}(\mathsf{O}+\mathsf{F})^{-1}\mathsf{A}. \tag{13}$$

Inserting (13) into (12) results in (11), and we have shown that the cost function is the same in the two cases. Therefore, the assimilation of the refractivity via refractivity mapping and the assimilation of the quasi excess phase – if the two observation operators were implemented in the same data assimilation system with consistent specifications of error covariances – should in principle lead to identical assimilation results.

As mentioned, the calculation of the refractivity via FARGO-N is a nonlinear operation, whereas both refractivity mapping and quasi excess phase modeling are linear operations because the trajectories are predefined and does not depend on the model refractivity. This is a computational advantage, because then the determination of the trajectories does not have to be repeated in the minimization procedure of the cost function in a variational data assimilation system. FARGO-N can be easily linearized with respect to refractivity by integrating $\Delta\alpha$ along predefined trajectories instead of along trajectories determined by the model refractivity. In particular, if the trajectories are chosen to be straight lines and we only consider integration along lines below the uppermost model level at $r_{\rm max}$, then (7), in combination with the expression for $\Delta\alpha$ in (6), can be written

$$\bar{N}(r) = N_{\rm local}(r) - \frac{1}{\pi}\int_r^{r_{\rm max}} \frac{1}{\sqrt{y^2-r^2}}\left\{\int_{-L}^{L}\left[\left(\frac{\mathrm{d}N}{\mathrm{d}y}\right)_x - \left(\frac{\mathrm{d}N}{\mathrm{d}y}\right)_{x=0}\right]\mathrm{d}x\right\}\mathrm{d}y. \tag{14}$$

As in Sect. 5.3, x and y are Cartesian coordinates in the occultation plane for a given straight line "ray", and $\bar{N}$ on the left-hand-side is introduced to distinguish the thus obtained refractivity profile from the model refractivity, $N = N(x, y)$. On the other hand, if $\mathbf{N}_{\rm SS}$ symbolizes the spherically symmetrical atmosphere corresponding to the local model refractivity profile, $\mathbf{N}_{\rm local}$, then $\mathsf{AV}\mathbf{N}_{\rm SS} = \mathbf{N}_{\rm local}$, and we can write refractivity mapping as

$$\bar{\mathbf{N}} = \mathbf{N}_{\rm local} + \mathsf{AV}(\mathbf{N} - \mathbf{N}_{\rm SS}). \tag{15}$$

With the aid of (9), which basically gives the AV operator on integral form, and noting that the derivative in the integrand in (9) can be put inside the first integral in (8) (because x and y are orthogonal coordinates), we can now see the similarity between a linearization of FARGO-N and refractivity mapping.

Just as FARGO-N basically is the *inverse* Abel transform applied to FARGO-α, a new bending angle observation operator could be defined by applying the *forward* Abel transform to the result of refractivity mapping. As mentioned in Sect. 3.3, the forward Abel transform introduces a significant amount of non-linearity into the observation operator, and the errors of representativeness in the lower troposphere may become difficult to describe. It may, however, be an advantage to apply the forward Abel transform to refractivity mapping in the upper part of the atmosphere if the uppermost level of the NWP model is at a high enough altitude.

6 Summary and Prospects

In this review we have given a brief account of the many efforts made over the past decade to develop efficient observation operators for the assimilation of GPS radio occultation data into atmospheric models. We divided the observation operators into three categories: 1) simple observation operators; 2) 2D ray tracing operators; 3) advanced alternatives to ray tracing.

The simple operators are characterized by being fast and relatively easy to implement, but also by large representativeness errors because the atmospheric horizontal gradients are not (or only partially) accounted for in the modeling of the measurements. Observation operators based on 2D ray tracing takes into account most of the influence from the horizontal gradients, but they are much more computationally expensive. Alternatives to ray tracing have been developed in attempts to speed up the computations, while still accounting for the horizontal gradients to a large degree. We have shown that some of these alternatives are closely related. However, it has yet to be verified how fast they actually are in comparison with existing 2D ray tracing operators implemented in some data assimilation systems.

Future research with regard to recently developed operators should address the estimation and characterization of the errors of representativeness. This should include both the error magnitude as well as the vertical error correlations. Although the representativeness errors are expected to be small for these operators, they may still be dominant in the lower troposphere (anticipating improved performance of future GPS radio occultation receivers [51]) where moisture can vary appreciably on horizontal scales of a few tens of kilometers. The representativeness errors may also depend on latitude and season. However, it should be noted that the errors due to limited NWP model resolution (not addressed elsewhere in this paper) may dominate if the model horizontal resolution is too coarse [50].

For variational data assimilation systems, the development of the adjoint of an observation operator is also necessary. Adjoint operators for some of the observation operators mentioned in this paper have been developed and/or described in [13, 56, 62].

Finally, it should be emphasized that the principles of refractivity mapping and quasi excess phase modeling are quite general, and that these operators could become useful in the future for the assimilation of all kinds of occultation data where the observations basically are integrals of one or more atmospheric constituents along ray paths.

Acknowledgements. The authors gratefully acknowledge Sergey Sokolovskiy (UCAR, Boulder, CO) and Andrea Hahmann (NCAR, Boulder, CO) for valuable comments on the manuscript.

References

[1] Ahmad B (1998) Accuracy and resolution of atmospheric profiles obtained from radio occultation measurements. PhD thesis, Center for Radar Astronomy, Stanford University, California

[2] Ahmad B, Tyler GL (1998) The two-dimensional resolution kernel associated with retrieval of ionospheric and atmospheric refractivity profiles by Abelian inversion of radio occultation phase data. Radio Sci 33:129–142

[3] Amstrup B, Mogensen KS, Huang XY (2000) Use of GPS observations in an optimum interpolation based data assimilation system. Scientific Report 00-14, Danish Meteorological Institute, Copenhagen, Denmark

[4] Collard AD, Healy SB (2003) The combined impact of future space-based atmospheric sounding instruments on numerical weather-prediction analysis fields: A simulation study. Quart J Roy Meteorol Soc 129:2741–2760

[5] Danmarks Meteorologiske Institut (2003) SAF Training Workshop: 2nd GRAS SAF User Workshop, EUMETSAT, Darmstadt, Germany

[6] Eyre JR (1994) Assimilation of radio occultation measurements into a numerical weather prediction system. Technical Memorandum No 199, European Centre for Medium-Range Weather Forecasts

[7] Eyre JR (1997) Variational assimilation of remotely-sensed observations of the atmosphere. J Meteorol Soc Japan 75:331–338

[8] Fjeldbo G, Kliore AJ, Eshleman VR (1971) The neutral atmosphere of Venus as studied with the Mariner V radio occultation experiments. Astron J 76:123–140

[9] Foelsche U, Kirchengast G (2004) Sensitivity of GNSS occultation profiles to horizontal variability in the troposphere: A simulation study. In: Kirchengast G, Foelsche U, Steiner AK (eds) Occultations for Probing Atmosphere and Climate, Springer, Berlin, pp:127–136

[10] Gobiet A, Kirchengast G (2004) Advancements of Global Navigation Satellite System radio occultation retrieval in the upper stratosphere for optimal climate monitoring utility. J Geophys Res 109:D24110, doi:10.1029/2004JD005117

[11] Gorbunov ME (1996) Three-dimensional satellite refractive tomography of the atmosphere: Numerical simulation. Radio Sci 31:95–104

[12] Gorbunov ME (2002) Ionospheric correction and statistical optimization of radio occultation data. Radio Sci 37:1084, doi:10.1029/2000RS002370

[13] Gorbunov ME, Kornblueh L (2003) Principles of variational assimilation of GNSS radio occultation data. Report No 350, Max-Planck-Institute for Meteorology, Hamburg, Germany
[14] Gorbunov ME, Sokolovskiy SV (1993) Remote sensing of refractivity from space for global observations of atmospheric parameters. Report No 119, Max-Planck-Institute for Meteorology, Hamburg, Germany
[15] Gorbunov ME, Gurvich AS, Bengtson L (1996) Advanced algorithms of inversion of GPS/MET satellite data and their application to reconstruction of temperature and humidity. Report No 211, Max-Planck-Institute for Meteorology, Hamburg, Germany
[16] Gorbunov ME, Sokolovsky SV, Bengtsson L (1996) Space refractive tomography of the atmosphere: Modeling of direct and inverse problems. Report No 210, Max-Planck-Institute for Meteorology, Hamburg, Germany
[17] Healy SB (1998) A statistical comparison of GPS/MET radio occultation data with numerical weather prediction analyses. Technical Report No 247, UK Meteorological Office, Bracknell, UK
[18] Healy SB (2001) Radio occultation bending angle and impact parameter errors caused by horizontal refractive index gradients in the troposphere: A simulation study. J Geophys Res 106:11875–11889
[19] Healy SB (2001) Smoothing radio occultation bending angles above 40 km. Ann Geophys 19:459–468
[20] Healy SB, Eyre JR (2000) Retrieving temperature, water vapour and surface pressure information from refractive-index profiles derived by radio occultation: A simulation study. Quart J Roy Meteorol Soc 126:1661–1683
[21] Healy SB, Thépaut JN (2005) Assimilation experiments with CHAMP GPS radio occultation measurements. Quart J Roy Meteorol Soc (in press)
[22] Healy SB, Jupp AM, Offiler D, Eyre J (2003) The assimilation of radio occultation measurements. In: Reigber C, Lühr H, Schwintzer P (eds) First CHAMP Mission Results for Gravity, Magnetic and Atmospheric Studies, Springer, Berlin pp:453–461
[23] Healy SB, Jupp AM, Marquardt C (2005) Forecast impact experiment with GPS radio occultation measurements. Geophys Res Lett 32:L03804, doi:10.1029/2004GL020806
[24] Huang XY (2000) Variational analysis using spatial filters. Mon Weather Rev 128:2588–2600
[25] Huang XY, Vedel H (2003) An introduction to data assimilation. In: SAF Training Workshop: 2nd GRAS SAF User Workshop, EUMETSAT, Darmstadt, Germany, 25–37
[26] Kuo YH, Zou X, Huang W (1997) The impact of Global Positioning System data on the prediction of an extratropical cyclone: an observing system simulation experiment. Dynam Atmos Oceans 27:439–470
[27] Kuo YH, Sokolovskiy SV, Anthes RA, Vandenberghe F (2000) Assimilation of GPS radio occultation data for numerical weather prediction. Terrestrial, Atmospheric and Oceanic Sciences 11:157–186
[28] Kuo YH, Wee TK, Sokolovskiy S, Rocken C, Schreiner W, Hunt D, Anthes RA (2004) Inversion and error estimation of GPS radio occultation data. J Meteorol Soc Japan 82:507–531

[29] Kursinski ER, Hajj GA, Schofield JT, Linfield RP, Hardy KR (1997) Observing Earth's atmosphere with radio occultation measurements using the Global Positioning System. J Geophys Res 102:23429–23465

[30] Kursinski ER, Healy SB, Romans LJ (2000) Initial results of combining GPS occultations with ECMWF global analyses within a 1DVar framework. Earth Planets Space 52:885–892

[31] Lee LC, Rocken C, Kursinski R (eds) (2000) Applications of Constellation Observing System for Meteorology, Ionosphere & Climate. Springer, Hong Kong

[32] Liu H, Zou X, Shao H, Anthes RA, Chang JC, Tseng JH, Wang B (2001) Impact of 837 GPS/MET bending angle profiles on assimilation and forecasts for the period June 20–30, 1995. J Geophys Res 106:31771–31786

[33] Liu H, Zou X (2003) Improvements to a GPS radio occultation ray-tracing model and their impacts on assimilation of bending angle. J Geophys Res 108:4548, doi:10.1029/2002JD003160

[34] Lohmann MS (2005) Application of dynamical error estimation for statistical optimization of radio occultation bending angles. Radio Sci 40:RS3011, doi:10.1029/2004RS003117

[35] Lorenc AC (1986) Analysis methods for numerical weather prediction. Quart J Roy Meteorol Soc 112:1177–1194

[36] Marquardt C, Eyre JR, Healy SB, Jupp A, Offiler D (2003) Use of GPS radio occultation data in meteorological services. In: Stauning P, Lühr H, Ultré-Guérard P, LaBrecque J, Purucker M, Primdahl F, Jørgensen JL, Christiansen F, Høeg P, Lauritsen KB (eds) OIST-4 Proceedings, 4'th Oersted International Science Team Conference, Narayana Press, Denmark, 261–268

[37] Palmer PI, Barnett JJ, Eyre JR, Healy SB (2000) A nonlinear optimal, estimation inverse method for radio occultation measurements of temperature, humidity, and surface pressure. J Geophys Res 105:17513–17526

[38] Poli P (2003) A fast forward model for simulating GPS radio occultation bending angles and refractivity in a two dimensional plane: Implementation and simulations. In: SAF Training Workshop: 2nd GRAS SAF User Workshop, EUMETSAT, Darmstadt, Germany, 38–48

[39] Poli P (2004) Effects of horizontal gradients on GPS radio occultation observation operators. II: A Fast Atmospheric Refractivity Gradient Operator (FARGO). Quart J Roy Meteorol Soc 130:2807–2825

[40] Poli P, Joiner J (2003) Assimilation experiments of one-dimensional variational analyses with GPS/MET refractivity. In: Reigber C, Lühr H, Schwintzer P (eds) First CHAMP Mission Results for Gravity, Magnetic and Atmospheric Studies, Springer, Berlin, pp:515–520

[41] Poli P, Joiner J (2004) Effects of horizontal gradients on GPS radio occultation observation operators. I: Ray tracing. Quart J Roy Meteorol Soc 130:2787–2805

[42] Poli P, Joiner J, Kursinski ER (2002) 1DVAR analysis of temperature and humidity using GPS radio occultation refractivity data. J Geophys Res 107:4448, doi:10.1029/2001JD000935

[43] Rhodin A (2003) Assimilation of radio occultation data at DWD. In: SAF Training Workshop: 2nd GRAS SAF User Workshop, EUMETSAT, Darmstadt, Germany, 80–84

[44] Rieder MJ, Kirchengast G (2001) Error analysis and characterization of atmospheric profiles retrieved from GNSS occultation data. J Geophys Res 106:31755–31770

[45] Rubek F, Sørensen MB (2003) GRAS-SAF system, products, and archive. In: SAF Training Workshop: 2nd GRAS SAF User Workshop, EUMETSAT, Darmstadt, Germany, 60–72

[46] Schlatter TW (2000) Variational assimilation of meteorological observations in the lower atmosphere: a tutorial on how it works. J Atmos Solar-Terr Phys 62:1057–1070

[47] Shao H, Zou X (2002) The impact of observational weighting on the assimilation of GPS/MET bending angle. J Geophys Res 107:4717, doi:10.1029/2001JD001552

[48] Smith EK, Weintraub S (1953) The constants in the equation for atmospheric refractive index at radio frequencies. Journal of Research of the National Bureau of Standards 50:39–41

[49] Sokolovskiy S, Kuo YH, Wang W (2004) Evaluation of a linear phase observation operator with CHAMP radio occultation data and high-resolution regional analysis. Mon Weather Rev (in press)

[50] Sokolovskiy S, Kuo YH, Wang W (2005) Assessing the accuracy of a linearized observation operator for assimilation of radio occultation data: case simulations with a high-resolution weather model. Mon Weather Rev 133:2200–2212

[51] Sokolovskiy SV (2001) Tracking tropospheric radio occultation signals from low Earth orbit. Radio Sci 36:483–498

[52] Steiner AK, Kirchengast G (2004) Ensemble-based analysis of errors in atmospheric profiles retrieved from GNSS occultation data. In: Kirchengast G, Foelsche U, Steiner AK (eds) Occultations for Probing Atmosphere and Climate, Springer, Berlin, pp:149–160

[53] Syndergaard S (1999) Retrieval analysis and methodologies in atmospheric limb sounding using the GNSS radio occultation technique. Scientific Report 99-6, Danish Meteorological Institute, Copenhagen, Denmark

[54] Syndergaard S, Flittner D, Kursinski R, Herman B (2003) Simulating the influence of horizontal gradients on refractivity profiles from radio occultations. In: Stauning P, Lühr H, Ultré-Guérard P, LaBrecque J, Purucker M, Primdahl F, Jørgensen JL, Christiansen F, Høeg P, Lauritsen KB (eds) OIST-4 Proceedings, 4'th Oersted International Science Team Conference, Narayana Press, Denmark, 245–250

[55] Syndergaard S, Flittner DE, Kursinski ER, Feng DD, Herman BM, Ward DM (2004) Simulating the influence of horizontal gradients on retrieved profiles from ATOMS occultation measurements—a promising approach for data assimilation. In: Kirchengast G, Foelsche U, Steiner AK (eds) Occultations for Probing Atmosphere and Climate, Springer, 221–232

[56] Syndergaard S, Kursinski ER, Herman BM, Lane EM, Flittner DE (2005) A refractive index mapping operator for assimilation of occultation data. Mon Weather Rev 133:2650–2668

[57] von Engeln A, Nedoluha G (2005) Retrieval of temperature and water vapor profiles from radio occultation refractivity and bending angle measurements using an optimal estimation approach: a simulation study. Atmospheric Chemistry and Physics 5:1665–1677

[58] von Engeln A, Nedoluha G, Kirchengast G, Bühler S (2003) One-dimensional variational (1-D Var) retrieval of temperature, water vapor, and a reference pressure from radio occultation measurements: A sensitivity analysis. J Geophys Res 108:4337, doi:10.1029/2002JD002908

[59] Wee TK, Kuo YH (2004) Impact of a digital filter as a weak constraint in MM5 4DVAR: An observing system simulation experiment. Mon Weather Rev 132:543–559
[60] Zhang X, Liu Y, Wang B, Ji Z (2004) Parallel computing of a variational data assimilation model for GPS/MET observation using the ray-tracing method. Advances in Atmospheric Sciences 21:220–226
[61] Zou X, Kuo YH, Guo YR (1995) Assimilation of atmospheric radio refractivity using a nonhydrostatic adjoint model. Mon Weather Rev 123:2229–2249
[62] Zou X, Vandenberghe F, Wang B, Gorbunov ME, Kuo YH, Sokolovskiy S, Chang JC, Sela JG, Anthes RA (1999) A raytracing operator and its adjoint for the use of GPS/MET refraction angle measurements. J Geophys Res 104:22301–22318
[63] Zou X, Wang B, Liu H, Anthes RA, Matsumura T, Zhu YJ (2000) Use of GPS/MET refraction angles in three-dimensional variational analysis. Quart J Roy Meteorol Soc 126:3013–3040
[64] Zou X, Liu H, Anthes RA (2002) A statistical estimate of errors in the calculation of radio-occultation bending angles caused by a 2D approximation of ray tracing and the assumption of spherical symmetry of the atmosphere. J Atmos Ocean Technol 19:51–64
[65] Zou X, Liu H, Anthes RA, Shao H, Chang JC, Zhu YJ (2004) Impact of CHAMP radio occultation observations on global analysis and forecasts in the absence of AMSU radiance data. J Meteorol Soc Japan 82:533–549

Analysis of Atmospheric and Ionospheric Wave Structures Using the CHAMP and GPS/MET Radio Occultation Database

A. G. Pavelyev[1], J. Wickert[2], Y. A. Liou[3], A. A. Pavelyev[1], and C. Jacobi[4]

[1] Institute of Radio Engineering and Electronics RAS, Moscow, Russia
pvlv@ms.ire.rssi.ru
[2] GeoForschungsZentrum Potsdam, Germany
[3] CSRSR/National Central University, Taipei, Taiwan
[4] Institute for Meteorology, University of Leipzig, Germany

Abstract. In this paper new directions of application of GPS radio occultation (RO) method are presented: (1) measuring the vertical gradients of the refractivity in the atmosphere and electron density in the lower ionosphere, (2) study of the ionospheric disturbances on a global scale, (3) investigation of the internal wave activity in the atmosphere. New directions may be informative for investigations of the connections between processes in the atmosphere and ionosphere, for study of the thermal regime in the intermediate height interval upper stratosphere to lower mesosphere, and for analysis of influence of space weather phenomena on the lower ionosphere.

1 Introduction

Significant refinements and modernization in the RO technique have been introduced in recent years (Hocke et al. 1999; Igarashi et al. 2000; Gorbunov 2002; Hajj et al. 2002; Pavelyev et al. 2004). New application of RO method for investigation of gravity wave (GW) activity using the temperature variations in the 5 km to 40 km interval found from the phase part of GPS RO radio holograms has been indicated by Steiner and Kirchengast (2000); Tsuda et al. (2000); Tsuda and Hocke (2002). The importance of the amplitude channel of GPS radio holograms for RO investigation of the atmosphere and ionosphere has been noted by Igarashi et al. (2000, 2001); Sokolovskiy (2000, 2002); Pavelyev et al. (2002, 2003); Liou et al. (2002, 2003). The different sensitivity of the amplitude and phase of RO signals to wave structures in the atmosphere and ionosphere has been established formerly (Igarashi et al. 2001; Pavelyev et al. 2004). Amplitude scintillations of RO signals can be considered as a radio holographic image of internal waves containing important information on the amplitude and phase of the wave structures in the atmosphere and ionosphere. The aim of this paper consists in application of the RO amplitude radio holographic method to investigate the atmosphere and ionosphere. In Section 2 the seasonal, geographical and temporal distributions of the ionospheric disturbances are found with global coverage from amplitude scintillations in the RO signals. The local mechanism of the quasi-regular amplitude scintillations caused by inclined ionospheric layers is introduced, and example of estimation of the electron density distribution and its gradient using the RO amplitude data is considered. In Section 3 the connection between the phase and amplitude variations in the RO signals is analyzed and the relationships for the retrieval of the refractivity and vertical temperature gradient are introduced. It is pointed out, that the amplitude of RO signals is more sensitive than the phase to wave structures in the atmosphere and ionosphere. In Section 4 possibilities for measuring the internal GW parameters (e.g., horizontal wind perturbations) is discussed using the GW polarization and dispersion relationships.

2 Local Mechanism of RO Amplitude Variations

The main contribution to the amplitude variations of the RO signal in the case of the quiet ionosphere is introduced by relatively small area along the ray GTL with center at the tangent point T (Fig. 1), where the ray trajectory GTL is perpendicular to the local gradient of the refractivity. If precise orbital data are given, the height h of point T can be evaluated under assumption of the spherical symmetry of the ionosphere and atmosphere with center at point O (Fig. 1). For some RO experiments strong amplitude scintillations have been observed at the height h of point T in the 40 km to 80 km interval (Sokolovskiy et al. 2002). These scintillations might be associated with plasma disturbances in the E- and/or F-layers of the ionosphere. Also sporadic amplitude scintillations have been indicated at GPS frequencies in satellite-earth links (Yeh and Liu 1982; Karasawa et al. 1985). Karasawa et al. (1985) showed that sporadic amplitude scintillations can be caused by plasma disturbances in the ionospheric E- and F-layers. In the absence of the global spherical symmetry a new tangent point 1 can appear in the ionosphere, where a sharp gradient of the electron density in an inclined plasma layer is perpendicular to the ray trajectory GTL (Fig. 1). This point can appear on both sides of the ray GTL (GT or TL) depending on which part of the ray trajectory is occupied by an inclined sporadic plasma layer. In any case the height h(T) of the observed RO amplitude variations depends on inclination δ of the inclined plasma layer relative to the local horizontal direction (Fig. 1). As a consequence the apparent height displacement Δh can arise in the estimated value of the altitude of the inclined plasma layer (Fig. 1). The layer's inclination δ and its horizontal displacement d relative to point T can be evaluated using Fig. 1:

$$\delta = (2\Delta h / r)^{1/2}, \; d = (2\Delta h r)^{1/2}, \; \Delta h = d^2/(2r), \tag{1}$$

where r is the distance OT.

Below the amplitude variations of the RO signal will be described by magnitude of the S_4 scintillation index (Yeh and Liu 1982):

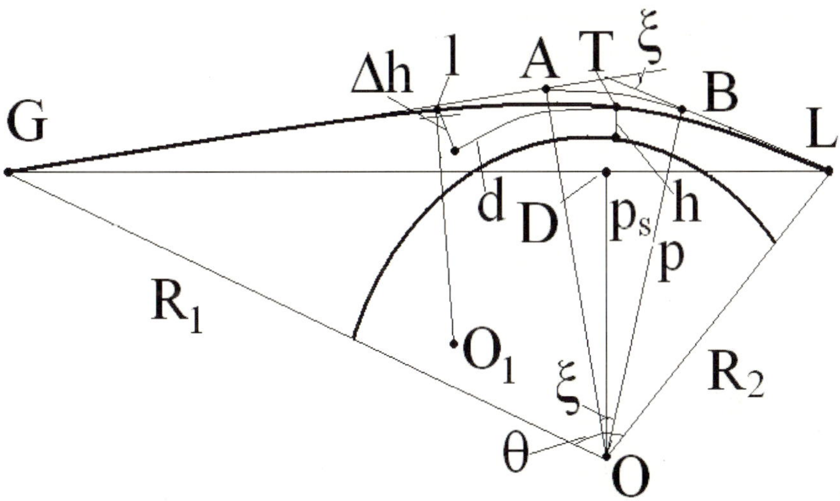

Fig. 1. Key geometrical parameters in the RO experiment. A GPS satellite emitting the radio waves is located at point G, a LEO satellite registering the amplitude and phase of RO signal at two GPS frequencies is located at point L. Line GTL indicates the curved RO signal trajectory in the atmosphere and ionosphere. Inclined plasma layer 1 in the ionosphere, where gradient of the electron density is perpendicular to RO ray, may be the cause of the sporadical amplitude and phase scintillations measured at point L. Apparent horizontal and height displacements of the inclined ionospheric layer are Δh and d, respectively. O_1 is the center of the local spherical symmetry of the inclined plasma layer.

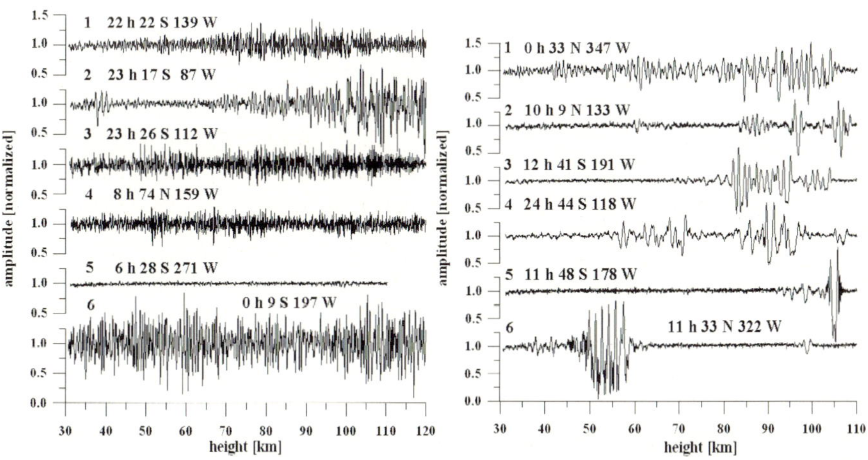

Fig. 2. Left panel: C-type noisy amplitude scintillations of RO signal. Right panel: S-type quasi-regular amplitude variation. Legends indicate the local time (LT) and the geographical coordinates of RO experiments.

$$S_4 = 2\left[\left\langle \left(A(t) - \langle A \rangle\right)^2 \right\rangle \Big/ \langle A \rangle^2 \right]^{1/2} \quad (2)$$

where $\langle\ \rangle$ is the average relevant to the height interval h(T), 40 km to 80 km; $A(t)$ is the amplitude of the RO signal. For analysis we used the amplitude data obtained during CHAMP RO experiment described by Wickert et al. (2001). The CHAMP RO amplitude variations, which can be recognized as C- and S-types of the amplitude scintillations observed previously (Karasawa et al. 1985) in the communication Inmarsat link at frequency 1.5 GHz, are shown in Fig. 2. The noisy C-type amplitude variations in the RO signals are shown in Fig. 2, left panel. These scintillations have been observed in the equatorial region at local evening during RO events No. 0051, November 19; No. 0053, July 05; No. 0069, February 24; No. 0135, July 04, 2003. Average values of the S_4 index are equal to 0.19, 0.32, 0.20, 0.5 (curves 1,2,3,6, respectively). Event 0159 corresponds to local morning (May 03), in the polar ionosphere near the north geomagnetic pole with $S_4 = 0.16$ (curve 4). Curve 5 relates to quiet nighttime ionosphere with $S_4 = 0.03$ (event No. 0198, May 03), when the amplitude fluctuations in the height interval h(T) 30 km to 110 km were caused mostly by random noise of the receiver. The geographical position and local time of the noisy RO events correspond to the same parameters of the noisy amplitude scintillations observed previously in satellite-earth links (e.g., Yeh and Liu 1982). Noisy scintillations can be associated with small-scale plasma irregularities in the F- or E-layers of the ionosphere (Yeh and Liu 1982; Karasawa et al. 1985).

Quasi-regular amplitude variations are shown in Fig. 2, right panel. Curves 1, 2, 4 correspond to the CHAMP RO events No. 0093, 0239, 0010, February 24 (mid-latitude nighttime and equatorial daytime ionosphere); curves 3, 5 relates to events No. 0050, 0171, November 19; and curve 6 corresponds to event No. 0246, July 05, 2003 (daytime mid-latitude ionosphere). The quasi-regular amplitude variations can be caused by inclined plasma layers in the E- or F-region of the ionosphere. Apparent displacement in height of a plasma layer Δh from its normal location can be applied to estimate the inclination δ and horizontal displacement d of the layer relative to point T (Fig. 1) by using equation (1). However, the uncertainty in the sign of the horizontal displacement d and inclination δ exists because the ionospheric disturbance can be located at the same distance d from point T on the part GT or LT of the ray trajectory GTL (Fig. 1). Note that application of BP method as shown by Sokolovskiy et al. (2002), gives a promise to solve in special cases the task of localization of the ionospheric disturbances. Quasi-regular amplitude variations can be used for restoration of the electron density distribution and its vertical gradient under assumption of the local spherical symmetry of the inclined ionospheric layer by method described formerly by Pavelyev et al. (2002). For demonstration the CHAMP RO event (January 14, 2003, 0 h 56 min LT, 76.4°N, 172.7°W) with strong quasi-regular wave-like amplitude variations will be used. In Fig. 3, left panel, the amplitude variations in the CHAMP RO signal (curve 1) are compared with the result of the ionospheric correction by means of

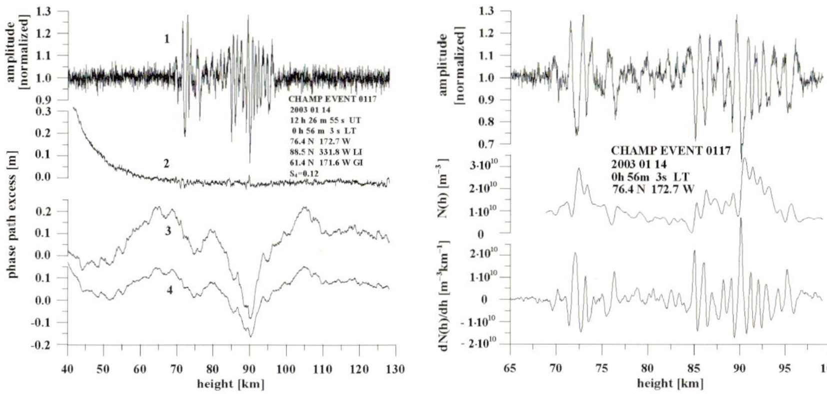

Fig. 3. Left panel: amplitude of the CHAMP RO signal (curve 1), result of ionospheric correction by means of the linear combination of the excess phase measured at frequencies F1 and F2 at the same time sample (curve 2), the excess phase at frequency F1 (curve 3) and F2 (curve 4) as functions of the height of the tangent point T (Fig. 1). The curves 3 and 4 have been obtained after subtracting the trend described by the polynomial of the second order. S_4 index was equal to 0.12, which corresponds to the disturbed ionosphere. Right panel: The amplitude variations (top curve), the electron density distribution $N(h)$ (middle curve) and the gradient of the electron density distribution $dN(h)/dh$ (bottom curve) as functions of height h of the tangent point T (Fig. 1).

linear combination of the phase excesses F0 (curve 2) and the phase excess at frequencies F1 and F2 (curves 3 and 4, respectively). The form of the amplitude variations indicates that the ionospheric disturbance consists of two connected patches which are responsible for the maximums in the amplitude changes in the 72 km to 76 km and 84 km to 96 km intervals h(T). In the 76 km to 84 km interval the amplitude variations are not so strong. Connection between the amplitude variations in two height intervals indicates that the ionospheric disturbance is a unified structure distributed in the horizontal and vertical directions. The corrected phase variations F0 are small (Fig. 3, left panel). This indicates regular layered structure of the plasma disturbance. One can retrieve the electron density distribution and its gradient from amplitude variations of the RO signal by using method described by Liou et al. (2002). The results of restoration of the electron density distribution and its gradient are shown in Fig. 3, right panel. The electron density variations are concentrated in the interval $0 < N(h) < 3.5{\cdot}10^{10}$ [electrons/m^{-3}]. These magnitudes of $N(h)$ are somewhat below the usual values of $N(h)$ for sporadic E-layers. The height interval of the amplitude variations is nearly equal to height interval of the variations in the electron density and its gradient. As follows from estimation of the possible location of the both patches, the first patch is located on line GT at a distance 300 km from point T. It is concentrated in the 92 km to 104 km interval with inclination to horizontal direction δ of about 3°. The second patch is located on the line GT in the 94 km to 100 km interval at the

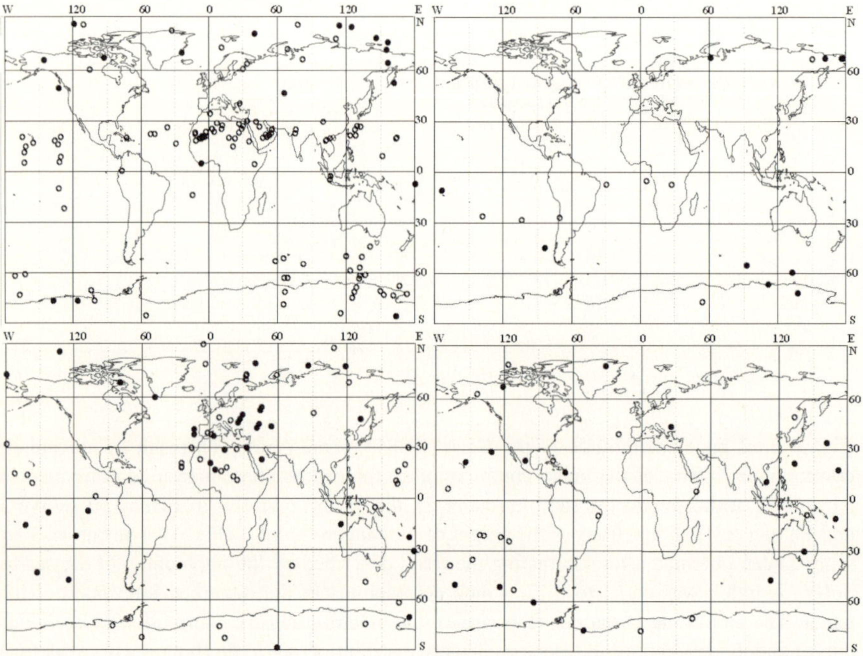

Fig. 4. Maps of strong ionospheric events, with S_4 index greater than 0.2, for September 2001 (left top panel), September 2002 (right top panel), September 26 – October 24, 2003 (left bottom panel), January 2004 (right bottom panel). The circles show the geographical position of the tangent point T (Fig. 1).

distance 500 km relative to the tangent point T (Fig. 1) with inclination of about 5°.

The geographical distribution of the strong ionospheric events (with a S_4 index greater than 0.2), for all types of amplitude scintillations, in the CHAMP RO signals at 1575.42 MHz is demonstrated in Figs. 4, 5. The distribution of the ionospheric events indicates that they are concentrated in some regions (e.g., the equatorial and geomagnetic north and south polar zones in Figs. 4 and 5).Strong activity in some equatorial regions may be connected with the evening ionospheric disturbances that arise after sunset, 20–24 hours of local time, in accordance with earth-based measurements reviewed earlier (e.g., Yeh and Liu 1982). As seen in Figs. 4, 5, the number of strong ionospheric events and their intensity decreases with time from September 2001 to January 2004. This may correspond to decrease in solar activity. The seasonal displacement of the regions with intense ionospheric events in the south and north directions during the periods from May 14 – July 14, 2001, November–December 2001, (Fig. 5, top panels), October 28 – November 26, 2003 and April 2004 (Fig. 5, bottom panels) can be noted. The number of intense ionospheric events increases in the North Polar Region with time from May – July 2001 to November – December 2001. Also the number of strong ionospheric events increases in the south equatorial region for the same time interval.

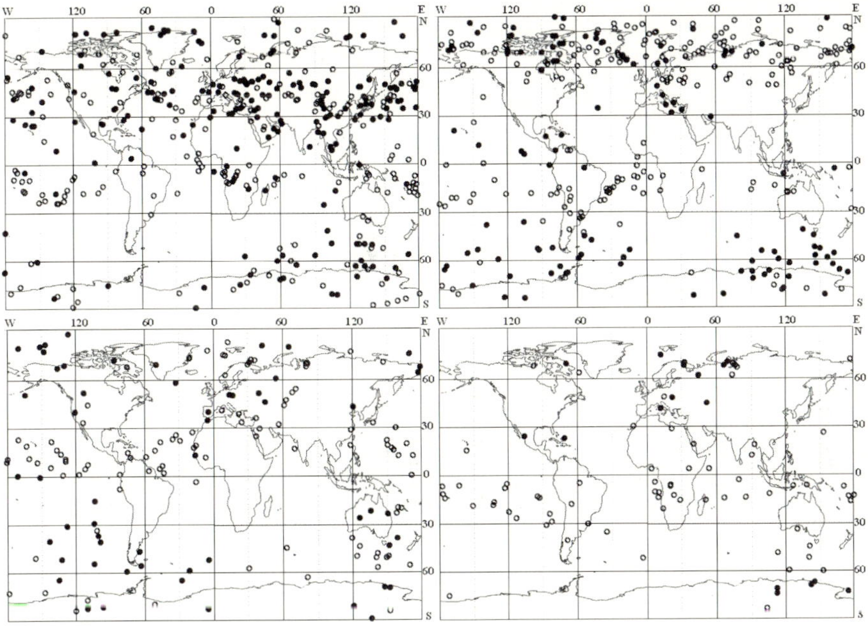

Fig. 5. Seasonal dependence of the global distribution of CHAMP RO events with strong amplitude variations (with magnitude of S_4 index greater than 0.2) for periods May 14 – July 14, 2001, (left top panel), November-December 2001 (right top panel), October 28 – November 26, 2003 (left bottom panel), April 2004 (right bottom panel). The closed circles (local time 08 – 20 hours) indicate the day time events, the open circles (local time 20 – 08 hours) indicate the night-time events.

These changes indicate two important mechanisms governing the ionospheric disturbances. The first one is connected with processes of ionizations caused by energetic electrons in the polar regions; the second is due to solar radiation. The influence of solar radiation has a seasonal character because ionization in the ionosphere follows annual and diurnal motion of the ionospheric sub-sun point. As follows from these considerations the amplitude part of RO radio holograms is appropriate for finding the spatial distribution of the electron density and its gradient, and for localization of the ionospheric disturbances.

3 Connection Between the Amplitude and Phase Variations

The phase and amplitude of signals propagating along the ray GTL (Fig. 1) can be expressed for the case of spherical symmetry (Pavelyev et al. 2004)

$$\Phi(p) = L(p) + \kappa(p),\ \theta = \pi + \xi(p) - a(p),$$
$$L(p) = \left(R_2^2 - p^2\right)^{1/2} + \left(R_1^2 - p^2\right)^{1/2} - \left[\left(R_2^2 - p_s^2\right)^{1/2} + \left(R_1^2 - p_s^2\right)^{1/2}\right] + p\xi(p) \tag{3}$$

$$\xi(p) = a(p) - a(p_s);$$
$$a(p) = \sin^{-1}(p/R_1) + \sin^{-1}(p/R_2);$$
$$-\frac{\mathrm{d}\kappa(p)}{\mathrm{d}p} = \xi(p). \tag{4}$$

The relationships (3), (4) follow from the eikonal equation and connect the phase excess $\Phi(p)$ with refraction angle $\xi(p)$, impact parameter p, the main refractivity part of the phase excess $\kappa(p)$. Equations (3) and (4) are valid also for each ray in multipath situation under assumption of the spherical symmetry. In the lower troposphere where the horizontal gradients of the refractivity may be significant the relationships (3) and (4) can be used as an approximation to real conditions. The refraction attenuation has been derived using geometrical optics under assumption of spherical symmetry by Pavelyev and Kucherjavenkov (1978)

$$X(p) = pR_0^2\left[R_1R_2\left(R_2^2 - p^2\right)^{1/2}\left(R_1^2 - p^2\right)^{1/2}\sin\theta\right]^{-1}\left|\frac{\partial\theta}{\partial p}\right|^{-1}, \tag{5}$$

where R_0 is the distance GDL (Fig. 1). The refraction attenuation $X(p)$ depends mainly on the second derivative $\mathrm{d}^2\kappa/\mathrm{d}p^2$:

$$X(p) \approx \frac{b(p_s)}{\left|\mathrm{d}^2\kappa/\mathrm{d}p^2 + b(p)\right|},$$
$$b(p) = \frac{1}{\left(R_2^2 - p^2\right)^{1/2}} + \frac{1}{\left(R_1^2 - p^2\right)^{1/2}}. \tag{6}$$

As a consequence, the amplitude variations are more sensitive than the phase excess to wave structures in the atmosphere and ionosphere. The amplitude variations may be used for obtaining the height distribution of the vertical gradient of the refractivity. The solution of the inverse RO problem for the amplitude channel of RO signal has been given earlier by Kalashnikov et al. (1986). For the case of circular orbit of LEO and GPS satellites this solution is given by equations

$$\frac{\mathrm{d}p}{\mathrm{d}t} = X[t(p_s)]\frac{\mathrm{d}p_s}{\mathrm{d}t} \quad \text{and} \quad p - p_0 = \int X[t(p_s)]\,\mathrm{d}p_s \tag{7}$$

Equations (6) and (7) can be used to find the vertical distribution of the vertical gradient of the refractivity d$N(h)$/dh (Liou et al. 2002):

$$\frac{\mathrm{d}N(h)}{\mathrm{d}h}=-\frac{n^2(h)J(p)}{p[1+J(p)]},\ J(p)=\frac{1}{\pi}\int\frac{\mathrm{d}^2\xi(x)}{\mathrm{d}x^2}\left(x^2-p^2\right)^{1/2}\mathrm{d}x,$$
$$P=n(h)(a+h),\ n(h)=1+N(h)\cdot 10^{-6},\ \frac{\mathrm{d}\xi(p)}{\mathrm{d}p}=b(p)\frac{(X-1)}{X}. \tag{8}$$

Integration in (8) is performed from p up to infinity.

The amplitude information may be used to retrieve the vertical gradient of the temperature profile (Liou et al. 2002):

$$\frac{\mathrm{d}T^*(h)/\mathrm{d}h}{T^*(h)}=-\frac{\mathrm{d}N(h)/\mathrm{d}h}{N(h)}-\frac{T_x}{T_a(h)}$$
$$T^*(h)=\frac{T(h)}{1+4810\,e(h)/[P(h)T(h)]},\ T_x\approx 34.16\ \text{K/km} \tag{9}$$
$$T_a(h)=T(h)\left[1+\frac{0.378\,e(h)}{P(h)}\right],\ N(h)=77.6\frac{P}{T^*}$$

where $T(h)$ is the temperature of the atmosphere [K], $T^*(h)$ is the "wet" temperature of the atmosphere depending on the water vapor pressure $e(h)$ and atmospheric pressure $P(h)$ [hPa], respectively. The first equation (9) connects the vertical gradient of the logarithm of the refractivity with the vertical gradient of the logarithm of the "wet" temperature $T^*(h)$. At the height above 10 km, equations (9) may be used to find the vertical gradient of the temperature profile $T(h)$ if the refractivity gradient is known. Equations (7) to (9) were applied to find the vertical gradients of the refractivity and temperature $T(h)$ using the amplitude variations of the RO signal and dependence of the free space impact parameter $p_s(t)$ on time.

4 Amplitude Variations and Wave Structures in the Atmosphere

Below we will consider perturbations in the vertical gradient of the refractivity and temperature found from the amplitude variations of the RO signals with the aim to establish parameters of wave structures in the atmosphere. The amplitudes of the RO signal are shown in Fig. 6, left panel, for the CHAMP RO event 0005 (curve A indicates amplitude variations at frequency F1) and GPS/MET RO event 0316 (curves A1, A2 indicate amplitude variations at frequencies F1 and F2, respectively). Results of simulations of the amplitude dependence on height are shown in Fig. 6, left panel, by the curves M0 and M. For calculating M0 we used the refractivity model $N(h)=N_o\exp(-h/H)$ with $N_o=340$ (N-units) and

$H = 6.4$ km, and the analytical connections between the refraction angle $\xi(p)$, the refraction attenuation, and the refractivity gradient described above. We applied the same method to obtain the curves M in Fig. 6, left panel, but the refractivity model was a sum of damped complex exponentials $N(h) = \mathrm{Re}\,[\Sigma Nj \exp(-\alpha_j h)]$ with real and complex N_j and α_j. The parameters N_j and α_j and number of the terms in the sum (23 for the GPS/MET event and 11 for the CHAMP event) have been determined by an iterative procedure such as to give the amplitude variations that are better coinciding with the experimental data. To obtain the vertical profiles of the refractivity, temperature and their gradients we use expressions (7) to (9). The vertical gradients of the temperature retrieved from the amplitude data are indicated for CHAMP (the bottom curves A and M) and GPS/MET RO (the upper curves A1, A2 and M) events 0005 and 0316 in Fig. 6, right panel. Note that in the CHAMP RO experiments only the amplitude variations at the first GPS frequency F1 has been measured. The curves M in Fig. 6, right panel, indicate the simulation results relating to the CHAMP (the second curve from bottom in Fig. 6, right panel) and GPS/MET (the second upper curve in Fig. 6, right panel) RO events, the curves A1, A2, and A describe the vertical temperature gradient variations restored from the initial amplitude changes shown in Fig. 6, left panel. It is evident from Fig. 6, right panel, that the wave activity in the atmosphere is a function of height. Maximum of the wave activity is observed near the tropopause region in the 15 km to 22 km interval. Quasi-regular wave structures with vertical wavelength $\lambda_v \sim 0.8$ km to $\lambda_v \sim 2$ km are clearly seen in both experimental and model data. The observed waves in the altitude distribution of the amplitude and vertical temperature gradient can correspond to the GW activity. If the observed wave structures are caused by GW activity then the vertical temperature gradients can be related with horizontal wind perturbations. The magnitude of the horizontal wind perturbations can be estimated using the polarization relationships, which are valid for the medium-frequency case, when the intrinsic frequency of the GW is greater than the inertial frequency f, but is well below the buoyancy frequency ω_b. Note, that from only RO profiles, we may not be sure whether the temperature wave structures observed correspond to low, medium or high intrinsic frequencies. To establish origin of the wave structures one must use additional observations provided by different methods (e.g., radiosondes measurements).

The GW dispersion relation has the form (Fritts and Alexander 2003; Eckermann et al. 1995)

$$\lambda_v = \frac{2\pi\left|c - U\cos\varphi\right|}{\omega_b} \tag{10}$$

where λ_v is the vertical wavelength of the GW, U is the background wind speed, c is the ground-based GW horizontal phase speed, and φ is the azimuth angle between the background wind and the GW propagation vectors. Equation (10) connects the vertical wavelength λ_v with the intrinsic phase speed of GW $v_i = \left|c - U\cos\varphi\right|$, which can be measured by an observer moving with the background wind velocity (Eckermann et al. 1995). A GW polarization relation has

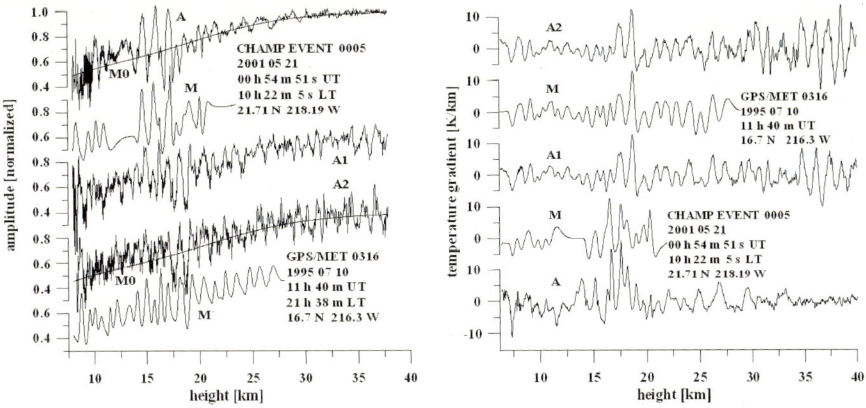

Fig. 6. Left panel: wave trains in the GPS/MET (curves A1 and A2) and CHAMP (curve A) RO amplitude data. Legend indicates the time of the RO experiments and the geographical co-ordinates of the RO regions. Right panel: Temperature gradient variations found from amplitude variations for the GPS/MET (curves A1 and A2) and CHAMP (curve A) RO events. Curves M0 and M describe results of simulation.

been published previously by Lindzen (1981). It connects the complex amplitude of the temperature variation, $t(h)$, with the horizontal wind perturbations $v(h)$, corresponding to the GW influence

$$v(h) = \mathrm{Re}\left[\frac{ig}{T_b\omega_b}t(h)\right] \tag{11}$$

where $\omega_b^2 = g/T_b\Gamma$, $\Gamma = \partial T_b/\partial h + 9.8°/\mathrm{km}$, g is the gravity acceleration, and T_b is the background temperature. Eckermann et al. (1995) used the polarization relationships for statistical analysis of the rocketsonde data. One can obtain from (11) a connection between the vertical gradients $\mathrm{d}v(h)/\mathrm{d}h$ and $\mathrm{d}t(h)/\mathrm{d}h$

$$\frac{\mathrm{d}v(h)}{\mathrm{d}h} = \frac{\mathrm{d}\,\mathrm{Re}\left[ig/\left(T_b\omega_b\right)t(h)\right]}{\mathrm{d}h} \approx \mathrm{Re}\left[\frac{ig}{T_b\omega_b}\frac{\mathrm{d}t(h)}{\mathrm{d}h}\right]. \tag{12}$$

Equation (12) is valid assuming that $T_b(h)$ and $\omega_b(h)$ are slowly changing at the vertical scales $\sim\lambda_v$. The functions T_b and ω_b are known from the model (M0) of the atmosphere used for the calculation of the refraction attenuation and refractivity in the RO region. To find the function $\mathrm{d}v(h)/\mathrm{d}h$ from the second equation (12) one can implement the radio holographic analysis by applying the Hilbert transform (Rabiner and Gold 1978). Implementation of the Hilbert transform gives the analytic presentation consisting of the amplitude $a_t(h)$ and phase $\Phi_t(h)$ of the real function $\mathrm{d}t(h)/\mathrm{d}h$ found from the amplitude variations of the RO signals:

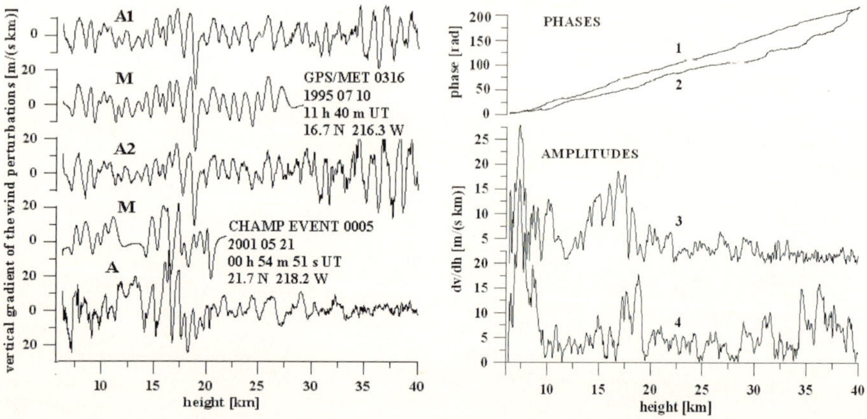

Fig. 7. Left panel: vertical gradient of the horizontal wind perturbations retrieved on the variations of the vertical temperature gradient for GPS/MET (three upper curves) and CHAMP (two bottom curves) RO events. Right panel: phase (curve 1 and 2) and amplitude (curves 3 and 4) of the GW as function of height.

$$\frac{\mathrm{d}t(h)}{\mathrm{d}h} = \mathrm{Re}\{a_t(h)\exp[i\Phi_t(h)]\}. \tag{13}$$

Then the vertical gradients of the horizontal wind perturbations function d$v(h)$/dh can be restored from (12). The vertical gradients of the horizontal wind perturbations can presented in the form d$v(h)$/dh = $a(h)\cos\Phi(h)$, where $a(h)$ and $\Phi(h)$ are the amplitude and phase of the analytic signal relevant to d$v(h)$/dh. The functions $a(h)$ and $\Phi(h)$ together present a GW portrait. The results of restoration of the vertical gradients of the horizontal wind perturbations are indicated in Fig. 7 (left panel). The upper curves (A1, A2, M) in Fig. 7 (left panel) are corresponding to the vertical gradient of the horizontal wind perturbations restored for the GPS/MET RO event 0316, while the bottom two curves (M, A) are relating with the CHAMP RO event 0005. The quasi-regular modulation of dv/dh by wave structures in the atmosphere is clearly seen both in the experimental and model data. It is important that the vertical period of this modulation is practically the same as seen in the amplitude variations in Fig. 7, left panel. The height dependence of the GW phases $\Phi(h)$ and amplitudes $a(h)$ are shown in Fig. 7 (right panel) by curves 1, 2 and 3, 4, respectively, for the GPS/MET and CHAMP RO events 0316 and 0005.

Comparison of the GW portraits found from the wave trains in the amplitude indicates increasing of the spatial frequency of the GW with height (Fig. 7, right panel). The phase in the GPS/MET event (curve 1) changes monotonically, in average, as a function of the height h, thus corresponding to quasi-monochromatic GW. For the CHAMP RO event (curve 3) the amplitude changes are concentrated mainly below 20 km. The amplitude, relevant to the GPS/MET event (curve 4), demonstrates essential changes in the interval 0.5...16 m/(s km) above 17 km.

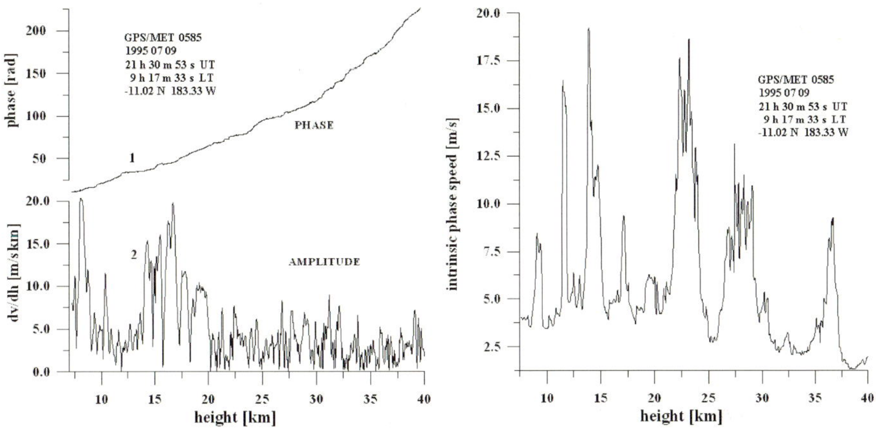

Fig. 8. Left panel: GW portrait for the GPS/MET RO event 0585. Right panel: intrinsic phase speed of the GW for the GPS/MET RO event 0585.

For this event one can see altitudes with high (17 km to 19 km, 29 km to 31 km, 35 km to 38 km) and low (32 km to 34 km) GW activity.

Another example of the GW activity in the atmosphere is given in Fig. 8 for the GPS/ MET RO event 0585 (July 07, 1995, 09 h 17 min LT, 11°S 183°W). The upper and bottom curves 1, 2 (left panel in Fig. 8) demonstrate the height dependence of the phase and amplitude of the GW. The vertical profile |d*v(h)*/d*h*| (curve 2 in Fig. 8) indicates the region where the amplitude of the GW is high: at the heights 14 km to 20 km, 23 km to 25 km, 28 km to 29 km, 31 km to 32 km and low: at the heights 20 km to 22 km, 25 km to 26 km, 33 km to 35 km. This indicates the layers in the atmosphere with different levels of the wave activity and is coinciding with the previously satellite observations of the waves activity provided by Eckermann and Preusse (1999). The phase curve 1 in the left panel in Fig. 8 reveals in the interval 10 km to 30 km the quasi-linear dependence on the height thus indicating quasi-monochromatic character of the observed wave structure. The detailed information on the height distribution of the spatial frequency of the GW can be obtained by differentiating the phase $\Phi(h)$ (curve 1 in the left panel in Fig. 8). After differentiating one can obtain the spatial frequency f_v and the vertical wavelength $\lambda_v = 1/f_v$ as functions of height and then estimate the intrinsic phase speed of the GW v_i using the relation (10). As seen in Fig. 8 (right panel) the value $v_i(h)$ changes in the range 2 m/s to 18 m/s. These values are similar to the intrinsic phase speed observed by rocketsondes (Eckermann et al. 1995) and satellite (Eckermann and Preusse 1999).

Integration of the vertical gradient d*v(h)*/d*h* can reveal the horizontal wind perturbations *v*(*h*) as a function of height. The function *v*(*h*) is depicted in Fig. 9, left panel, for two RO events: GPS/MET 0316 (part a)) and CHAMP 0005 (part b)). The curve A in Fig. 9, left panel, part a) has been obtained by integrating the average wind speed gradient $\mathrm{d}v/\mathrm{d}h = [\mathrm{d}v_1(h)/\mathrm{d}h + \mathrm{d}v_2(h)/\mathrm{d}h]/2$ (curves A1, A2 in Fig. 7 (left panel)) restored from the GPS/MET RO amplitude data. The curve A in

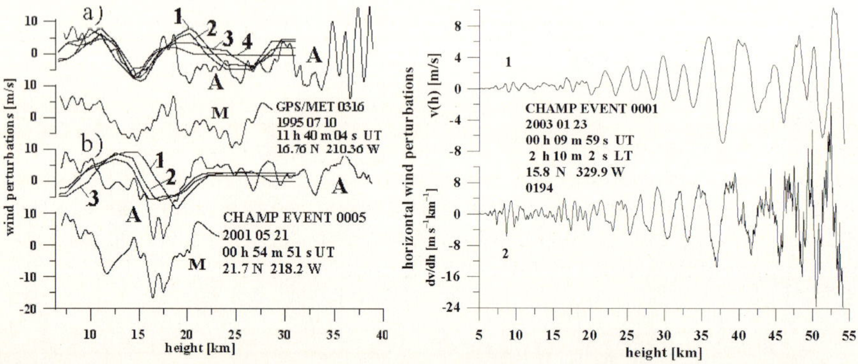

Fig. 9. Left panel: comparison of the wind perturbations found from GPS/MET and CHAMP RO events with Earth-based radiosondes data. Right panel: horizontal wind perturbations (curve 1) and its vertical gradient (curve 2) for CHAMP RO event No. 001, January 23, 2003.

Fig. 9, left panel, part b) has been obtained by integrating the wind speed gradient d*v*/d*h* corresponding to the bottom curve A in Fig. 7 (left panel) restored from the CHAMP RO amplitude data. The curves M (Fig. 9, left panel, parts a), b)) describes the results of the simulation. The curves 1 – 4 in Fig. 9, left panel, part a) indicate the radiosondes (RS) data corresponding to two stations in Taiwan: Hualian (1, 4) (24.0°N, 238.4°W) and Taipei (2, 3) (25.0°N, 238.5°W), obtained on July 15, 1995 at 00 h UT (1, 2) and 12 h UT (3, 4), respectively. The curves 1 – 3 in Fig. 9, left panel, part b) correspond to the same stations and indicate the RS data, obtained on May 30, 2001 at 00 and 12 h UT (1, 2) (Taipei) and 12 h UT (3) (Hualian), respectively. The difference between the Taiwan Earth-based stations and the GPS/MET RO region latitudes for the event 0316 is about 8°, the corresponding value for longitudes is about 28°. The difference between the Taiwan and the CHAMP RO region latitudes for the event 0005 is of about 3°, the corresponding value for longitudes is about 20°. The time for comparison between the RO events and RS observations has been chosen according to the average background velocity of the global equatorial westerly air flow about 6 m/s to 10 m/s in the height interval 8 km to 30 km. The RS wind perturbations (curves 1 – 4 in Fig. 9, left panel, part a), curves 1 – 3 in Fig. 9, left panel, part b)) have been obtained by subtracting the polynomial approximation of the fifth power from the experimental vertical profiles of the horizontal wind speed. As follows from Fig. 9, left panel, part a) the RS data (1 – 4) are in agreement with the results found from the amplitude of the GPS/MET RO signal and simulation (curves A, M in Fig. 9, left panel, part a), respectively). Some discrepancies of about ±2 m/s to ±4 m/s exist in the height interval 10 km to 20 km. The discrepancies of the CHAMP data (curve A in Fig. 9, left panel, part b) from the RS data (curves 1 – 3 in Fig. 9, left panel, part b)) is greater ±4 m/s to ±6 m/s. Note that the RS data do not reveal high-spatial frequencies observed in the RO results. It appears this is due to possible smoothing effects of the RS measurements. The RO values of $v(h)$ (curves A in Fig. 9, left panel) are variable from ±1 m/s to ±12 m/s at the height interval 10 km to 35 km and indicate some higher magnitudes for the $v(h)$ as compared with the model's results (curves M in Fig. 9, left panel). The difference illustrates a current state of the inversion accuracy.

Example of horizontal wind perturbations in the 5 km to 55 km interval is shown in Fig. 9, right panel (curve 1), with its vertical gradient (curve 2). An increase of the vertical wavelength (from 1 km up to 3 km – 4 km) and amplitude (from 0.5 m/s up to 8 m/s) of the horizontal wind perturbations in the 10 km to 40 km interval is connected with GW propagation in the stratosphere. This example corresponds to increasing of the intrinsic phase speed of GW with height.

5 Conclusions

The amplitude of GPS signal is a radio holographic indicator of the ionospheric disturbances in the trans-ionospheric RO links. The classification of the amplitude variations of GPS signals is introduced. We showed that scintillations of the GPS RO signals can be classified into two types of amplitude variations also previously observed in satellite-Earth trans-ionospheric links. Quasi-regular amplitude variations can be inverted to obtain estimation of the electron density distribution and its gradient in inclined ionospheric layers.

The geographical and temporal distributions of the CHAMP RO events with high S_4 index values observed during 2001 – 2004 showed dependence on season, geographical location and solar activity. This dependence indicated two mechanisms of ionization: (1) ionization owing to the fast electron moving in downward direction from the magnetosphere and (2) solar radiation. In the equatorial regions the most of the CHAMP RO events with high S_4 index values have been observed during local evening and night that corresponds to specifical mechanism of the plasma perturbations in the lower ionosphere.

The amplitude variations can be used to measure the perturbations in the vertical gradient of the refractivity and temperature. The GW polarization relationships can be applied to find the horizontal wind perturbations and its vertical gradient as function of height. Comparison of the horizontal wind perturbations obtained from the amplitude data analysis with radiosondes data revealed good agreement.

The amplitude part of GPS RO radio hologram is important for application of the RO method to analysis of the wave phenomena in the atmosphere and ionosphere, and for experimental studies of processes in the atmosphere, mesosphere and ionosphere and solar activity.

Acknowledgments. The authors gratefully acknowledge GFZ-Potsdam for delivering the CHAMP and GPS/MET RO data for analysis. The work is partly supported by Russian Fund of Basic Research, grant No. 03 02 17814.

References

Eckermann SD, Preusse P (1999) Global Measurements of Stratospheric Mountain Waves from Space. Science 286:1534–1537

Eckermann SD, Hirota I, Hocking WK (1995) Gravity wave and equatorial wave morphology of the stratosphere derived from long-term rocket soundings. Q J R Meteorol Soc 121:149–186

Fritts DC, Alexander MJ (2003) Gravity wave dynamics and effects in the middle atmosphere. Rev Geophys 41(1), 1003, doi:10.1029/2001RG000106

Gorbunov ME (2002) Ionospheric correction and statistical optimization of radio occultation data. Radio Sci 37(5):17-1–17-9, doi:10.1029/2000RS002370

Hajj GA, Kursinski ER, Romans LJ, Bertiger WI, Leroy SS (2002) A technical description of atmospheric sounding by GPS occultation. J of Atmos Solar-Terr Phys 64:451–469

Hocke K, Pavelyev A, Yakovlev O, Barthes L, Jakowski N (1999) RO data analysis by radio holographic method. J of Atmos Solar-Terr Phys 61:1169–1177

Igarashi K, Pavelyev A, Hocke K, Pavelyev D, Kucherjavenkov IA, Matugov S, Zakharov A, Yakovlev O (2000) Radio holographic principle for observing natural processes in the atmosphere and retrieving meteorological parameters from radio occultation data. Earth Planets Space 52(11):893–899

Igarashi K, Pavelyev A, Hocke K, Pavelyev D, Wickert J (2001) Observation of wave structures in the upper atmosphere by means of radio holographic analysis of the radio occultation data. Adv Space Res 27(6/7):1321–1327

Kalashnikov I, Matugov S, Pavelyev A, Yakovlev O (1986) Analysis of the features of RO method for the Earth's atmosphere study. In: Electromagnetic waves in the atmosphere and space. 208–218. "Nauka" Ed Moscow (in Russian)

Karasawa Y, Yasukawa K, Yamada M (1985) Ionospheric scintillation measurement at 1.5 GHz in mid-latitude region. Radio Sci 20(3):643–651

Lindzen RS (1981) Turbulence and stress owing to gravity wave and tidal breakdown. J Geophys Res 86:9707–9714

Liou YA, Pavelyev AG, Huang CY, Igarashi K, Hocke K (2002) Simultaneous observation of the vertical gradients of refractivity in the atmosphere and electron density in the lower ionosphere by radio occultation amplitude method. Geophys Res Lett 29(19), 1937, doi:10.1029/2002GL015155

Liou YA, Pavelyev AG, Huang C-Y, Igarashi K, Hocke K, Yan SK (2003) Analytic method for observation of the gravity waves using radio occultation data. Geophys Res Lett 30(20), 2021, doi:10.1029/2003GL017818

Pavelyev AG, Kucherjavenkov AI (1978) Refraction attenuation in the planetary atmosphes. Radio Eng and Electron Phys 23(7):13–19

Pavelyev A, Igarashi K, Reigber C, Hocke K, Wickert J, Beyerle G, Matyugov S, Kucherjavenkov A, Pavelyev D, Yakovlev O (2002) First application of radioholographic method to wave observations in the upper atmosphere. Radio Sci 37(3), 1043, doi:10.1029/2000RS002501

Pavelyev AG, Liou YA, Wickert J (2004) Diffractive vector and scalar integrals for bistatic radio holographic remote sensing. Radio Sci 39, RS4011, doi:10.1029/2003RS002935

Rabiner L, Gold B (1978) Theory and Application of Digital Signal Processing. Prentis Hall

Steiner AK, Kirchengast G (2000) Gravity wave spectra from GPS/MET occultation observations. J Atmos Ocean Tech 17(4):495–503

Sokolovskiy SV (2000) Inversion of radio occultation amplitude data. Radio Sci 35(1):97–105

Sokolovskiy SV, Schreiner W, Rocken C, Hunt D (2002) Detection of high-altitude ionospheric irregularities with GPS/MET. Geophys Res Lett 29(3), 1033, doi:10.1029/2001GL013398

Tsuda T, Hocke K (2002) Vertical wave number spectrum of temperature fluctuations in the stratosphere using GPS occultation data. J Meteorol Soc Japan 80(4B):925–938

Tsuda T, Nishida M, Rocken C, Ware RH (2000) A global morphology of gravity wave activity in the stratosphere revealed by the GPS occultation data (GPS/MET). J Geophys Res 105:7257–7273

Yeh KC, Liu CH (1982) Radio wave scintillations in the ionosphere. Proc IEEE 70:324–360

Wickert J et al. (2001) Atmosphere sounding by GPS ratio occultation: First results from CHAMP. Geophys Res Lett 28(17):3263–3266

Are we Observing Mountain Waves Above the Andes Range from GPS Occultation Profiles?

A. de la Torre[1], P. Alexander[1], and C. G. Menéndez[2]

[1] Departamento de Fisica, FCEN, Universidad de Buenos Aires, Pab I, Ciudad Universitaria, 1428 Buenos Aires, Argentina
delatorr@df.uba.ar

[2] Centro de Investigaciones del Mar y de la Atmosfera, Pab II, Ciudad Universitaria, 1428, Buenos Aires, Argentina

Abstract. GPS radio occultation temperature profiles retrieved from SAC-C and CHAMP exhibit a significant wave activity in the troposphere and lower stratosphere at midlatitudes (30°S to 40°S) above the Andes Range. Large amplitude structures, with long vertical wavelength, have been repeatedly reported in this region, as detected from other experimental devices and attributed to mountain waves. The possibility to associate the observed enhancements in wave activity to mountain forcing, or instead, to other significant sources in the region considered is discussed. The generation in the vicinity to a permanent jet situated above the mountains of inertio gravity waves by geostrophic adjustment, with longer horizontal and perhaps shorter vertical wavelengths than those expected from mountain waves, seems to be the main source. These waves could be more easily detected from GPS profiles than mountain waves. As it is known, inertio gravity waves are the means by which mass and momentum are redistributed so as to ultimately achieve geostrophic balance from an initially unbalanced state.

1 Introduction

As recently pointed out by [8], one main observational limitation to identify GW sources in the lower and middle stratosphere is given by the difficulty in separating wave structures from background conditions. As it is generally accepted, the main GW sources are: topographic, convective and shear generation, geostrophic adjustment and wave-wave interaction. The observation of wave structures in the lower and middle atmosphere are performed by means of various ground-based and spaceborne techniques. In the last years, the use of occultation measurement principle for observing the Earth's atmosphere and climate exploit solar, lunar, stellar, navigation and satellite-crosslink signals. In particular, atmosphere parameters as temperature, pressure, water vapor and geopotential have been obtained since 2001 from LEO (Low Earth Orbit) satellites CHAMP and SAC-C. The simultaneous global coverage, sub-Kelvin vertical temperature resolution, long-term stability and the absence of limitations imposed by weather conditions make this technique unique (see e.g., [9], [12] and references therein). Recently, an analysis of global distribution of GW activity in the higher troposphere and lower stratosphere between June 2001 and March 2003 was performed, using temperature profiles retrieved from GPS RO (Global Positioning System Radio Occultation) experiments on board the SAC-C and CHAMP satellites ([4], [5]). Particular attention was put on equatorial and midlatitude Andes regions. It was mainly observed, for short vertical wavelengths, a considerable wave activity in equatorial areas above Indonesia, Brazil and India, well correlated with outgoing longwave radiation, and systematic enhancements in the winter hemisphere. At long vertical wavelengths, a particular attention was given to signals above Central Andes, and the possibility to associate them to the presence of mountain waves (MWs) was suggested. Previously to these results, limb viewing satellite data from CRISTA (Cryogenic Infrared Spectrometers and Telescopes for the Atmosphere) were used to infer that topography was the source for waves observed in the stratosphere from space ([7], [16]). At the same time, [15] and [11], from MLS (Microwave Limb Sounder), concluded that the source giving rise to the strong radiance variances observed above the Andes Range should be MWs. Similar conclusion was suggested by [10] from GPS/MET RO data. In the present work, we discuss about possible origins of GW signatures observed above Central Andes, from CHAMP and SAC-C RO data.

2 GPS RO Data and Analysis

After subdividing both hemispheres in independent, adjacent cells of 5 per 10 degrees in longitude and latitude respectively, normalized temperature fluctuations $(T'/T_b)^2$ were calculated, where T' is the difference between the retrieved temperature T, splined with a 200 m step, and its lowpass filtered profile T_b. In the data set here considered and previously processed at JPL (Jet Propulsion Laboratory), bias in the temperature fluctuations greater than 3 K are frequently observed above and below 29 km and 8 km, respectively, and sometimes even within this range. As our main interest here is the upper troposphere and lower stratosphere regions, we considered the altitude interval 10 km to 27 km only. The T' cutoffs were chosen at 3 km and 9 km. The filter applied is non-recursive, and to avoid Gibbs effects, a Kaiser window was used.

We are interested in the region defined by 70°W–65°W and 30°S–40°S, situated in the Province of Mendoza, Argentina, at the eastern side of the highest Andes Mountains (> 7000 m). [5] detected in this specific region, and during few months, a signature of wave activity considerably intense respect to other non-equatorial regions in the Southern Hemisphere. From the almost 150 000 RO profiles globally retrieved between June 1, 2001 and March 31, 2003, a total of 89 events fell within this region. In Figure 1, the relative variance $(T'/T_b)^2$ of these events is shown as a function of height, chronologically ordered in the horizontal axis. Note the non uniform space and time distribution of occultations (Figure 2).

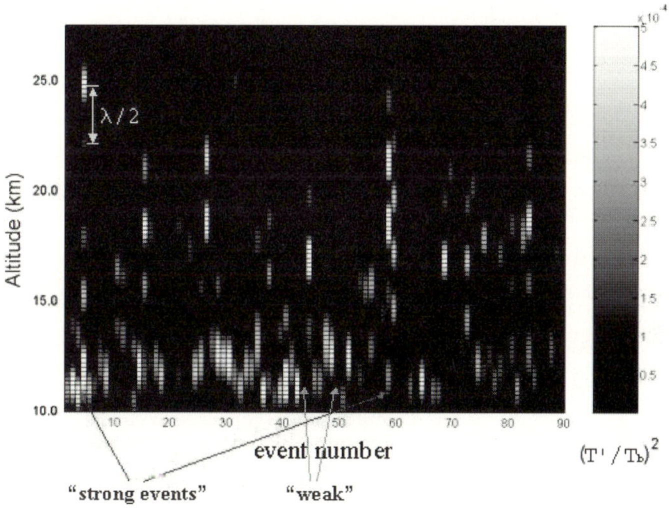

Fig. 1. Occultation events-height view of relative temperature variance, $(T'/T_b)^2$, in the region defined by 70°W–65°W and 30°S–40°S, between June 1, 2001 and March 31, 2003 (89 RO events).

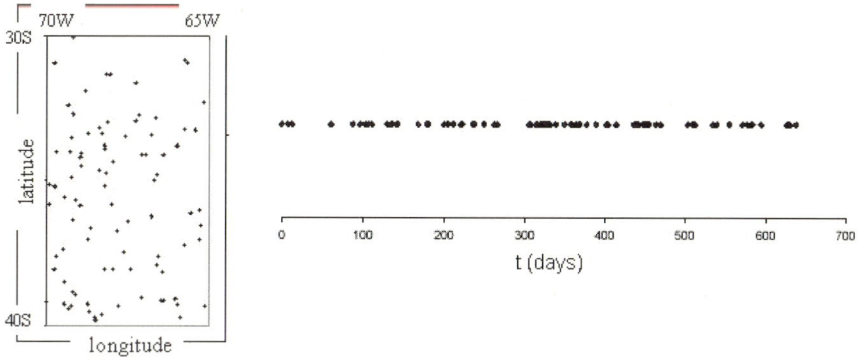

Fig. 2. Space (left) and time (right) distribution of the 89 RO occultations shown in Figure 1.

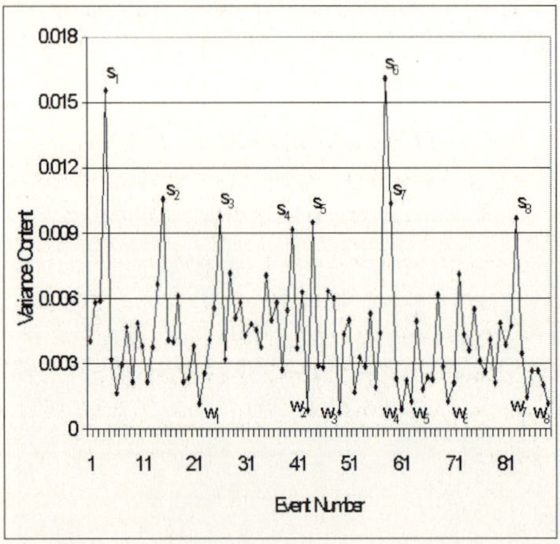

Fig. 3. Relative variance content $(T'/T_b)^2$ between 10 km and 27 km. The 8 strongest events are noted with a "s" letter and the 8 weakest events with a "w".

Each event in Figure 1 is represented by a single vertical column of pixels between 10 km and 27 km, and with a horizontal width of one pixel. The vertical extension of each pixel equals the spline step of 200 m chosen. Periodic enhancements of wave activity may be observed, and the distance between two consecutive maxima correspond to half a vertical wavelength. As 3D limb sounding paths lie, by definition, far from the vertical direction, the sloping effect of phase surfaces introduce a systematic discrepancy between "real" and "apparent" vertical wavelengths. As the duration of each occultation event is very short, this sloping effect will be more or less important, depending on the intrinsic frequencies of the wave structures observed [3]. In particular, this difference is expected to be more significative for MWs than for inertio-gravity waves. Some features can be remarked from the Figure 1: i) the relative intensity of wave activity shows "stronger" and "weaker" events; ii) a predominant apparent vertical wavelength, situated between 5 km and 6 km is clear; iii) minima around 15.0 km, 17.5 km and 20.0 km slowly oscillate in time. Similar plots belonging to adjacent cells (70°W–65°W, 40°S–50°S and 20°S–30°S, not shown here for brevity) exhibit a relative smaller wave activity than in Figure 1.

In Figure 3, the relative variance content $(T'/T_b)^2$ between 10 km and 27 km (the integral between both altitudes divided by 17 km) is shown.

Two particularly intense events emerge from the rest, and a total of 8 strong events, noted with a "s" letter may be differentiated. For comparison, we marked also the 8 weakest events (letter "w" in the figure). At first glance, one is tempted to state that the enhancements here observed correspond to MWs, taking into account the strong "potential" forcing at these latitudes, able to be attributed to the highest Andes Mountains. Let us consider the strongest case, observed on August 30, 2001. For steady forcing of a stratified fluid, such as flow over orography, the Scorer parameter L^2, is defined by $L^2 = (N(z)^2/U(z)^2) - U_{zz}/U$, where N and U are the buoyancy frequency and the wind in the direction of wave propagation, both depending on the altitude z (see e.g., [1]). For propagating waves, the relation $(2\pi/\lambda_v)^2 = L^2 - (2\pi/\lambda)^2 > 0$ must be verified, where λ_v and λ are the vertical and horizontal wavelengths. As the direction of possible generated MWs in the region is unknown, the minimum λ value required was estimated for all leewards directions between north and south every 15 degrees, using GPS RO temperature data and horizontal wind obtained from NCEP reanalyses. In Figure 4, we observe that free propagation was allowed for horizontal wavelengths longer than 25 km.

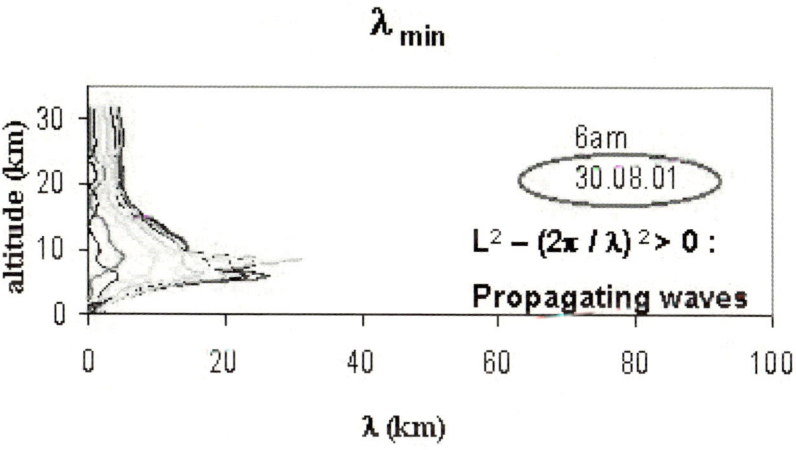

Fig. 4. Minimum horizontal wavelength allowing MWs to propagate to the upper troposphere, for all leewards directions between north and south every 15 degrees.

Indeed, this is not a severe restriction for typical MWs (see [8]). MWs would be free to propagate from the troposphere to the lower stratosphere, as from NCEP horizontal wind data no critical levels seem to be present (as well as no veering of mean wind with altitude). Nevertheless, we calculated a very low correlation between mean low level westerlies at the West of the Andes Range (at 925 hPa, 850 hPa and 700 hPa) and wave activity, this last quantified through the relative variance content of the 89 RO events shown in Figure 3.

We wonder about the capability of GPS–LEO RO observations to detect short horizontal wavelength MWs. The vertical resolution available from this technique depends on different factors: large vertical refractivity gradients, measurement precision, inversion scheme and diffraction limit [13]. We may agree that, if a sounding path was situated along the vertical direction, no restrictions would exist to detect usual large amplitude wavelike structures, mostly situated between 3 km and 15 km. According to this, currently observed large amplitude waves, with intrinsic frequencies between inertial and buoyancy limits could be, in principle, detected. Nevertheless, the available resolution due to integration effects along the line-of-sight (LOS) during the occultation event, is similar to the horizontal resolution, which is limited to around 150 km. This restriction prevents for the observation of short horizontal wavelengths, typically generated by mountains. In addition, depending on the angle defined between LOS and the wave vector, the wave signal detected and the resolution attained may even render worse [11]. If this is true, GPS RO vertical profiles might not include small gravity waves, and the wave activity calculated from the T variance used to described climatologies of GWs must be expected to be underestimated.

At this point, we remark that the region selected in the Province of Mendoza, represents a natural laboratory where the simultaneous competition between at least 4 different typical sources of GWs generation may be distinguished: i) topographic forcing, ii) geostrophic adjustment, iii) shear and convective generation. The relative weight of each of them is expected to follow a seasonal variability, and may respond to local atmospheric conditions, too. Possible selection mechanisms of wave propagation into the lower and middle atmosphere seem to be related to these sources.

In addition to the obvious association of the observed wave activity with the direct forcing of the Andes Range (accepting that possibly an important short-scale orographic contribution may remain invisible to RO GPS data), we recall that a westerly jet is currently situated above central Andes, centered around 30°S, and subject to an annual variability. A typical representation of it, is illustrated in Figure 5a, and its daily variability in the time interval here considered, is shown in Figure 5b.

The generation near to the jet of inertio-GWs by geostrophic adjustment, with longer horizontal and perhaps shorter vertical wavelengths than those expected from MWs, should be considered. These waves could be more easily detected from GPS-RO profiles than MWs. As it is known, inertio-GWs are the means by which mass and momentum are redistributed so as to ultimately achieve geostrophic balance from an initially unbalanced state ([2], [17]). These last authors found, from 13 case studies in Northern Hemisphere, that inertio-GWs waves are mainly situated in the exit region of the jet streak, and preferentially on its anticyclonic shear side (left side in Southern Hemisphere). In Figure 6, the position corresponding to the two "strongest" GPS RO events observed (August 30, 2001, 04:09 UTC, and September 9, 2002, 15:20 UTC) is shown, in zonal velocity 6-hours-composite charts from NCEP reanalyses, at 250 hPa. In both cases, the events were situated within the jet streak. In Figure 7, two typical "weak" events, are shown for comparison, both situated far from the jet streak, where almost no vertical shear is observed.

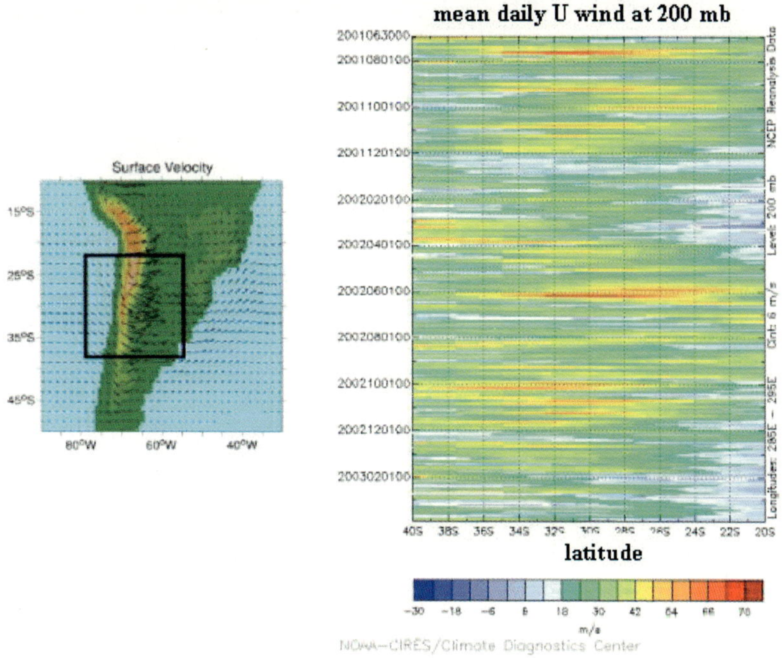

Fig. 5. a) (left) Typical qualitative representation of the westerly jet currently situated above central Andes, and centered around 30°S, and b) (right) its daily variability in the time interval here considered

A systematic correlation is observed between wave activity content for the 16 selected strong/weak events, and their respective close/far localization respect to the jet. This suggests about the significance of the jet in the generation of waves in this region.

In addition to the horizontal shear, a considerable vertical shear is associated with the jet too. However, most theoretical studies of shear instability wave generation coincide in the prediction of wavelengths much smaller than those observed in mesoscale wave events ([14], [8]). Accepting an important contribution of MWs to the GW spectrum in this region, from Figure 6 no MW critical levels serving as "generation levels", from which MW perturbations could extract energy from the mean flow and thus grow rapidly, are observed (zero mean flow in the direction of MW propagation). This mechanism of generation of large amplitude waves with medium horizontal scales, seems to be less possible here. We recall that even accepting that MWs are invisible to limb sounding observations, the presence of the jet provides dynamic conditions that are favorable for the upward propagation of MWs [18]. Finally, we know that wave generation by deep convection is characterized by its intermittence and correspondence with convective events. Usually there is

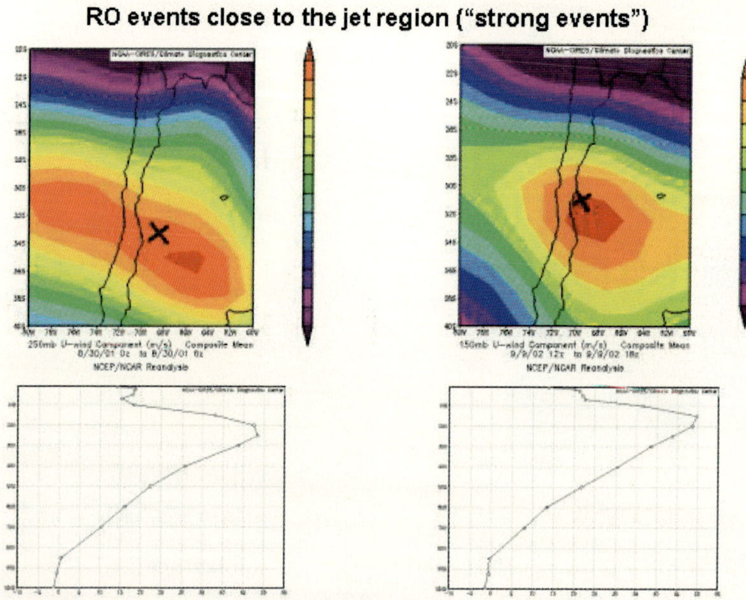

Fig. 6. Geographic position (cross) corresponding to the two "strongest" GPS RO events observed (August 30, 2001, 4:09 UTC, and September 9, 2002, 15:20 UTC), in zonal velocity 6-hours-composite charts from NCEP reanalyses, at 250 hPa.

not one prominent single phase speed, wavelength and frequency as in the case of MWs. Instead, a full range is usually observed [8]. In the two cases here shown, no deep convection was observed, as expected. It is known to be usually absent during August and early September. Typically, frequent storms (sometimes very intense) in the region are daily developed only during summer between late October and March [6]. Finally, we draw attention in Figure 8 to the long term modulation in the nodes observed in Figure 1, now plotted linearly in time.

RO low horizontal resolution prevents for the observation of horizontal wavelengths typically generated by mountains. In addition, a very low correlation between low level westerlies and wave activity is found. Inertio-GWs generated by geostrophic adjustment possess longer horizontal wavelengths, whereby RO strong events were found close to a nearly permanent jet at 30°S that may be producing them. We will further explore these ideas through mesoscale model simulations.

Acknowledgements. AT, PA and CGM are members of CONICET. This work was supported by UBA and CONICET grants X021 and PEI 6373. Figures 5b, 6 and 7 provided by web page of NOAA/CIRES CDC, Boulder, CO.

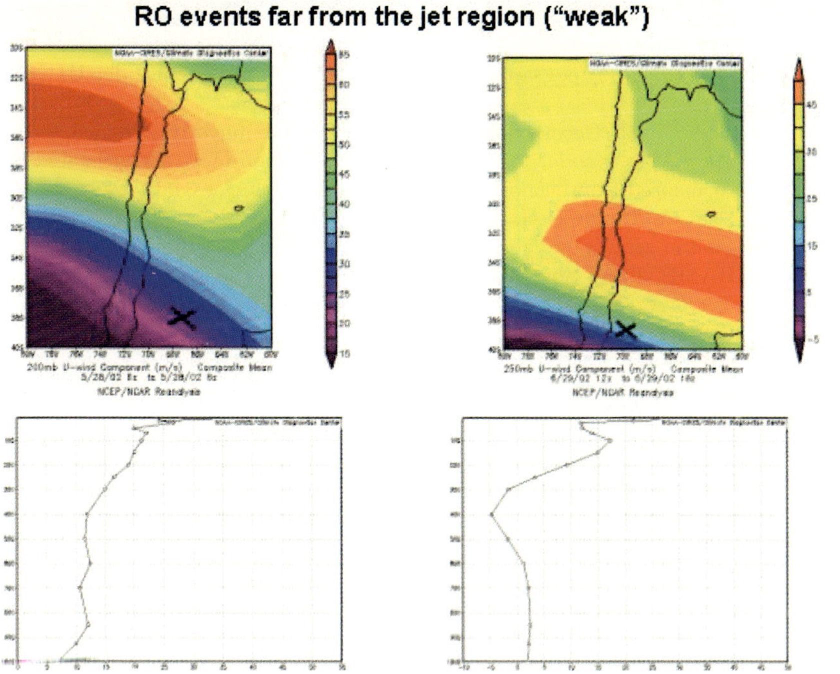

Fig. 7. For comparison with Figure 6, two typical “weak” events are shown.

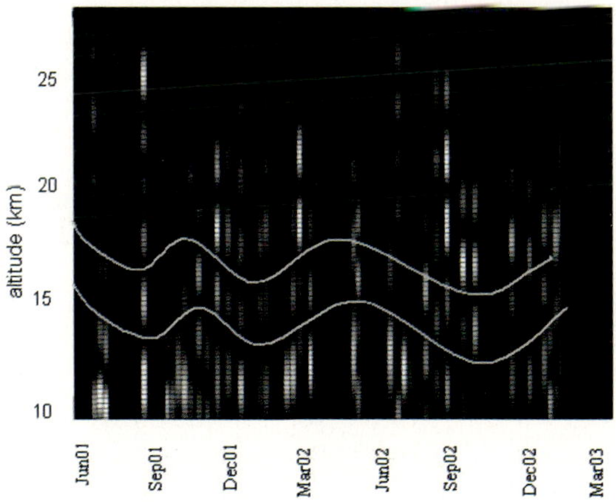

Fig. 8. Long term modulation in the nodes observed in Figure 1.

References

[1] Baines PG (1995) Topographic effects in stratified fluids. Cambridge University Press

[2] Blumen W (1972) Geostrophic adjustment. Rev Geophys 10(2):485–528

[3] de la Torre A, Alexander P (1995) The interpretation of wavelengths and periods as measured from atmospheric balloons. J Appl Met 34(12):2747–2754

[4] de la Torre A, Tsuda T, Hajj G, Wickert J (2004) A global distribution of the stratospheric gravity wave activity from GPS occultation profiles with SAC-C and CHAMP. J Met Soc Japan 82(1B):407–417

[5] de la Torre A, Tsuda T, Hocke K, Giraldez A (2004) Stratospheric gravity wave fluctuations and sporadic E at mid-latitudes with focus on possible effect of the Andes. In: Kirchengast G, Foelsche U, Steiner AK (eds) Occultations for Probing Atmosphere and Climate. Springer, Berlin Heidelberg New York, pp 353–364

[6] de la Torre A, Vincent D, Tailleaux R, Teitelbaum H (2004) A Deep convection event above the Tunuyn Valley near the Andes Mountains. Mon Weather Rev 132(9):2259–2268

[7] Eckermann SD, Preusse P (1999) Global measurements of stratospheric mountain waves from space. Science 286(5444):1534–1537

[8] Fritts DC, Alexander MJ (2003) Gravity wave dynamics and effects in the middle atmosphere. Rev Geophys 41(1):1003, doi:10.1029/2001RG000106

[9] Hajj GA et al. (2004) CHAMP and SAC-C atmospheric occultation results and intercomparisons. J Geophys Res 109:D06109, doi:10.1029/2003JD003909

[10] Hocke K, Tsuda T, de la Torre A (2002) A study of stratospheric GW fluctuations and sporadic E at midlatitudes with focus on possible orographic effect of Andes. J Geophys Res 107(D20):4428

[11] Jiang JH, Wu DL, Eckerman SD (2002) Upper atmosphere research satellite (UARS) MLS observation of mountain waves over the Andes. J Geophys Res 107(D20):8273, doi:10.1029/2002JD002091

[12] Kirchengast G (2004) Occultations for probing atmosphere and climate: Setting the scene. In: Kirchengast G, Foelsche U, Steiner AK (eds) Occultations for Probing Atmosphere and Climate. Springer, Berlin Heidelberg New York, pp 1–8

[13] Kursinski ER et al. (1997) Observing Earth's atmosphere with radio occultation measurement using the Global Positioning System. J Geophys Res 102(D19):23429–23465

[14] Lalas DP, Einaudi F (1976) On the characteristics of gravity waves generated by atmospheric shear layers. J Atmos Sci 33(7):1248–1259

[15] McLandress C, Alexander MJ, Wu DL (2000) Microwave Limb Sounder observations of gravity waves in the stratosphere: A climatology and interpretation. J Geophys Res 105(D9):11947–11967

[16] Preusse PA et al. (2002) Space-based measurements of stratospheric mountain waves by CRISTA 1. Sensitivity, analysis method, and a case study. J Geophys Res 107(D23):8178, doi:10.1029/2001JD000699

[17] Uccellini LW, Koch SE (1987) The synoptic settings and possible energy sources for mesoscale wave disturbances. Mon Weather Rev 115(3):721–729

[18] Whiteway JA, Duck TJ (1999) Enhanced arctic stratospheric gravity wave activity above a tropospheric jet. Geophys Res Lett 26(16):2453–2456

Analysis of Seasonal and Daily Mid-Latitude Tropopause Pressure Using GPS Radio Occultation Data and NCEP-NCAR Reanalyses

B. Bizzarri[1], I. Bordi[1], A. Dell'Aquila[2], M. Petitta[1,4], T. Schmidt[3], A. Sutera[1], and J. Wickert[3]

[1] Department of Physics, University of Rome "La Sapienza", P.le Aldo Moro 2 00185 Rome, Italy

[2] Ente per le Nuove tecnologie, l'Energia e l'Ambiente, ENEA, Casaccia, Via Anguillarese 301, Rome, Italy

[3] GeoForschungsZentrum Potsdam (GFZ), Department 1, Geodesy & Remote Sensing, Potsdam, Germany

[4] Corresponding author marcello.petitta@roma1.infn.it

Abstract. In the last decades the tropopause has received renewed attention for the analysis of atmospheric variability and, more in general, for climate studies. GPS (Global Positioning System) radio occultation measurements can provide accurate values of temperature and height of the tropopause, as recently shown in several papers, due to the high vertical resolution of this observing technique and the limited disturbance from water vapor, scarse in the tropopause region. The relatively high density of radio occultation measurements in the mid and high latitude regions enables to evaluate the horizontal scale of variability of the tropopause with sufficient accuracy. Data from GPS/MET (GPS Meteorology) and CHAMP (Challenging Minisatellite Payload) have been recently used to study the relationship between tropopause variability in tropical regions and convective structures. In this paper we focus on the analysis of variability of the tropopause structure in mid and high latitudes by comparing NCEP-NCAR (National Centers for Environmental Prediction – National Center for Atmospheric Research) reanalyses and radio occultation data from the CHAMP satellite. The comparison shows good agreement of the tropopause pressure fields both in total and seasonal means. The comparison between two single days shows that the current number of CHAMP measurements is insufficient to correctly retrieve the global tropopause field. In the next years, when more satellites will be equipped with GPS radio occultation payloads, the daily tropopause field will be better retrieved.

1 Introduction

An increasing amount of interest and resources are being dedicated by the scientific community to the study of climate and to the analysis of a possible global change of the climatic equilibrium. In this context, the tropopause has received increasing interest during the last three decades. A precise knowledge of the temporal and spatial structure of the tropopause, in particular in the extra-tropics, allows to evaluate the exchange of mass, water, and chemical constituents between the troposphere and the stratosphere [6].

Elementary models of climate change induced by external factors demonstrate the existence of a linkage between modifications of the planetary surface temperature and the tropopause height. In fact, the tropospheric lapse rate is fixed by radiative-convective processes. In the lower stratosphere the temperature depends on the radiative equilibrium and on the possible presence of absorbers. As shown by Manabe and Wetherald [10], in a radiative-convective equilibrium hypothesis a rise of the mean tropopause height corresponds to a rise of the mean surface temperature.

Recent studies ([5] and [14]) that used the tropopause pressure as an indicator of climate change were able to detect some trends on global and regional means, especially in the last two decades. These signals seem to be due to human-induced changes in ozone and well-mixed greenhouse gases, as simulated in GCM (General Circulation Model) runs by Santer et al. [14]. The detected trend is about 5 hPa/decade that roughly corresponds, in a standard atmosphere, to a rise of about 250 m in the mean tropopause height.

Several processes and mechanisms concur to determine and modify this atmospheric feature. For example, in the extra-tropics, different authors have investigated the effects of baroclinic activity on the tropopause height by means of conceptual models [9], GCM simulations [1] and global reanalyses [4]. In particular, Dell'Aquila [4] has estimated the mean effects of baroclinic disturbances on the mean observed pressure tropopause as about 10 hPa, that in a standard atmosphere corresponds to about 500 m.

It is noted that figures such as 250 m or 500 m at tropopause level constitute a challenge for current observing systems, both ground- and satellite-based, thus also for analyses and reanalyses. Current satellite operational sounders exploiting infrared or microwave cross-nadir scanning do not have sufficiently high vertical resolution whereas radiosondes, that could provide tropopause detection with the accuracy comparable with the magnitude of the variability, are too sparse.

GPS (Global Positioning System) radio occultation (RO) measurements of temperature and pressure have sufficiently high vertical resolution (e.g., 200 m) and may have good spatial density, depending on the number of properly equipped satellites. As shown by Bizzarri et al. [2], the vertical resolution data obtained by GPS RO measurements is highly relevant for atmospheric study, specifically for the detection of tropopause temperature and height.

The disturbing impact of water vapor is negligible in the tropopause region, due to scarse amount.

In recent papers some relevant studies on the tropical tropopause retrieved by RO data have been reported. These works analyzed and compared the tropopause field derived by RO with the one obtained by other sources. Randel et al. [11] have shown that the tropopause retrieved from data from the GPS/MET (GPS Meteorology) mission can be related to the convective structure in the tropics. Schmidt et al. [15] analyzed the tropical tropopause from May 2001 to November 2003 by using data from CHAMP (Challenging Minisatellite Payload). They showed that the temperature bias between the RO data and the reanalyses from ECMWF (European Centre for Medium-Range Weather Forecasts) is less than 0.5 K in the layer 300 hPa to 30 hPa. In the tropopause structure the authors have been able to recognize the most important features of the tropical atmosphere such as the annual cycle and spatial variability, and also some signature of the Quasi Biennial Oscillation.

Due to the interest for the tropopause height in mid and high latitude regions, and as a follow-on of previous studies, we focused on the 40°N to 90°N region. In this area, we have compared the seasonal structure of tropopause pressure obtained from CHAMP data in years 2001 – 2004 with the one obtained from the reanalyses by NCEP (National Centers for Environmental Prediction) and NCAR (National Center for Atmospheric Research). In addition, we analyzed daily data to explore the possibility of measuring high frequency variability of tropopause structures.

This paper is organized as follow: in Section 2 data and method used to evaluate tropopause height are presented; Section 3 shows the results obtained, and, finally, some conclusions are traced out in the final section.

2 Data and Method

We have used vertical temperature profiles derived in dry air conditions from GPS occultation measurements from the German-American CHAMP satellite [12] in the period May 2001 to July 2004. 160 000 soundings were available. The data were analyzed using the operational occultation processing software from GeoForschungsZentrum (GFZ), Potsdam, Germany. Details of the retrieval, results of validation studies and relevant references are given in [16]. The analysis software is subject of continuous improvement. In this paper the recent version (005) from GFZ [17] has been used.

For the comparison with global reanalyses we used the tropopause pressure by the NCEP-NCAR archive, that freely provides daily data on a 2.5° × 2.5° latitude-longitude grid [7].

Starting from the total number of globally distributed vertical profiles, we evaluated the tropopause height, pressure and temperature defined according to WMO (World Meteorological Organization) [18], that fixes the tropopause height as the lower boundary of the layer where the temperature lapse rate is less than 2 K/km for a depth of at least 2 km. The tropopause so defined is called *thermal tropopause*.

This definition is difficult to be applied in polar regions, where several stable stratified layers could be present; and also in the tropics, where the presence of a double tropopause (*hydro tropopause* and *thermal tropopause*) might be observed. In our study we preferred to remove these cases from the dataset; however only 1 % of the total number of profiles have been rejected.

Monthly and seasonal tropopause pressure fields were retrieved with excellent accuracy from the GPS RO in [11] and [15]. Moreover, Leroy [8] showed that the thermal atmospheric structure can be retrieved with sufficient accuracy by using weekly data.

Our aim is to evaluate the possibility to use RO data to describe tropopause variability at different time and spatial scales. To do that, we used daily sparse occultations for the Northern Hemisphere and interpolated the tropopause values to a grid of 1×1 degree. The accuracy of the resulting interpolation depends on the number of occultations available in each day. If the number of daily occultations is low, the reconstructed field does not fit the observed one because it is undersampled and some structures could be missed, in particular between 40 and 60 degrees latitude, where the wave activity is high and the wave patterns and the baroclinic disturbances influence the tropopause height. A higher number of measurements would be necessary to correctly sample the field in these conditions.

Due to the geometry of the CHAMP orbit, there is a different spatial density, for a given number of occultations, as a function of latitude. In tropical regions, where the number of daily occultations is low, the interpolation does not provide good results. In Fig. 1 the histogram shows the normalized total number of measurements provided by CHAMP, latitudinally weighed, function of latitude. Due to the spherical shape of the Earth and to the geometry of GPS and CHAMP satellite orbits, the highest density of occultations is between mid-latitudes and the pole. Our analysis is focused on detecting the main structures of the mid and high latitude tropopause field and, in this region (40°N to 90°N), the density of occultations is higher than in the tropics and the daily interpolation conveys more reliable results.

Another problem consists of a cold bias between the tropopause retrieved by CHAMP data and the one provided by NCEP-NCAR reanalyses. Comparing the total zonal mean (not shown) there is a significant cold bias between the two datasets. This bias was previously found by Schmidt et al. [15] in tropical regions by comparing CHAMP data and ECMWF reanalyses, and is probably due to the different algorithm used in the evaluation of the tropopause layer.

In this preliminary study we compared the RO measurements with the reanalyses using three different time periods: one covering the entire available CHAMP dataset, one covering the winter (December, January and February: DJF) and summer (June, July and August: JJA) seasons, and one covering 2 individual days (January 31, 2004 and July 7, 2001).

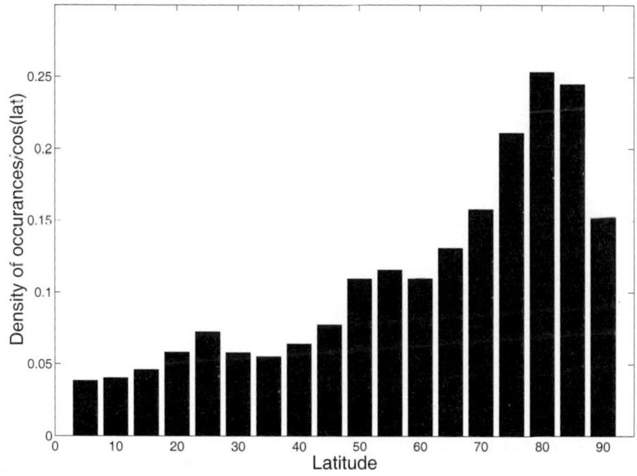

Fig. 1. Histogram of the density of the CHAMP occultations weighed for the area as a function of latitude. The maximum density of the occultations is between mid-latitude and polar regions.

3 Results

In Fig. 2 the mean tropopause pressure eddy fields evaluated for the entire period by CHAMP (Fig. 2a) and by the reanalysis (Fig. 2b) are shown. The difference is shown in Fig. 2c.

We calculated the correlation between the fields derived from CHAMP and the NCEP-NCAR reanalysis to have a measure of the agreement of the two datasets. Table 1 reports the values of the correlations for the total mean, the seasonal means and the individual daily means. Moreover we calculated the correlation between CHAMP and reanalysis data for each latitude to show the meridional variability of the two fields. The results are show in Fig. 3 for all the cases.

From the comparison of the mean fields for the entire period some remarkable features emerge:

- The correlation between the two maps (Fig. 2a and 2b) is 86 % and the latitudinal correlation shows very high values (between 0.8 and 0.9) for latitudes higher than 45 degrees (Fig. 3).
- The characteristic wavenumber 1-2 structure can be observed with the maxima located behind the main mountain chains and the minima in front of them. This pattern could be related to the stationary tropospheric orographic waves that, according to Bordi et al. [3], are troposphere-filling and modify the mean tropopause structure.
- The main maxima and minima are located in the same positions, but numerical values are quite different (as shown in Fig. 2C), probably due to the different vertical resolution of the datasets, to the algorithm for tropopause

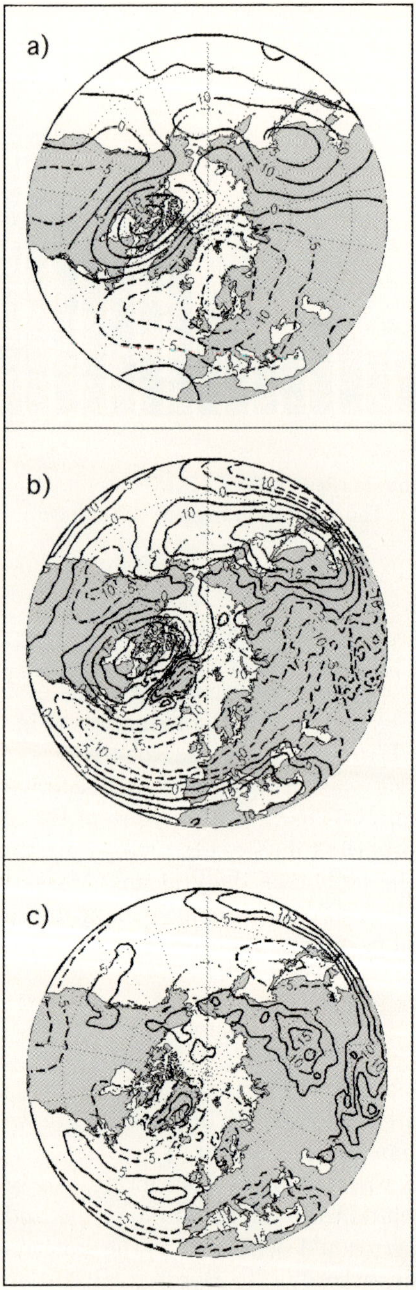

Fig. 2. Mean eddy field of tropopause pressure for the period from May 2001 to July 2004. a) CHAMP data. b) NCEP-NCAR reanalysis. c) Difference between CHAMP and NCEP/NCAR fields.

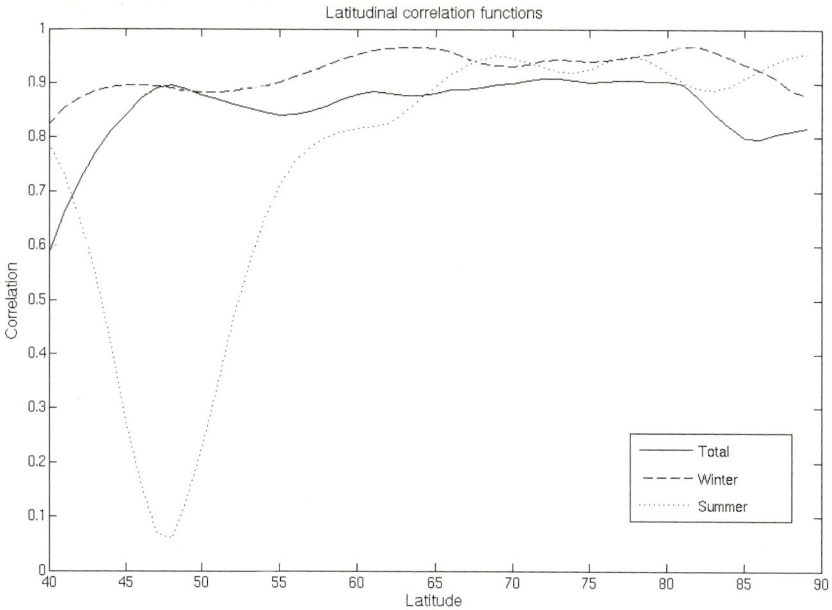

Fig. 3. Eddy tropopause pressure correlation between CHAMP and NCEP-NCAR for each latitude. Solid line: total period. Dashed line: DJF period. Dotted line: JJA period.

detection and to the problem related to the daily mean interpolation of CHAMP data.

- In the lower latitudes, near tropics, the density of CHAMP occultations is more sparse and the reconstructed field is less reliable.

Analyzing the winter seasons and the summer seasons (here not shown) some features can be outlined:

- In the winter case (DJF) there is a good accordance between the two fields showing a correlation of 89 % which is higher than in the total case. The features induced by the orography over the Pacific region, the Rocky Mountains and the Hudson bay have been correctly reconstructed by CHAMP occultations. Over Northern Europe and Asia a weaker agreement is observed in respect of the total case. The latitudinal correlation (Fig. 3 dashed line) is high for all the latitudes.
- For the summer period (JJA) the correlation between the two fields is weaker than in the other cases (64 %), and the meridional correlation in Fig. 3 (dotted line) shows a minimum between 45° and 50°. In the CHAMP field the dipole structure over the North American continent is captured but in other regions there are no significant similarities.

To understand the ability of RO data to capture detailed structures we have also compared the tropopause obtained by CHAMP RO with the reanalyses in two days when the number of RO is particularly high. We selected a winter and a summer day. In both cases the accordance is poor, missing small areas and features.

On January 31, 2004 (see Fig. 4) the correlation between the two datasets is very low (22 %); the interpolation of CHAMP data over the entire hemisphere does not provide a good estimation of the tropopause pressure. In particular, the maximum observed in the reanalysis dataset over the Atlantic Ocean between its center and the USA coast is not detected, because of the absence of measurements in that region. However some accordance can be observed over the polar regions and in the south of the USA.

Table 1. Correlation between the NCEP-NCAR and CHAMP tropopause pressure fields for different periods.

Total	86 %
Winter	89 %
Summer	64 %
January 31, 2004	22 %
July 7, 2001	12 %

For the summer day, July 7, 2001 (not shown), the eddy field is less enhanced and the main maxima observed by the reanalysis are only partially captured by RO data, as well as the feature over Iceland. In addition, the minimum near Greenland is estimated to be deeper than in the reanalysis data. In this case the correlation is 12 %.

From these two particular days some considerations can be outlined: as a matter of fact, the number of occultations during a day is not sufficient to accurately describe the fine structure of the tropopause. A better estimation of daily atmospheric patterns to reveal small scale structures requires higher number of occultations, i.e., more satellites. Alternatively, more accurate retrieval of the tropopause field can be obtained by averaging over more days (5 or 6).

4 Conclusions and Comments

The importance of the tropopause was emphasized by several works that associated its variability to various phenomena such as baroclinic adjustment, planetary waves, Quasi Biennal Oscillation, climate change. Therefore, accurate measurements of this atmospheric parameter are requested. The RO technique can lead to a significant improvement in tropopause studies. Our study investigated the capability of CHAMP data to measure fine structures of the tropopause field.

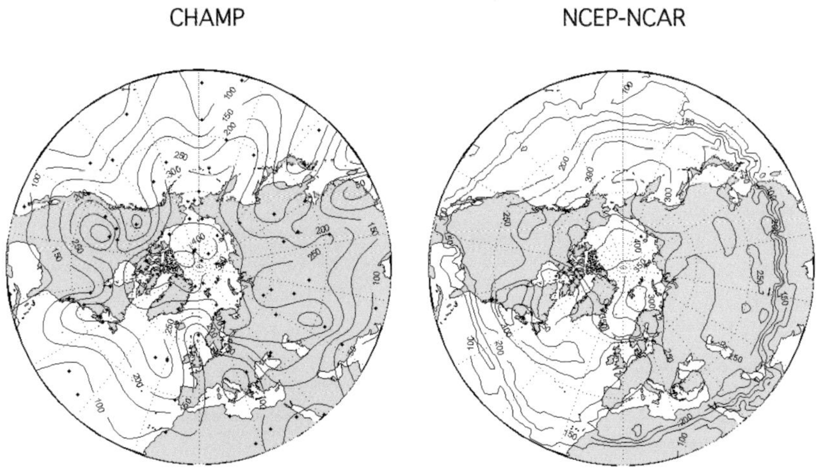

Fig. 4. Tropopause height for January 31, 2004. a) CHAMP data. b) NCEP-NCAR reanalysis.

160 000 CHAMP occultations from May 2001 to July 2004 were used. The thermal tropopause as defined by WMO [18] was evaluated. After tropopause detection, daily interpolation was performed to obtain a field with a resolution of 1×1 degrees. From these fields, time mean eddy fields on the entire period, on the winter and summer seasons, and on two single days, were computed and compared with the ones obtained from NCEP-NCAR reanalyses.

The main results can be summarized as follows:

- A cold bias between CHAMP and reanalyses is found as concerns the zonal mean of the tropopause pressure.
- The low number of measurements in the tropical region leads to inaccurate estimation of the tropopause height and to inconsistency in daily interpolation.
- The comparison between the RO and the reanalysis data for total and seasonal means shows good agreement and enable detection of main features, with high correlation coefficients between the two datasets. However finest structures cannot be observed with adequate detail.
- Higher correlation observed in winter can be related to higher variance of the field in this season, characterized by stronger disturbances and larger amplitudes waves. In summer, due to lower variability, denser sampling is necessary if small scale features are to be detected.
- Single-day analyses have poor correlation, particularly when sampling is poor as compared with the complexity of the meteorological situation.

- To avoid possible diurnal bias, means should be separately performed for four times each day; however, the number of available occultations was not sufficient for this purpose (interpolation would have become unreliable).

The follow-on of this study could consider:

- Evaluation of the minimum number of occultation events needed to obtain precise information on the fine structure of the tropopause field.
- Assessment of the impact of including RO data from SAC-C (Satlite Argentino Cientfico-C) and GRACE (Gravity Recovery And Climate Experiment) [17].
- Use ECMWF data for comparing with RO data.
- Use of RO data to attempt detecting baroclinic and synoptic wave patterns that modify the extratropical tropopause structure.

This preliminary study shows that RO data could improve the knowledge of the tropopause field and of the mechanisms that drive the variability of this atmospheric feature. In the next years, when additional data will be available from Metop and the COSMIC constellation [13], more detailed description of the tropopause field will be possible, enabling detection of its high frequency variability and fine spatial structures.

References

[1] Barry L, Craig GC, Thuburn J (2000) A GCM investigation into the Nature of baroclinic adjustment. J Atmos Sci 57(8):1141–1155

[2] Bizzarri B, Bordi I, Dell'Aquila A, Petitta M, Sutera A (2004) GPS radio occultation sounding to support General Circulation Models. Nuovo Cimento C27(1):59–71

[3] Bordi I, DellAquila A, Speranza A, Sutera A (2004) On the mid-latitude tropopause height and the orographic-baroclinic adjustment theory. Tellus 56A(4):278–286

[4] Dell'Aquila, A (2004) Midlatitude winter tropopause: the observed state and a theory of baroclinic adjustment. PhD Thesis, University of Genoa, Genoa

[5] Hoinka KP (1998) Statistics of the global tropopause pressure. Mon Wea Rev 126:3303–3325

[6] Holton JR, Haynes PH, McIntyre ME, Douglass AR, Rood RB, Pfister L (1995) Stratosphere-troposphere exchange. Rev Geophys 33:403–439

[7] Kalnay E, Kanamitsu M, Kistler R, Collins W, Deaven D, Gandin L, Iredell M, Saha S, White G, Woollen J, Zhu Y, Leetmaa A, Reynolds R, Chelliah M, Ebisuzaki W, Higgins W, Janowiak J, Mo KC, Ropelewski C, Wang J, Jenne R, Joseph D (1996) The NCEP/NCAR 40-Year Reanalysis Project. Bull Am Meteor Soc 77:437–471

[8] Leroy SS (1997) Measurement of geopotential heights by GPS radio occultation. J Geophys Res 102(D6):6971–6986

[9] Lindzen RS (1993) Baroclinic neutrality and the tropopause. J Atmos Sci 50(8):1148–1151

[10] Manabe S, Wetherald RT (1967) Thermal Equilibrium of the Atmosphere with a Given Distribution of Relative Humidity. J Atmos Sci 24(3):241–259.

[11] Randel WJ, Wu F, Rivera Rios W (2003) Thermal variability of the tropical tropopause region derived from GPS/MET observations. J Geophys Res 108(D1):4024–4031

[12] Reigber C, Schwintzer P, Lühr H, and Wickert J (eds) (2005) Earth Observation with CHAMP: Results from three years in orbit. Springer, Berlin

[13] Rocken C, Kuo Y-H, Schreiner W, Hunt D, Sokolovskiy S, McCormick C (2000) COSMIC System Description. Terrestrial, Atmospheric and Oceanic Science 11(1):21–52

[14] Santer BD, Sausen R, Wigley TML, Boyle JS, AchutaRao K, Doutriaux C, Hansen JE, Meehl GA, Roeckner E, Ruedy R, Schmidt G, Taylor KE (2003) Behavior of tropopause height and atmospheric temperature in models, reanalyses, and observations: Decadal changes. J Geophys Res 108(D1), doi:10.1029/2002JD002258

[15] Schmidt T, Wickert J, Beyerle G, Reigber C (2004) Tropical tropopause parameters derived from GPS radio occultation measurements with CHAMP. J Geophys Res 109:D13105, doi:10.1029/2004JD004566

[16] Wickert J, Schmidt T, Beyerle G, König R, Reigber C, Jakowski N (2004) The radio occultation experiment aboard CHAMP, Operational data analysis and validation of vertical atmospheric profiles. J Met Soc Japan 82(1B):381–395

[17] Wickert J, Beyerle G, König R, Heise S, Grunwaldt L, Michalak G, Reigber C, Schmidt T (2005) GPS radio occultation with CHAMP and GRACE: A first look at a new and promising satellite configuration for global atmospheric sounding. Ann Geophysicae 23:653–658.

[18] WMO (1957) Meteorology: A three dimensional science. WMO Bull 6:134–138

CHAMP Radio Occultation Detection of the Planetary Boundary Layer Top

A. von Engeln[1,2], J. Teixeira[3], J. Wickert[4], and S. A. Buehler[2]

[1] Satellite Applications, Met Office, Exeter, United Kingdom
engeln@uni-bremen.de
[2] Institute of Environmental Physics, University of Bremen, Germany
[3] UCAR/VSP at Marine Meteorology Division, Naval Research Laboratory, Monterey, CA, USA
[4] GeoForschungsZentrum Potsdam (GFZ), Department 1, Geodesy & Remote Sensing, Potsdam, Germany

Abstract. A simple approach to derive the Planetary Boundary Layer (PBL) top altitude from the CHAllenging Minisatellite Payload (CHAMP) radio occultation data is presented. The lowest sampled altitude is assumed to be determined by the top of the PBL. We average CHAMP measurements for the years 2002 and 2003 over 5 degree latitude longitude boxes and compare them to European Centre for Medium-Range Weather Forecasts (ECMWF) data. The ECMWF PBL top was calculated from the relative humidity gradient with respect to altitude. The altitude where this gradient has its minimum and temperatures are above 273 K is assumed to represent the PBL top. Agreement between the two datasets is good in terms of mean PBL height, especially over sea. The CHAMP data show the major features of PBL height with a realistic transition from stratocumulus regions to shallow and deep cumulus areas. CHAMP also shows a substantial amount of PBL height variability that might proof useful to study PBL dynamics.

1 Introduction

Since the 1850s when C. Piazzi-Smyth found the trade-wind inversion by measuring the temperature while climbing the peaks of Tenerife of the Canary Islands, the trade-wind inversion has been at the center of many research efforts in tropical and sub-tropical meteorology. Several campaigns followed these initial measurements. The work of von Ficker (1936) has helped to establish the trade-wind inversion as an ubiquitous part of the Hadley circulation and the tropical climate.

Recently scientist have become aware of just how important the dynamics of PBL inversions is to the overall climate system, e.g., Ma et al. (1996); Philander et al. (1996). The sub-tropics are the regions of the globe where persistent sheets of Stratocumulus (Sc) clouds cool the planet by reflecting a substantial portion of the downwelling shortwave radiation. The regions of the world where the cloud radiative forcing is largest are the ones associated with stratocumulus off the west coasts of continents. The amount of clouds in the sub-tropics is highly correlated with the PBL depth (the height of the PBL inversion). Shallower boundary layers are associated with larger cloud cover due to a moister PBL. Deeper boundary layers are associated with trade-wind cumulus and smaller values of cloud cover.

How the transition from shallower PBLs with Sc to deeper PBLs with cumulus happens, and the physical mechanisms involved is still a matter of research, although significant progress has been achieved during the last twenty years. Much of this progress has been achieved through a successful combination of theory and intensive observational campaigns in the sub-tropics (ASTEX, FIRE, DYCOMS), see e.g., Ackerman et al. (2004). But unfortunately, these experiments are always localized in space and time and do not provide a realistic global picture.

Satellite remote sensing in the visible and infrared has been relatively successful in measuring variables associated with clouds. However, direct measurements of the PBL inversion characteristics have been difficult to achieve mainly due to limited vertical resolution. Although recently some preliminary efforts have been made using AIRS data to study PBL inversions (Fetzer et al. 2004).

Radio occultation (RO) offers a promising alternative for global PBL inversion measurements. Probing of the PBL using RO instruments has already been discussed in e.g., Hajj et al. (2004) based on the signal reappearance after entering the PBL. In this study we use the lowest altitude sampled by CHAMP (Wickert et al. 2004) to derive the PBL height. We analyze these results by comparing them to values of the PBL height obtained from ECMWF data. A similar version of this article has also been published with Geophysical Research Letters (von Engeln et al. 2005).

2 Data Processing

The used ECMWF global analysis cover the years 2002/2003. They have a 1.5° resolution and 60 vertical levels (about 18 levels between 0 km and 3 km). ECMWF fields show generally good agreement with the PBL altitude as seen in radiosonde data (von Engeln et al. 2003; von Engeln and Teixeira 2004). The top of the PBL in the ECMWF dataset was derived by finding at each gridpoint the altitude where the decrease of relative humidity with height is largest, under the constraint that the temperature is above 273 K. This approach is a typical one for the determination of the sub-tropical PBL top (Beljaars and Betts 1992).

The used CHAMP data was processed with the Full Spectrum Inversion (FSI) technique (Jensen et al. 2003) at the GeoForschungsZentrum Potsdam,

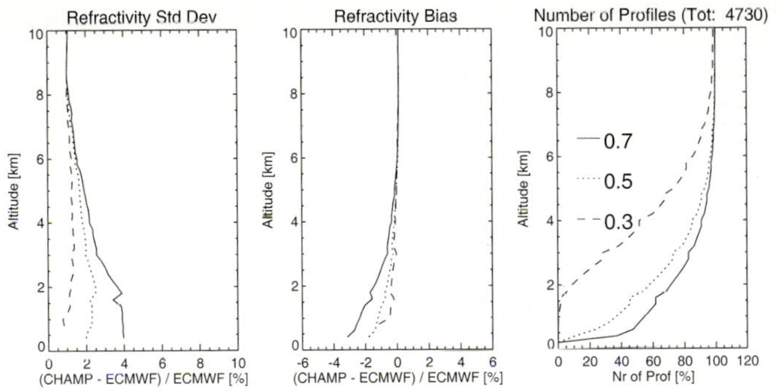

Fig. 1. Standard deviation (left) and bias (middle) of CHAMP profiles processed with different FSI threshold settings. Occultations are compared to co-located ECMWF profiles. Number of profiles is shown on the right.

Germany. Processing is stopped when the FSI amplitude of an occultation is reduced by 50 %. This threshold has been derived by a quality check (bias and standard deviation with respect to other available data and simulations) of CHAMP data and thus independently from the PBL top altitude. Here we use off-line produced CHAMP data, for the operational data stream several quality checks are additionally implemented leading to slightly higher altitudes for the cutoff.

The global impact of the FSI threshold setting on the data quality is shown for a subset of profiles in Figure 1. Threshold settings corresponding to an amplitude reduction by 30 % (factor 0.3) and 70 % (factor 0.7) are compared to the operational setting of 0.5. Refractivity standard deviation and bias are calculated with respect to the above mentioned ECMWF data at the mean location of the observation. The more conservative the threshold, the better the data quality but the lower the penetration into the lower atmospheric layers. A factor of 0.5 still allows a significant proportion of profiles entering the lowest 2 km.

A further comparison of the PBL top altitude and the 0.5 FSI amplitude threshold settings based on simulations in shown in von Engeln et al. (2005). Note that our PBL detection is independent on the actual quality of the refractivity profiles since it uses the FSI amplitude. It only requires that the measurement penetrates down to the PBL top.

In total there are 109 641 occultations available for the years 2002/03, out of which we removed about 20 000 that terminated above 3 km. This step was introduced since the PBL is usually below 3 km and the ECMWF resolution is only able to capture PBL inversions up to about 2.5 km (von Engeln and Teixeira 2004). In total, 87 598 occultations were left (38 879 over land). CHAMP does measure also within the PBL, although here we assume that the majority of the measurements is terminated by the PBL top. This is caused by the refractivity gradient present at the PBL, where the decreasing humidity

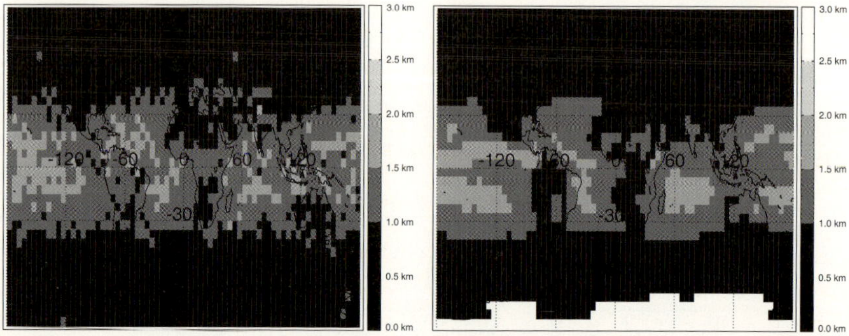

Fig. 2. Left: Mean minimum altitude found in CHAMP data. Right: Mean altitude of minimum relative humidity gradient calculated from ECMWF data. Data is averaged over a 5° latitude longitude grid. White areas indicate that temperatures were never above 273 K.

and possible clouds along with radiative and evaporative cooling contribute to the gradient. Although it is in general possible to use this changing refractivity gradient to identify the PBL top in ECMWF data, the use of a relative humidity change proves more robust and was thus applied here.

Note that the presented results depend on the tracking algorithm used onboard the receiver. Software updates to the tracking algorithm will modify the relation of the PBL top to the lowest altitude sampled by CHAMP, in particular the open loop implementation will probe more frequently into the PBL (Sokolovskiy 2001).

3 Results

Figure 2 shows the mean CHAMP and ECMWF PBL altitude above the Earth surface. On average, about 33 measurements enter each grid box. Both plots show similar features, especially over the Ocean. Mean altitudes around 0 km are found for polar latitudes. Also visible is the gradual increase in PBL height when moving from polar latitudes toward the equator.

In particular over the sub-tropics and tropics, the transition from a shallow stratocumulus PBL close to the west coasts, to a deeper cumulus PBL (Duynkerke et al. 1999; Stevens et al. 2001; Siebesma et al. 2003), is visible in both datasets. Although it is clear that in the Sc regions ECMWF gives values that are consistently lower. In the deep tropics the two datasets diverge in some regions: 1.) In the Eastern Pacific, around 10°N, the ECMWF data shows a peak value for the PBL height that presumably corresponds to the inter-tropical convergence zone (ITCZ); 2.) In the Western Pacific, a maximum is apparent in CHAMP at around 10°N–15°N.

Although there are differences between CHAMP and ECMWF, the subtropical PBL evolution is captured relatively well by the CHAMP data. For

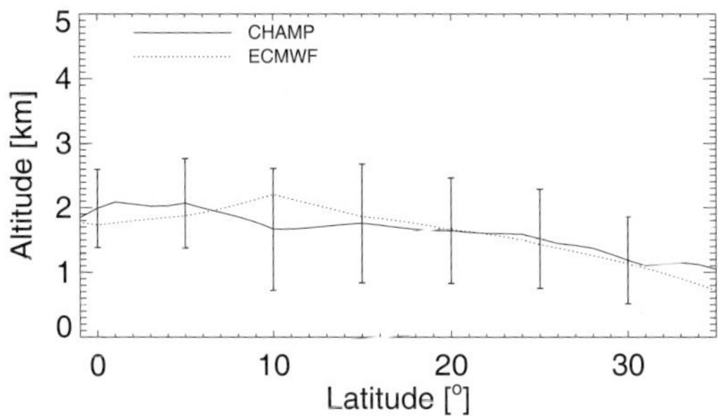

Fig. 3. Latitude altitude slice along 1°S, 173°W to 35°N, 125°W showing the mean PBL altitude from CHAMP data over a 5° latitude longitude grid. Vertical lines indicate the error bars. Also shown is the altitude where the minimum gradient of relative humidity was found in ECMWF data.

example the transition in PBL height from the coast of California to Hawaii and the equator is relatively similar in both datasets. In order to investigate such an evolution in some more detail, Figure 3 shows the mean PBL height along a cross-section from 1°S, 173°W to 35°N, 125°W. This particular cross-section was chosen since it is able to represent the climatology of a typical transition from stratocumulus, to cumulus and then to deep convection, and is has been used before in intercomparison studies of the parameterizations in atmospheric models (Siebesma et al. 2004). An increase in PBL height in both datasets from about 1 km at 30°N, to about 2 km south of 15°N is found. Also, the flatness of the curves in the tropics at about 2 km is found in both datasets. Close to the coast the two datasets diverge a little, possibly due to topography issues. Also shown in Figure 3 is the relatively large value of standard deviation (around 1 km) of the CHAMP dataset.

In order to get some more insight in the PBL variability, Figure 4 shows the histograms (occurrence in absolute numbers of observations) of PBL height for three boxes along the previous cross-section for altitudes up to 4 km.

In area A, closer to the Sc regions, we find two peaks in the histogram: one in the first 400 m above the surface (surface, low stratus and stratocumulus) and another one between 1.6 km and 2 km, with most of the observations in this range from the surface to 2 km. In area B, there is one dominant peak just above 2 km, but also several occultations terminating near the surface. The two-peak nature of the histograms in this stratocumulus and cumulus regions seem to point to a situation where two dominant regimes alternate: 1.) a low stratocumulus regime or a weak inversion with occultations reaching the surface; 2.) a fully developed trade cumulus regime. In area C, there is only one flat peak between about 2.5 km and 3.5 km presumably associated with deep moist convection.

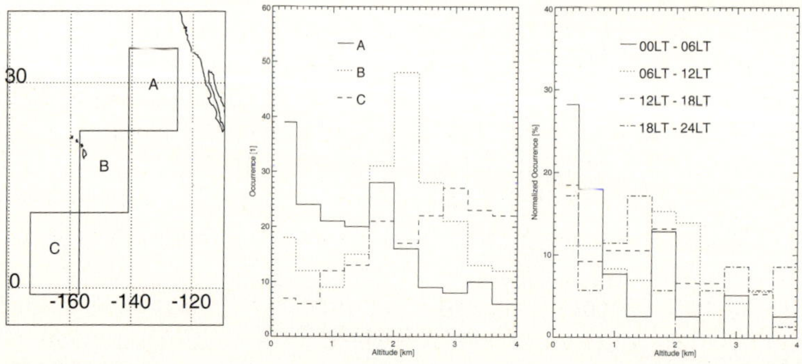

Fig. 4. Left: Map of Pacific Ocean and California showing 3 selected areas. Middle: Histogram of occultations reaching the given altitude in the selected area. Right: Histogram of normalized occurrence with respect to local time interval in area A. Bin size 0.4 km.

Also shown in Figure 4 is the normalized occurrence for four specific local time intervals in area A. The normalization is performed with respect to the local time interval since the coverage is not uniform (moving from 00LT to 24LT through the intervals there are 39, 72, 76, 35 occultations) and thus one should be cautious when interpreting the data. Although some features can identified: 1) At any time of the day there are always two peaks, one close to the surface and the other somewhere between 1 km and 2 km; 2) During local night and early morning (00LT to 06LT), when stratus and stratocumulus are deeper and more frequent due to a strong cloud top longwave cooling driving convection, there are more observations closer to the surface than close to 2 km; 3) During local morning and noon (06LT to 12LT) on the other hand, the most populated peak is closer to 2 km, with values close to the surface about one-third smaller than during the previous period. These features agree with what would be expected from previous studies. In fact, if the quasi-bimodal structure that is found in area A is actually reflecting the prominence of two specific regimes of the boundary layer, namely a low stratus/stratocumulus regime and cumulus regime, then the diurnal cycle results may well reflect the higher frequency of occurrence of stratus during local night and cumulus during local daytime.

4 Conclusion

We use FSI processed CHAMP RO data for the years 2002 and 2003 to estimate the PBL height. We use a simple approach by assuming that the lowest altitude sampled by the CHAMP instrument is representative of the PBL height. This altitude is compared to ECMWF analysis data for the same period. The top of the PBL in ECMWF data is calculated as the altitude where the gradient of relative humidity with respect to height shows its minimum. Maps of mean PBL height show good agreement between CHAMP and ECMWF, in particular over the Ocean. The transition in mean PBL height from the stratocumulus regions to the cumulus and deep cumulus areas is well captured by the CHAMP data. Moreover, the CHAMP data exhibits a fair amount of variability in PBL height. Although this variability looks reasonable it is still unclear how much of it is actually representing the dynamics of the PBL height.

This and other issues will need to be investigated further in order to precisely define the usefulness of the CHAMP dataset as an instrument to understand the dynamics of the PBL height in a global perspective. Future work will also focus on more sophisticated algorithms to derive the PBL height from RO data and on how to use the amplitude to derive further information such as inversion depth. It is anticipated that this data could be assimilated into weather prediction models.

Acknowledgements. A. von Engeln was partly funded by the German Federal Ministry of Education and Research (BMBF), within the AFO2000 project UTH-MOS (Grant 07ATC04), and during his visit at the Naval Research Laboratory, Monterey, USA by the Visitor Support Program of the Office of Naval Research International Field Office in London (Grant Number: N00014-04-1-4020). J. Teixeira acknowledges the support of the Office of Naval Research under Program Element 062345N. The authors would like to thank G. Beyerle (GeoForschungsZentrum Potsdam, Germany) for software development and data processing.

References

Ackerman A, Kirkpatrick M, Stevens D, Toon O (2004) The impact of humidity above stratiform clouds on indirect aerosol climate forcing. Nature 432:1014–1017, DOI 10.1038/nature03174

Beljaars A, Betts A (1992) Validation of the boundary layer representation in the ECMWF model. ECMWF Seminar Proc. Validation of models over Europe, ECMWF, Reading, United Kingdom

Duynkerke P, Jonker P, Chlond A, Zanten MV, Cuxart J, Clark P, Sanchez E, Martin G, Lenderink G, Teixeira J (1999) Intercomparison of three- and one-dimensional model simulations and aircraft observations of stratocumulus. Bound-Layer Meteor 92(3):453–487

von Engeln A, Teixeira J (2004) A ducting climatology derived from ECMWF global analysis fields. J Geophys Res 109(D18):D18104, DOI 10.1029/2003JD004380

von Engeln A, Nedoluha G, Teixeira J (2003) An analysis of the frequency and distribution of ducting events in simulated radio occultation measurements based on ECMWF fields. J Geophys Res 108(D21):4669, DOI 10.1029/2002JD003170

von Engeln A, Teixeira J, Wickert J, Buehler SA (2005) Using champ radio occultation data to determine the top altitude of the planetary boundary layer. Geophys Res Lett 32(6):L06815, DOI 10.1029/2004GL022168

Fetzer E, Teixeira J, Olsen E, Fishbein E (2004) Satellite remote sounding of atmospheric boundary layer temperature inversions over the subtropical eastern pacific. Geophys Res Lett 31(17):L17102, DOI 10.1029/2004GL020174

von Ficker H (1936) Die Passatinversion. Veröffentlichungen Meteor. Institut, Universität Berlin, 33 pp.

Hajj G, Ao C, Iijima B, Kuang D, Kursinski E, Mannucci A, Meehan T, Romans L, de la Torre Juarez M, Yunck T (2004) CHAMP and SAC-C atmospheric occultation results and intercomparisons. J Geophys Res 109:D06109, DOI 10.1029/2003JD003909

Jensen A, Lohmann M, Benzon H, Nielsen A (2003) Full spectrum inversion of radio occultation signals. Radio Sci 38(3):1040, DOI 10.1029/2002RS002763

Ma CC, Mechoso C, Robertson A, Arakawa A (1996) Peruvian stratus clouds and the tropical pacific circulation: a coupled ocean-atmosphere GCM study. J Climate 9(7):1635–1645

Philander S, Gu D, Halpern D, Lambert G, Lau NC, Li T, Pacanowski R (1996) Why the ITCZ is mostly north of the equator. J Climate 9(12):2958–2972

Siebesma A, et al. (2003) A large eddy simulation intercomparison study of shallow cumulus convection. J Atmos Sci 60(10):1201–1219

Siebesma A, et al. (2004) Cloud representation in general circulation models over the northern pacific ocean: A EUROCS intercomparison study. Quart J Roy Meteor Soc 130(604):3245–3267

Sokolovskiy S (2001) Tracking tropospheric radio occultation signals from low Earth orbit. Radio Sci 36(3):483–498

Stevens B, Ackerman A, Albrecht B, Brown A, Chlond A, Cuxart J, Duynkerke P, Lewellen D, Macvean M, Neggers R, Sanchez E, Siebesma A, Stevens D (2001) Simulations of trade wind cumuli under a strong inversion. J Atmos Sci 58(14):1870–1891

Wickert J, Schmidt T, Beyerle G, König R, Reigber C, Jakowski N (2004) The radio occultation experiment aboard CHAMP: Operational data analysis and validation of vertical atmospheric profiles. J Meteorol Soc Jpn 82(1B):381–395

Use of GNSS Occultation Data for Climate Monitoring and Climate Change Studies

Monitoring Climate Variability and Change by Means of GNSS Data

M. Stendel

Danish Meteorological Institute, Copenhagen, Denmark
mas@dmi.dk

Abstract. Although temperature and water vapor are key variables for the detection of climate change, the direct observation of these quantities in the free atmosphere is difficult using conventional techniques like radiosondes or satellites. The main problems are due to inhomogeneities, instrumental changes, sensor drift and lack of global coverage. More reliable information about climate variability and change can be gained from GNSS (Global Navigation Satellite System)-based vertical profiles of temperature and humidity. GNSS observations are particularly well suited for climate monitoring, since they provide high precision and vertical resolution data under all weather conditions and with time and space independent sampling characteristics, cover the whole globe, have virtually no instrumental bias and are intrinsically self-calibrating.

1 Introduction

The variations of temperature and water vapor within the atmosphere are key parameters for the determination of climate change. Water vapor is the dominating greenhouse gas, and its role in the hydrological cycle of the atmosphere is crucial for understanding greenhouse warming, since models generally show a rapid increase in water vapor with a warming climate (Fig. 1), as a consequence of the Clausius-Clapeyron relation. Moreover, models that incorporate the effects of anthropogenic activities predict a warming in the troposphere and a cooling in the lower stratosphere. At present, the causes and magnitudes of these effects are not understood well enough, especially the role of water vapor.

Despite considerable effort these basic results have proven difficult to confirm from observations. The reason is simply the lack of suitable observing techniques. For many decades radiosondes have been the only device to obtain information about the free atmosphere. Since the late 1970s these observations have been augmented by measurements from satellites, using mainly passive sensors. However, these observing systems have limitations, and estimated temperature trends in the free atmosphere differ significantly even when using the best available observations for the last few decades, while little is known about trends in water vapor. The main reasons for the difficulty in determining trends for temperature and water vapor are related to inhomogeneities in observing systems, changes of in-

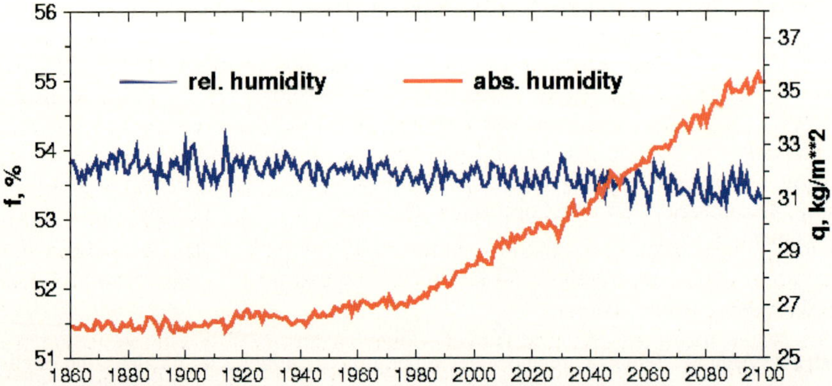

Fig. 1. Annual mean global values of relative humidity f (in %) vertically averaged for 850 hPa to 300 hPa and vertically integrated absolute humidity q (in kg/m^2). From Semenov and Bengtsson (2002).

struments and to instrumental drift problems, in particular in the satellite observations.

2 Conventional Observations of Temperature and Water Vapor

Tropospheric temperature is generally much better observed than water vapor. However, there are still surprisingly large discrepancies between estimates based on different data sets. An example is the recent and rather strong difference between deep-layer average lower and middle tropospheric temperature (TLT and TMT, respectively) trends found between data from microwave sounding units (MSU) on board NOAA polar orbiting satellites and from radiosondes (Christy et al. 2001; Vinnikov and Grody 2003). From the same satellite data set, Christy et al. (2003) found slightly negative trends in TLT and TMT, while other researchers (Prabhakara et al. 2000; Fu et al. 2004) concluded that these temperatures had risen almost as much as the surface temperature (Folland et al. 2001). The reason for this ongoing controversy is related to the inherent drift of the MSU and AMSU instruments (which were never designed for climate monitoring purposes), changing of satellites without sufficiently long overlap periods (Hurrell and Trenberth 1998), orbit variations ("falling satellite effect", Wentz and Schabel 1998) and its correction (Mears and Wentz 2005) as well as diurnal sampling problems.

Climate models are able to reproduce the vertical structure of temperature trends for the last decades (Santer et al. 2005), while observational data sets do not. This may partly be caused by incomplete spatial coverage of radiosonde data (only the Northern Hemisphere extratropics and a few regions on the Southern Hemisphere are adequately covered), but lower tropospheric and lower stratospheric temperatures obtained from the MSU/AMSU instruments, which have so far been the only observational time series with a global coverage, also fail to show recent temperature increase. The situation becomes even more complicated if one compares observed trends from MSU and radiosondes to those obtained from recent reanalysis projects (NCEP/NCAR: Kalnay et al. 1996; ERA40: ECMWF 2002): NCEP/NCAR and radiosonde data show a strong stratospheric cooling after 1991, whereas ERA40 and MSU/AMSU data do not. Sherwood et al. (2005) demonstrated that radiosondes suffer from temporally inhomogeneous radiation shielding, while Mears and Wentz (2005) discuss MSU problems introduced by the correction of orbital variations. In summary, it has so far been impossible to determine trends in atmospheric temperature from observational data sets with the required accuracy.

For water vapor, the situation is even worse, as current observing systems have severe limitations with respect to measurement accuracy and sampling in space and time, and are therefore not good enough to show whether systematic changes exist. Radiosonde measurements suffer from numerous instrumental changes (Gaffen et al. 2000), poor accuracy at low temperatures and insufficient global coverage and therefore cannot be used to assess the quality of global scale trends derived from satellite data. Furthermore, due to the high operational cost, the radiosonde network is gradually reduced, especially in remote regions. Satellite-based passive sensors, such as SSM/I and HIRS channels 11 and 12, have biases that cannot easily be determined, making them less useful for climate monitoring purposes (Bengtsson et al. 2004a). Attempts to inter-calibrate from one satellite to the next have been hampered by insufficient length of overlapping periods and temporal drifts in both orbit (Wentz and Schabel 1998) and instrument sensitivity. Only microwave sounders have been thoroughly calibrated (Christy et al. 2003), although many problems remain (see e.g., Hurrell and Trenberth 1998). Furthermore, MSU data has only a poor vertical resolution (7 km to 10 km for individual spectral channels and 2 km to 3 km for combined products such as TLT). Future sounders (e.g., SEVIRI on Meteosat's second generation satellites) are intended to increase the global coverage and temporal sampling rate rather than the vertical resolution.

Results from reanalyses cannot be used as well, as has recently been demonstrated by Bengtsson et al. (2004b), due to model and data-assimilation biases in determining the hydrological cycle from reanalysed data sets.

3 Water Vapor Retrieval by Surface-Based GNSS Observations

Ground-based observations can be used to measure vertically integrated water vapor (IWV) in the atmosphere. Microwave signals broadcasted by the GNSS satellites are delayed by the ionosphere and the neutral atmosphere on their way to the ground receiver. Correcting for the dispersive ionospheric contribution, we obtain the nondispersive contribution from the electric neutral atmosphere which is mainly induced by dry air, water vapor, clouds and rain, and is proportional to the masses of the specific components along the ray path. The delay is smallest in the direction of the zenith and increases approximately with the reciprocal of sine of the elevation angle.

The elevation dependence can be described by a mapping function where the total atmospheric delay is partitioned into the sum of the "wet" (contribution of water vapor; the effect of clouds and rain is normally marginal (Solheim et al. 1998)) and the "dry" delay (hydrostatic component). The latter is proportional to the air masses and amounts to about 2.3 m for measurements in the zenith. The delay due to water vapor is considerably more variable and ranges from a few mm in arid regions to more than 400 mm in humid regions. Ground-based GNSS receivers use the elevation dependence of the signal arrival time which is introduced by the refractive index in the atmosphere, to estimate an equivalent zenith total delay (ZTD). Then, from the ZTD, precipitable water vapor can be derived by means of measured surface pressure and temperature and the hydrostatic assumption. An uncertainty in the ground pressure of 1 hPa corresponds to an uncertainty of 2.3 mm in the ZHD, which will introduce an uncertainty of 0.4 mm in the precipitable water. Even though it seems rather trivial to obtain a 1 hPa accuracy for pressure measurements, this has caused problems since few of the GNSS stations have been equipped with barometers so far.

Recent studies by Gradinarsky et al. (2002), monitoring integrated water vapor for a seven year period over northern Europe, and by Hagemann et al. (2003), developing a global monitoring concept using ZTD measurements in combination with operational analyses from ECMWF, have proven very promising. In particular, it appears feasible from the Hagemann et al. study to combine reanalyses of water vapor with in-situ GPS measurements to arrive at consistently checked comprehensive global fields. At the same time the observed trend from the directly measured ZTD provide fully independent data. Therefore, the reanalysis data sets can be used to identify gross systematic errors of the ZTD measurements and thereby exercise a continuous control of these data.

Ground-based RO observations can also be used to obtain information about recent trends in integrated water vapor, another climate change key observable. Figure 2 shows IWV trends from radiosondes and GNSS as well as from two independent techniques (Very Long Baseline Interferometry, VLBI, and microwave radiometry) obtained at the Onsala Space Observatory in Sweden.

4 Temperature and Water Vapor Retrieval by Space-Based GNSS Observations

Using the radio occultation (RO) technique we can obtain high vertical resolution measurements of temperature, geopotential and refractivity in the upper troposphere and the stratosphere. By further combining RO and other types of atmospheric mass data (pressure and temperature), also observations of vertical humidity profiles can be obtained with high accuracy and high vertical resolution for most of the troposphere. In space-based GNSS atmospheric sounding the phase

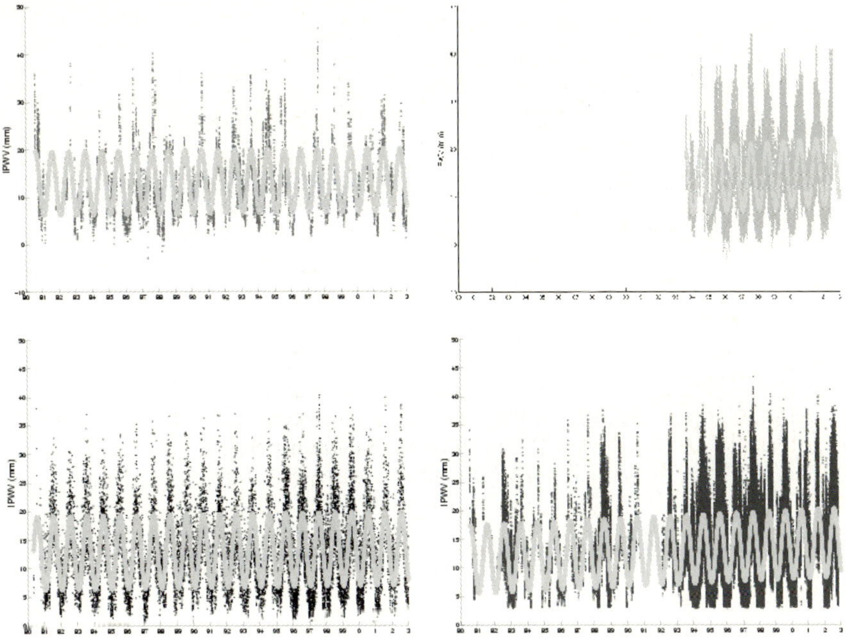

Fig. 2. Time series of IWV at Onsala Space Observatory from three independent techniques as well as from radiosondes launched from nearby Landvetter airport. GPS (upper right panel) and microwave radiometer data (lower right panel) are mainly from the last ten years. For this period the linear trend in IWV is larger (0.13 mm/yr to 0.24 mm/yr), compared to 1980 – 2003 from which VLBI (upper left panel) and radiosonde (lower left panel) data with trends less than 0.05 mm/yr are available. From Haas et al. (2003).

delay can be derived from occultation measurements of Low Earth Orbiting (LEO) satellites. From a time series of delays during an occultation, one can derive bending angles of the ray path. Using an Abel transform these can be converted into a vertical profile of refractive indices of the atmosphere (Kursinski et al. 1997). When the satellites of the Global Positioning System (GPS) are used as the transmitters, each LEO satellite can observe up to 500 globally distributed occultations (setting and rising) per day.

The refractivity N of the atmosphere is related to temperature, pressure and humidity by (Smith and Weintraub 1953)

$$N = (n-1)\times 10^6 \approx 77.6\frac{p}{T} + 3.73\times 10^5 \frac{p_\mathrm{w}}{T^2} \tag{1}$$

where n is the refractive index, p is pressure, p_w is the partial pressure of water vapor (both in hPa) and T is temperature (in K). From this equation one can see that the refractivity is highly useful in climate research, since it is related to key quantities in the atmosphere – pressure, temperature and humidity.

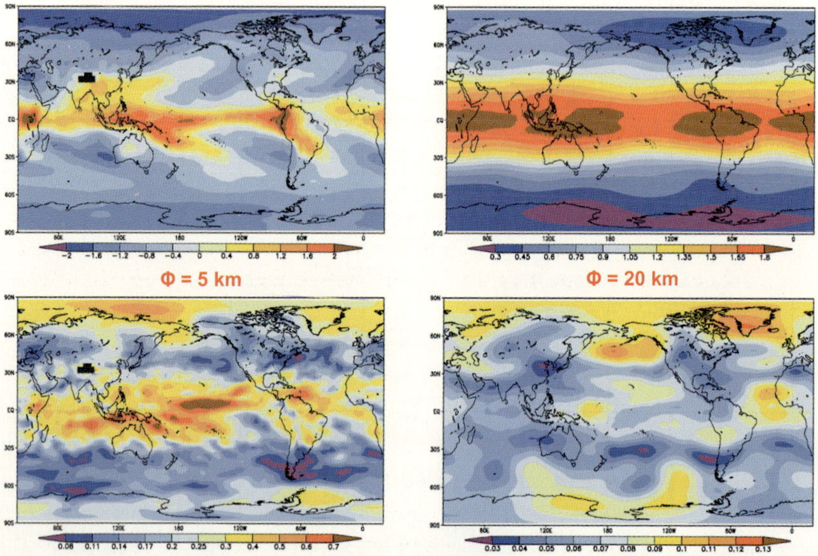

Fig. 3. Climate change signal (top row; 2091 – 2100 minus 1991 – 2000) and interannual variability (bottom row) of refractivity on geopotential levels of 5 and 20 kilometers in units of 10^{-6}. From Vedel and Stendel (2004).

The first term (contribution of O_2 and N_2 molecules) is referred to as the "dry" term and the second (contribution of water vapor) as the "wet" term. This implies that a single refractivity measurement, even if hydrostatic balance is assumed, is not sufficient to retrieve pressure, temperature and vapor pressure – the so-called "wet-dry ambiguity". However, as shown by Vedel and Stendel (2004), refractivity is as sensitive to global warming as geopotential thickness (i.e., atmospheric temperature) – in the stratosphere, it is even more sensitive. This means that refractivity itself is an ideal observable for climate monitoring and detection of global warming without having to convert to temperature and humidity profiles first (Fig. 3).

For comparing the GNSS occultation analysis results with independent meteorological data as e.g., reanalyses or radiosonde data, it is necessary to process the refractivity into other variables, such as temperature and geopotential heights. In the upper troposphere and stratosphere, the wet term in equation 1 is very small and can be neglected so that temperature is directly related to refractivity. Using a model for gravity and integrating the hydrostatic equation downwards to a given level from a high altitude we can obtain a series of geopotential heights and corresponding geopotential thicknesses, i.e., temperatures (Leroy 1997; Schrøder et al. 2003). Upper tropospheric geopotential height is very important in itself for climate monitoring, since, if surface pressure is known, it can be converted into a measure of mean tropospheric virtual temperature, which is one of the most fundamental parameters in climate monitoring.

On the other hand, when water vapor cannot be neglected, such as in the lower and middle troposphere, it is impossible to separate the effects from temperature and from water vapor. However, it is possible to calculate water vapor from refractivity when an independent measure of temperature is given. Also the other way (calculation of temperature from refractivity and an estimate of water vapor) is in principle possible. As shown by Ware et al. (1998), the errors in temperature and water vapor pressure relate to each other as $\Delta T \cong 4.35\ \Delta p_w$. In other words, if the temperature is known independently to an accuracy of 2 K, we can estimate water vapor to an accuracy of 0.5 hPa. Given that in many regions of the lower troposphere water vapor pressure usually lies between 5 hPa and 20 hPa and that there are no other measures of water vapor, this would be a very useful estimate.

5 GNSS Observations as a Climate Monitoring Tool

As shown above, GNSS observations, both ground-based and space-based, offer new and promising possibilities to monitor temperature and water vapor and have the potential to overcome all the problems conventional observations suffer of. The following properties make them particularly suitable for long-term climate monitoring:

- GNSS RO observations have a fairly uniform distribution around the world, both over land and ocean, in contrast to radiosondes and aircraft measurements. This is true for longer averaging periods (as is required for climate monitoring), whereas for shorter, e.g., daily, periods, observations tend to cluster due to the sun-synchronous orbit.
- GNSS RO observations are free of instrumental biases since the observation depends on a measurement of time, not of radiation intensities. If double-differencing is applied, the measurements are also essentially self-calibrating. These long-term stability properties imply that it is possible to compare two data sets separated by several years and taken by different sensors, which is not at all straightforward for present microwave sounders.
- They can be used in all-weather conditions, in particular in cloudy areas, which will not be covered adequately, even by future advanced infrared sounders. Furthermore, the observational quality and sampling characteristics are virtually the same in all geographical regions and at all times.
- They have a relatively high vertical resolution (1 km or better) compared to existing and planned passive infrared and microwave sounders, and thus addressing the main limitation of these systems. This is true for both temperature and water vapor measurements.
- Quantities that are key observables for climate change, such as near-tropopause geopotential height (Schrøder et al. 2003) or refractivity itself (Vedel and Stendel 2004) can be obtained almost totally independent of conventional measuring systems. Only surface pressure needs to be taken from other sources.

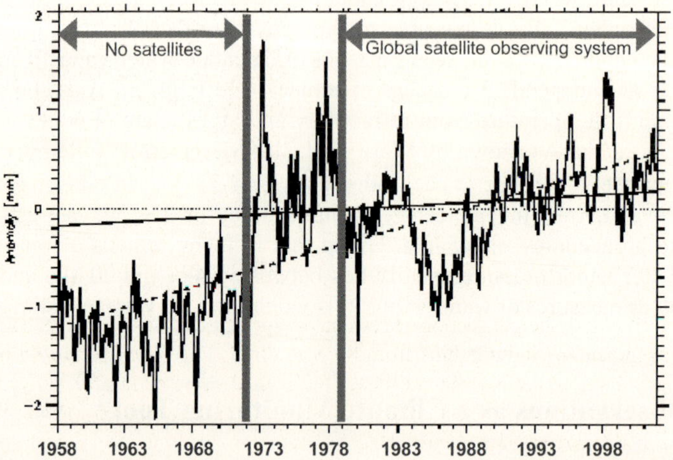

Fig. 4. ERA 40 integrated water vapor anomalies and trends for the period 1958 – 2001. From 1972 on, satellites were gradually introduced into the global observing system. The dashed line shows the linear trend which is twice as large as one would expect from the Clausius-Clapeyron equation, assuming conservation of relative humidity. The full line shows a corrected trend which is obtained by excluding all data from 1972 to 1978 and by adding a factor to the data prior to 1972 which is calculated from the difference of ERA 40 and an experiment excluding all satellite observations to mimic the observing system prior to 1979. From Bengtsson et al. (2004b).

- The unprecedented time stability, accuracy and resolution of GNSS RO data offers a new possibility to identify spurious trends in radiosonde, MSU/AMSU and reanalysis data sets. As an example, Fig. 4 demonstrates that excluding all satellite observations to mimic the observing system prior to 1979 leads to a gross underestimation of trends in water vapor compared to GPS observations.

The potential of RO data for climate change monitoring has been shown in simulation studies (e.g., Steiner et al. 2001; Foelsche et al. 2003) and is currently under investigation for data from CHAMP (Challenging Minisatellite Payload, see e.g., Foelsche et al. 2005a; Wickert et al. 2005). Figure 5 demonstrates that RO gives the opportunity to directly observe climate change on a timeframe of about 20 years. The signal to noise ratio (expected trend compared to temperature accuracy) is particularly large near the tropopause, which is where we expect the largest temperature changes from the Clausius-Clapeyron equation. In other words, due to accuracy, long-term stability and size of the climate change signal being sought for, we will be enabled to measure climate change by means of GNSS-LEO data thanks to their accuracy and long-term stability.

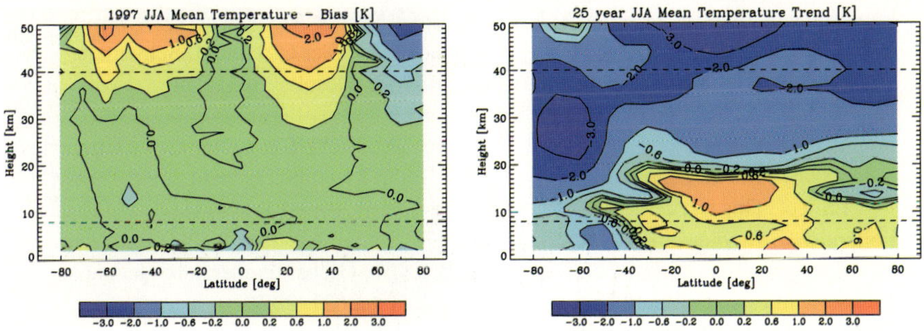

Fig. 5. Left panel: Latitude-height section of residual temperature biases for boreal summer (JJA) 1997 in seventeen 10° latitude bins from 80°S to 80°N. Each average profile in the left panel involves about 50 realistically simulated individual occultation profiles sampled from 6 LEOs for setting and rising occultations. From Foelsche et al. (2005). Right panel: ECHAM5 (T42L39) 25-yr boreal summer (JJA) temperature trends for the period 2001 – 2025. From Foelsche et al. (2006, this issue).

The joint US-Taiwanese COSMIC mission, scheduled for 2006, will use six satellites and therefore profoundly enhance data availability in near-real time. With roughly 2500 daily occultations, there will be almost an order of magnitude more data available than from the combined CHAMP and SAC-C missions. Figure 5 demonstrates that with such a constellation, we will be able to reliably measure climate trends which are expected for the next few decades, i.e., to observe climate change.

References

Bengtsson L, Hodges KI, Hagemann S (2004a) Sensitivity of the ERA40 reanalysis to the observing system: determination of the global atmospheric circulation from reduced observations. Tellus 56A:456–471

Bengtsson L, Hagemann S, Hodges K (2004b) Can climate trends be calculated from reanalysis data? J Geophys Res 109(D11111), doi:10.1029/2004JD004536

Christy JR, Parker DE, Stendel M, Norris WB (2001) Differential trends in tropical sea surface temperature and atmospheric temperatures since 1979. Geophys Res Lett 28:183–186

Christy JR, Spencer RW, Norris WB, Braswell WD, Parker DE (2003) Error estimates of version 5.0 of MSU-AMSU bulk atmospheric temperatures. J Atm Ocean Tech 20:613–629

ECMWF (2002) ERA workshop on reanalysis. ERA-40 Project Report Series No 3, ECMWF, Reading, UK. www.ecmwf.int/publications/library/do/references/list/192

Foelsche U, Kirchengast G, Steiner AK (2003) Global climate monitoring based on CHAMP/GPS radio occultation data. In: Reigber C, Lühr H, Schwintzer P (eds) First CHAMP mission results for gravity, magnetic and atmospheric studies. Springer, Berlin, pp 397–407

Foelsche U, Gobiet A, Löscher A, Kirchengast G, Steiner AK, Wickert J, Schmidt T (2005) The CHAMPCLIM project: An overview. In: Reigber C, Lühr H, Schwintzer P, Wickert J (eds) Earth Observation with CHAMP: Results from Three Years in Orbit. Springer, Berlin, pp 615–620

Foelsche U, Gobiet A, Steiner AK, Borsche M, Wickert J, Schmidt T, Kirchengast G (2006) Global Climatologies Based on Radio Occultation Data: The CHAMPCLIM Project. This issue

Folland CK, Rayner NA, Brown SJ, Smith TM, Shen SSP, Parker DE, Macadam I, Jones PD, Jones RN, Nicholls N, Sexton DMH (2001) Global temperature change and its uncertainties since 1861. Geophys Res Lett 28(13):2621–2624

Fu Q, Johanson CM, Warren SG, Seidel DJ (2004) Contribution of stratospheric cooling to satellite-inferred tropospheric temperature trends. Nature 429(6987):55–58

Gaffen DJ, Sargent MA, Habermann RE, Lanzante JR (2000) Sensitivity of tropospheric and stratospheric temperature trends to radiosonde data quality. J Climate 13(10):1776–1796

Gradinarsky LP, Johansson JM, Bouma HR, Scherneck HG, Elgered G (2002) Climate monitoring using GPS. Phys Chem Earth 27:335–340

Haas R, Elgered G, Gradinarsky L, Johansson J (2003) Assessing long term trends in the atmospheric water vapor content by combining data from VLBI, GPS, radiosondes and microwave radiometry. In: Schwegmann W, Thorandt V (eds) Proc. 16th Working Meeting on European VLBI for Geodesy and Astrometry, Leipzig, May 9-10, 2003, Bundesamt für Kartographie und Geodäsie, Frankfurt/Leipzig, pp 279–288

Hagemann S, Bengtsson L, Gendt G (2003) On the determination of atmospheric water vapour from GPS measurements. J Geophys Res 108, doi:10.1029/2002JD003235

Hurrell JW, Trenberth KE (1998) Difficulties in obtaining reliable temperature trends: Reconciling the surface and satellite microwave sounding unit records. J Climate 11(5):945–967

Kalnay E et al. (1996) The NCEP/NCAR 40-year reanalysis project. Bull Amer Met Soc 77(3):437–471

Kursinski ER, Hajj GA, Schofield JT, Linfield RP, Hardy KR (1997) Observing Earth's atmosphere with radio occultation measurements using the Global Positioning System. J Geophys Res 102:23429–23465

Leroy SS (1997) Measurements of geopotential heights by GPS radio occultation. J Geophys Res 102:6971–6986

Mears CA, Wentz FJ (2005) The effect of diurnal correction on satellite-derived lower tropospheric temperature. Science Express Reports, doi:10.1126/science.1114772

Prabhakara C, Iacovazzi Jr R, Yoo J-M, Dalu G (2000) Global warming: evidence from satellite observations. Geophys Res Lett 27:3517–3520

Santer BD, Wigley TML, Mears C, Wentz FJ, Klein SA, Seidel DJ, Taylor KE, Thorne PW, Wehner MF, Gleckler PJ, Boyle JS, Collins WD, Dixon KW, Doutriaux C, Free M, Fu Q, Hansen JE, Jones GS, Ruedy R, Karl TR, Lanzante JR, Meehl GA, Ramaswamy V, Russell G, Schmidt GA (2005) Amplification of surface temperature trends and variability in the tropical atmosphere. Science Express Reports, doi:10.1126/science.1114867

Schrøder T, Leroy S, Stendel M, Kaas E (2003) Stratospheric temperatures probed by Microwave Sounding Units or by occultation of the Global Positioning System. Geophys Res Lett 30, doi:10.1029/2003GL017588

Semenov VA, Bengtsson L (2002) Secular trends in daily precipitation characteristics: Greenhouse gas simulation with a coupled AOGCM. Clim Dyn 19(2):123–140

Sherwood S, Lanzante J, Meyer C (2005) Radiosonde daytime biases and late-20th century warming. Science Express Reports, doi:10.1126/science.1115640

Smith EK, Weintraub S (1953) The constants in the equation for atmospheric refractive index at radio frequencies. Proc IRE 41:1035–1037

Solheim F, Godwin JR, Westwater ER, Han Y, Keihm SJ, Marsh K, Ware R (1998) Radiometric profiling of temperature,water vapor and cloud liquid water using various inversion methods.Radio Sci 33:393–404

Steiner AK, Kirchengast G, Foelsche U, Kornblueh L, Manzini E, Bengtsson L (2001) GNSS occultation sounding for climate monitoring. Phys Chem Earth 26:113–124

Vedel H, Stendel M (2004) GPS RO refractivities as a climate change measure. Clim Change (submitted)

Vinnikov KY, Grody NC (2003) Global warming trend of mean tropospheric temperature observed by satellites. Science 302:269–272

Wentz FJ, Schabel M (1998) Effects of orbital decay on satellite-derived lower-tropospheric temperature trends. Nature 394:661–664

Wickert J, Gobiet A, Beyerle G, Steiner AK, Kirchengast G, Foelsche U, Schmidt T (2005) GPS radio occultation with CHAMP: Comparison of inversion results from GFZ Potsdam and IGAM Graz. In: Reigber C, Lühr H, Schwintzer P, Wickert J (eds) Earth observation with CHAMP: Results from three years in orbit. Springer, Berlin, pp 525–530

Climate Benchmarking Using GNSS Occultation

S. S. Leroy, J. A. Dykema, and J. G. Anderson

Harvard University, Cambridge, Massachusetts, USA
leroy@huarp.harvard.edu

Abstract. We put climate monitoring in a scientific context, which can be arrived at through a careful implementation of Bayesian inference. What we find is that a good climate monitoring tool must help address the physics of a climate model so as to make it better able to predict future climates. GNSS occultation is ideal because it offers sensitivity to improve the model physics which affects the stratospheric Brewer-Dobson circulation, the tropical tropospheric hydrological cycle, and the poleward migration of the mid-latitude storm track. Also, GNSS occultation is ideal because it can be readily made into a benchmark measurement provided clock calibration is always done by double-differencing, and measurements used to determine precise orbits and information on ionospheric activity are archived as auxiliary information. In doing so, GNSS occultation can be made S.I. traceable.

1 Introduction

The promise of radio occultation using the Global Navigation Satellite Systems (GNSS) as a climate monitoring technique has been recognized for a long time [14, 17, 21], but exactly how occultation data can be used for this purpose and why it might be appropriate to use radio occultation for climate monitoring has never been formalized. Certainly, radio occultation offers advantages over other sounding techniques which have been used in the past for climate monitoring, namely calibration using a time standard traceable to atomic clocks through double differencing (to be discussed later), but does radio occultation offer a guarantee of capturing the true state of the atmosphere? If GNSS occultation is needed, what are the actual requirements for making it an essential climate monitoring tool?

The above questions on climate monitoring by GNSS occultation can be answered by outlining a background to global climate change, describing how climate change research can be philosophically linked to climate monitoring, and defining the qualities of a measurement which makes it useful for climate monitoring. It is not in the scope of this work to discuss in any detail the background to climate change, and it is left to the reader to consult the

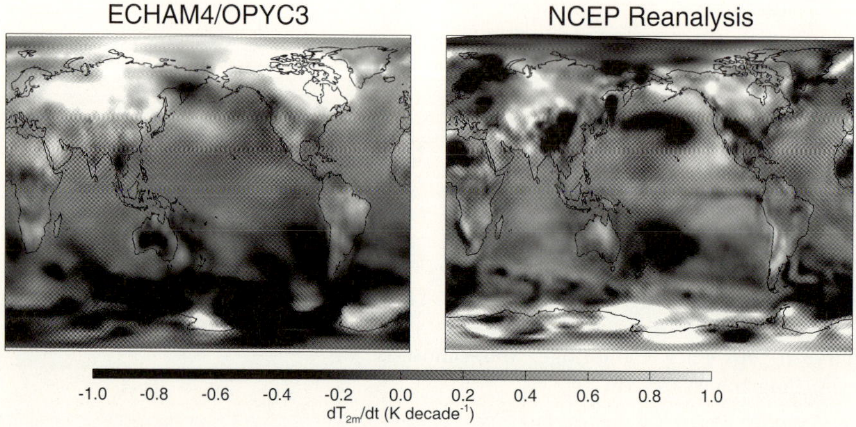

Fig. 1. Trend of 2-meter air temperature in ECHAM4/OPYC3 and the NCEP Reanalysis, 1970–1989. The left plot shows the temperature trend in the air temperature at 2 m above the surface as modeled by the ECHAM4/OPYC3 climate model with realistic greenhouse gas, sulfate aerosol, and ozone forcing. The plot on the right shows the same but taken from the NCEP Reanalysis.

Third Assessment Report of the Intergovernmental Panel on Climate Change (IPCC2001) [12] for a thorough review of observations used to probe climate change over the past century. We will, however, discuss a philosophy which puts climate monitoring in a scientific context and assess the accuracy with which current GNSS occultation must capture the true state of the climate.

2 An Overview of Climate Monitoring

The reason climate monitoring is of current interest is that global warming and the strong possibility that human activities are at least in part to blame has become a key scientific debate. In addition, any societal attempt to limit warming by greenhouse gas emissions in the future must be informed by reliable forecasts of climate change in the future.

That the surface air has been warming over the past century is clear [4]. The warming has occured primarily in two intervals: one from 1910 through 1940, and the second from 1970 to the present. The temporal pattern is consistent with an atmospheric greenhouse instigated by increasing carbon dioxide, methane, nitrous oxide, and chloro-fluorocarbons. The absence of a trend between 1940 and 1970 is thought to be due to the presence of sulfate aerosols, which reflect shortwave radiation.

While it is possible to construct models of warming over the past few centuries that can match the temporal pattern of observed trends, there remain substantial difficulties in modeling the spatial pattern of surface air temper-

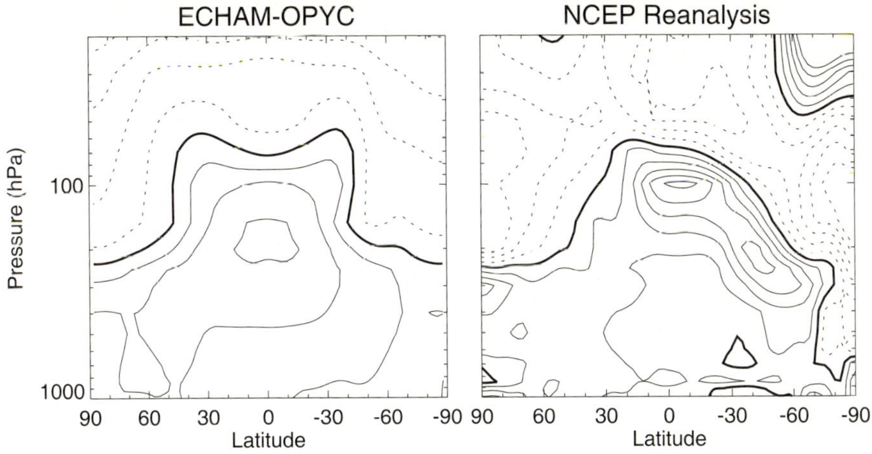

Fig. 2. Trends in upper air temperature in model and reanalysis due to increasing greenhouse gases, 1970–2000. The plot on the left shows the trend in upper air temperature as modeled by ECHAM4/OPYC3 when carbon dioxide increases at 1 % year^{-1}. The plot on right shows the trend in upper air temperature from the NCEP Reanalysis from 1970 through 2000. Each contour represents 0.2 K decade^{-1}. Solid (dashed) line contours are for positive (negative) values, and the thick line is the zero contour.

ature over recent decades. Fig. 1 shows the surface air warming trend in the climate model ECHAM4/OPYC3[1] for the period 1970 through 1989 in comparison with the NCEP Reanalysis [13]. Bulk features such as the relatively stronger warming over land versus over ocean are common to both model and reanalysis. Substantial differences over the ocean, especially the Pacific basin, however, reveal major difficulties in modeling change of the atmosphere's circulation. It is difficult to attribute the disagreement in southern high latitudes to the climate model or reanalysis, considering uncertainties in the ice-albedo feedback as implemented in climate models and the paucity of data available to steer reanalyses in southern high latitudes. Detailed differences in trends over land are a consequence of the simplicity of land surface/biosphere parameterizations. The discrepencies in the spatial patern of surface air warming point to inadequacies in our ability to physically model the climate system.

To further illustrate the point of difficulties in predicting atmospheric change, Fig. 2 shows patterns of upper air warming predicted to result from carbon dioxide increases. Upper air temperature trends predicted by the ECHAM4/OPYC3 model when it is subjected to a steady 1 % yr^{-1} carbon dioxide increase are compared to upper air temperature trends of the NCEP

[1] The 2-m surface air temperature output from ECHAM4/OPYC3 was obtained from the on-line data gateway for the Third Assessment Report of the Intergovernmental Panel for Climate Change.

Reanalysis from 1970 through 2000. The comparison is reasonable because greenhouse gas increases dominate the tropospheric temperature trend during this period. Because the model forcing is not absolutely realistic in that it does not contain the right carbon dioxide forcing, ozone forcing, volcanic forcing, etc., we only concern ourselves with the overall patterns of warming exhibited. Of particular note is the lack of upper tropical tropospheric warming in the reanalysis in comparison to the model. Moreover, the strong warming lobe at the tropopause at the southern subtropical front is completely absent in the model prediction. Such effects are indications of the model's physics lacking the ability to adequately predict dynamical trends in the atmosphere. The consequences for the hydrological cycle are substantial.

The inadequate physics of the climate model is related to the uncertain model parameterizations pertaining to the hydrological cycle. For example, the water cycling rate and precipitation efficiency, two key parameters of the hydrological cycle, exhibit strong control over precipitation patterns in the tropics [3]. Atmospheric dynamics is governed by the equations of motion, which are explicitly integrated in an atmospheric model. Atmospheric motion is ultimately forced by external and internal heating due to radiation and latent heating. Because the dynamical evolution of water and clouds contribute to these heating effects on scales much smaller than the model resolves, they are parameterized. Such parameterizations, especially those pertaining to the hydrological cycle, are notoriously inadequate in simulating a realistic hydrological cycle.

What is needed is a methodology which relates climate monitoring to the scientific process wherein theories, or models, are tested against empirical data. By monitoring climate change carefully it is possible to reduce uncertainties in climate models, especially those pertaining to parameterizations, and thus make them more useful tools in forecasting climate change.

3 Bayesian Inference and Climate Signal Detection

A climate monitoring system must be reliable enough to help refine our capability of predicting future climate change [8]. This is the result when detecting climate signals by optimal fingerprinting is put into its Bayesian context [15]. Bayesian statistics is useful when analysis involves data and model because it provides a rigorous framework for inference between theories (re: models) using observations [6].

In optimal fingerprinting, one searches for m climate signals with patterns $\mathbf{s}_i$, columns of $\mathbf{S}$, in an observational data set $\mathbf{d}$ with dimension n in the presence of naturally occuring variability $\mathbf{n}$. *A priori,* the amplitudes $\boldsymbol{\alpha}_i$ of the signals are considered unknown. The patterns of the climate signals are expected to uniquely identify, or "fingerprint" the forcing which causes them. As such, this technique has been used in climate signal detection and attribution studies [20]. The model for the data is

$$\mathbf{d} = \mathbf{S}\boldsymbol{\alpha} + \mathbf{n}. \tag{1}$$

In Bayesian statistics, the posterior knowledge for signal amplitudes given the data $p(\boldsymbol{\alpha}|\mathbf{d})$ is proportional to the evidence for the data $P(\mathbf{d}|\boldsymbol{\alpha})$ multiplied by the prior for the signals' amplitudes $p(\boldsymbol{\alpha})$. (A conditional probability $P(A|B)$ is the probability that event A is or will be "true" conditioned on the fact that event B is true or occurred.) All told,

$$\begin{aligned} &p(\boldsymbol{\alpha}|\mathbf{d}) \propto P(\mathbf{d}|\boldsymbol{\alpha})\; p(\boldsymbol{\alpha}) \\ &P(\mathbf{d}|\boldsymbol{\alpha}) = (2\pi)^{-n/2}\, |\mathbf{N}|^{-1/2}\, \exp\Big[-\frac{1}{2}(\mathbf{d}-\mathbf{S}\boldsymbol{\alpha})^{\mathrm{T}}\mathbf{N}^{-1}(\mathbf{d}-\mathbf{S}\boldsymbol{\alpha})\Big]. \end{aligned} \tag{2}$$

In optimal fingerprinting, the signal shapes, the interannual variability covariance $\mathbf{N}$ are considered known, no prior knowledge is considered, and the result is a probability distribution telling whether the consequences of specific forcings are detectable in the data. Gaussian statistics are assumed for natural variability.

This technique has been used to investigate the degree to which global warming can be attributed to human influences. Using models to formulate a distinctive pattern of temperature change caused by increasing greenhouse gases, it is possible to minimize the influence of natural variability in searching for the anthropogenic pattern of temperature change. If successful, then warming of the surface air can be attributed to human influence. Significant problems remain with such an interpretation, though. Primarily, the detection of a human influence on climate is not the same as an attribution of global warming to human influence. For example, the overwhelming influence of carbon dioxide on atmospheric temperature is its cooling of the stratoshere. A complete implementation of optimal fingerprinting would appropriately lead to the conclusion that stratospheric cooling is the result of a human influence on climate. Although tropospheric warming may be part of that influence, it is inappropriate to conclude that tropospheric warming is a reliably detected result of human influence. The reason such a conclusion cannot be drawn is that the physics of stratospheric cooling differs radically from the physics of tropospheric warming.

A better way to implement Bayesian statistics is to recognize that neither the signal patterns nor their amplitudes are known *a priori*. These unknowns can be bundled into a package of uncertain parameters of a climate model, the source of our uncertainties in predicting climate change. (We assume strong knowledge of all external forcing of the atmosphere.) The coefficients of the model are $\boldsymbol{\mu}$, and the data $\mathbf{d}$ is modeled as

$$\mathbf{d} = \mathbf{F}(\boldsymbol{\mu}) + \delta\boldsymbol{\mu} \cdot \nabla_{\boldsymbol{\mu}}\mathbf{F}(\boldsymbol{\mu}) + \mathbf{n} \tag{3}$$

where $\mathbf{F}$ is a kernel for how a signal is simulated by a model with parameters $\boldsymbol{\mu}$. For example, if the data represent the signal of the greenhouse effect as observed in radiosondes in the form of a trend, then $\mathbf{F}(\boldsymbol{\mu})$ is how a model predicts that trend would look given parameter values $\boldsymbol{\mu}$. The Jacobian

$\nabla_{\boldsymbol{\mu}}\mathbf{F}(\boldsymbol{\mu})$ tells how the trend would change by changing the parameters $\boldsymbol{\mu}$. If the prior for the model parameters is $\boldsymbol{\mu}_0$ with uncertainty covariance $\boldsymbol{\Sigma}_{\mu}$, i.e., $\boldsymbol{\mu} \sim \mathcal{N}(\boldsymbol{\mu}_0, \boldsymbol{\Sigma}_{\mu})$,[2] then the Bayesian solution is $\boldsymbol{\mu} \sim \mathcal{N}(\boldsymbol{\mu}_{\text{mp}}, \boldsymbol{\Sigma}_{\mu,\text{mp}})$ where the "most probable" model parameters $\boldsymbol{\mu}_{\text{mp}}$ and their error covariance $\boldsymbol{\Sigma}_{\mu,\text{mp}}$ are given by

$$\begin{aligned}\boldsymbol{\mu}_{\text{mp}} &= \boldsymbol{\mu}_0 + \left((\nabla_{\mu}\mathbf{F})^{\text{T}}\mathbf{N}^{-1}(\nabla_{\mu}\mathbf{F}) + \boldsymbol{\Sigma}_{\mu}^{-1} \right)^{-1} \\ &\qquad (\nabla_{\mu}\mathbf{F})^{\text{T}}\mathbf{N}^{-1}(\mathbf{d} - \mathbf{F}(\boldsymbol{\mu}_0)) \\ \boldsymbol{\Sigma}_{\mu,\text{mp}} &= \left((\nabla_{\mu}\mathbf{F})^{\text{T}}\mathbf{N}^{-1}(\nabla_{\mu}\mathbf{F}) + \boldsymbol{\Sigma}_{\mu}^{-1} \right)^{-1} . \end{aligned} \tag{4}$$

The Jacobian $\nabla_{\mu}\mathbf{F}$ is evaluated with parameters $\boldsymbol{\mu} = \boldsymbol{\mu}_0$ using several runs of a climate model, at least two runs for every parameter in $\boldsymbol{\mu}$. (Should a climate adjoint be constructed, wherein interdecadal trends of the model variables would be related to all parameters of a model run backwards in time, only two runs of the climate model would be necessary.)

Eqs. 4 show how to improve a model using Bayesian inference. Using a climate data set which contains information on how highly uncertain components of the climate system respond to a known forcing over the duration of the data, one can tune the parameters of the model so that it responds in the same way as observed. This type of analysis has been implemented using surface air temperature data over the past century to estimate oceanic parameters but with limited success [5].

4 Climate Model Uncertainties and How They are Realized in Microwave Refractivity

So what are the implications for GNSS radio occultation? That GNSS occultation can be used for climate monitoring has been mentioned elsewhere, but the requirements entailed have never been formalized. Climate model improvement by detecting the emergence of responses to external forcing provides the theoretical framework for deducing requirements for climate monitoring.

To implement Bayesian inference as described above, one must assess the influence on climatic trends as produced by changing each of a model's least constrained parameters. This is an expensive proposition (but one that has been undertaken by `climateprediction.net`). A simple way around this computational expense is to assume instead that the world's premier climate models differ in their responses to a prescribed forcing in ways that reflect what happens when the parameters within any given model are adjusted within reason. By examining the differences between different models'

[2] A normal distribution can be written as $\mathbf{y} \sim \mathcal{N}(\mathbf{y}_0, \mathbf{Y})$ where the probability distribution for the n-dimensional vector $\mathbf{y}$ is $p(\mathbf{y}) = (2\pi)^{-n/2} |\mathbf{Y}|^{-1/2} \exp\left[-\frac{1}{2}(\mathbf{y} - \mathbf{y}_0)^{\text{T}}\mathbf{Y}^{-1}(\mathbf{y} - \mathbf{y}_0)\right]$.

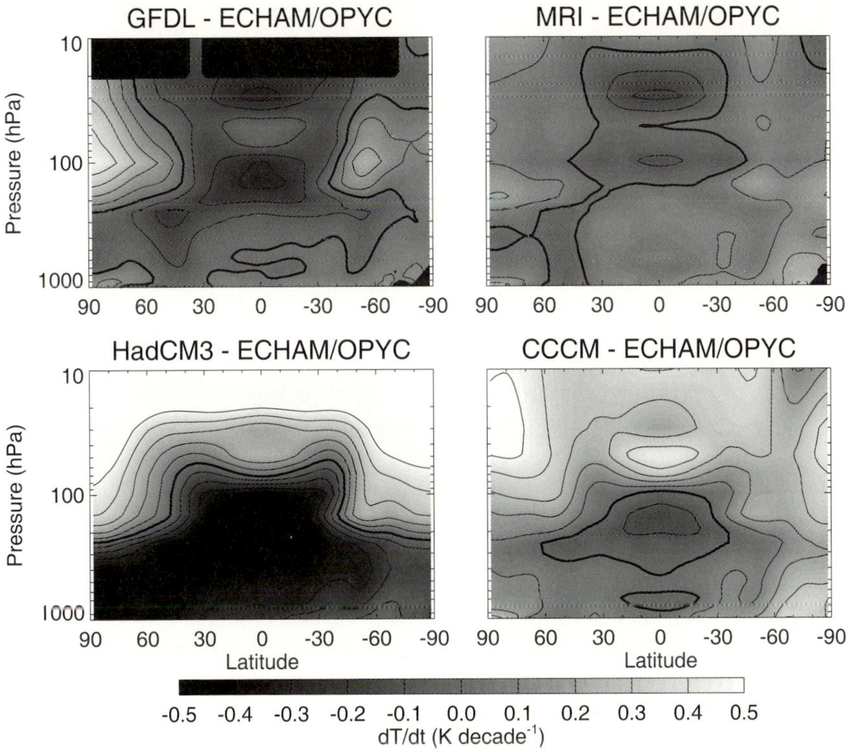

Fig. 3. Difference in zonal average temperature trends in five CMIP2+ models. These plots show the difference in zonal average temperature trends as simulated by five CMIP2+ models when subjected to increasing carbon dioxide at 1 % yr^{-1}. The reference model is ECHAM4/OPYC3. The other models are those of GFDL (upper left), MRI (upper right), HadCM3 (lower left), and CCCM (lower right). The zero contour is thick. See Fig. 2 for reference.

predictions of future climatic trends given a prescribed forcing, we can deduce what information GNSS occultation provides in improving model predictive capability.

We use the ensemble of climate models which contributed to the Coupled Model Intercomparison Project (CMIP2+) [2]. We chose to work with output of the models which provided temperature, humidity, and height on pressure surfaces, namely the German MPI Hamburg model (ECHAM4/OPYC3); the British Hadley Centre models (HadCM2, HadCM3); the Australian CSIRO model (CSIR Mk2); the Japanese Meteorological Research Institute (MRI, CGCM 2.3); the American Geophysical Fluid Dynamics Laboratory model (GFDL, version R30_c), NCAR's Climate System Model (CSM, version 1), and Parallel Computing Model (PCM, run B04.30); and Canadian Climate Centre Model (CCCM, version 2). Each of these models was subjected to increasing carbon dioxide at 1 % yr^{-1} for 70 years, about twice the rate of recent history.

The models of CMIP2+ show considerable disagreement in their predictions of atmospheric trends when subjected to the same forcing. Fig. 3 shows the zonal average difference of temperature trends of four models from ECHAM4/OPYC3. In all models, the stratosphere cools but with different patterns and in different amounts. The best agreement is between ECHAM4/OPYC3 and MRI, but even this comparison has interesting differences. While the amount of stratospheric cooling integrated over area is the same for these two models, the pattern shows enhanced cooling in MRI relative to ECHAM4/OPYC3 in the tropics and vice versa in the extratropics. This is the signature of the stratospheric Brewer-Dobson circulation increasing in strength more rapidly in the MRI model than in ECHAM4/OPYC3: upward (downward) motion in the tropics causes adiabatic expansion (compression) and cooling (warming) when the Brewer-Dobson circulation is strengthened. The GFDL model shows similar trends in the stratosphere with respect to ECHAM4/OPYC3 as does MRI but with a much greater difference in the Brewer-Dobson circulation strength. Both HadCM3 and CCCM show substantially less stratospheric cooling in an increasing carbon dioxide experiment than ECHAM4/OPYC3.

The Brewer-Dobson circulation is a residual mean circulation induced by the deposition of angular momentum by vertically propagating planetary waves originating in the troposphere [11]. If the Brewer-Dobson increases in strength, it necessarily means that more planetary waves propagate vertically from the troposphere into the stratosphere. In turn, these planetary waves are generated by flow over topography and baroclinic and barotropic instability in the troposphere. The stratospheric tropical-extratropical temperature patterns in Fig. 3 give information on the efficiency with which planetary waves are generated in the troposphere.

The differences in temperature trends in Fig. 3 also reveal fundamental tropospheric differences. The tropical tropospheric pattern of temperature trends of MRI less ECHAM4/OPYC3 reveals a Hadley circulation growing stronger with time in MRI than in ECHAM4/OPYC3. In the CCCM-ECHAM4/OPYC3 comparison, CCCM shows evidence of the tropopause increasing more rapidly in time than does ECHAM4/OPYC3. These differences can be related to the models' different mechanisms pertaining to the hydrological cycle.

The CMIP2+ models contributed output of temperature, geopotential height, and specific humidity, making it possible to simulate trends in microwave refractivity in a 1 % yr^{-1} carbon dioxide experiment. Recall that microwave refractivity N is empirically determined to be

$$N = (n-1) \times 10^6 = a \left(\frac{p}{T}\right) + b \left(\frac{p_\mathrm{W}}{T^2}\right) \tag{5}$$

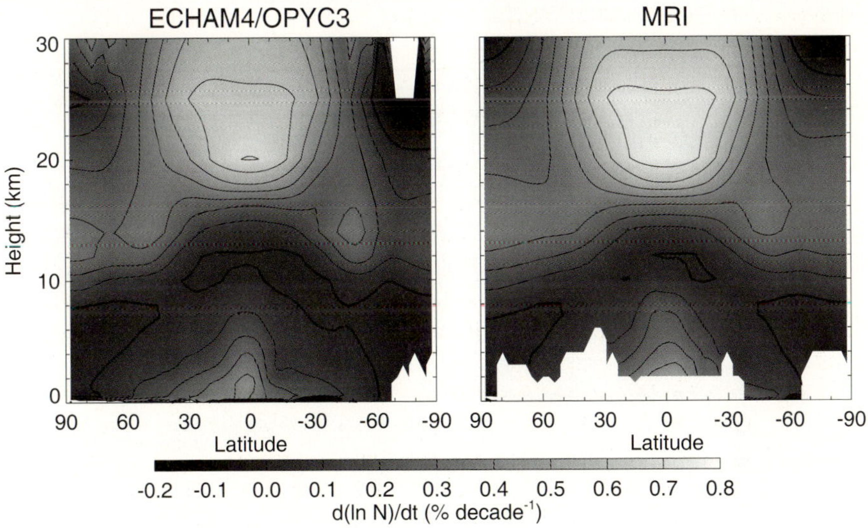

Fig. 4. Trends in microwave refractivity as simulated by the ECHAM4/OPYC3 and MRI climate models. The trends in zonal average microwave refractivity as a function of latitude and geopotential height as computed by 1 % yr^{-1} carbon dioxide forced runs of the ECHAM4/OPYC3 and MRI climate models. Contour intervals are 0.1 %. The thick line is the zero contour.

where n is the microwave index of refraction, p, T, and p_W are atmospheric pressure, temperature, and water vapor pressure, and the coefficients a and b are 77.6 K hPa^{-1} and 373.0×10^3 K^2 hPa^{-1} [19]. Microwave refractivity integrated vertically as a function of height is an ideal observeable for climate monitoring [16]. It can be thought of as a "dry" pressure: above the mid-troposphere, increasing dry pressure can be interpreted as thermal expansion of the underlying atmosphere, and below the mid-troposphere increasing dry pressure can be interpreted as a combination of thermal expansion of the underlying atmosphere and increasing precipitable water vapor in the overlying atmosphere.

First, Fig. 4 shows the trends in microwave refractivity as produced by forced runs of two of the CMIP2+ climate models. Except for the tropical lower troposphere, much of the troposphere shows little trend in microwave refractivity despite the presence of tropospheric warming. This is explained by the fact that, while temperatures are increasing, pressure as a function of geopotential height is also increasing. The overall effect on density is one of cancellation, and since the "dry" component of refractivity is proportional to density, only a trend of less than 0.1 % $decade^{-1}$ is seen throughout much of the troposphere.

The trend in integrated microwave refractivity conveys the notion of tropospheric warming better than trends in refractivity alone (see Fig. 5). As

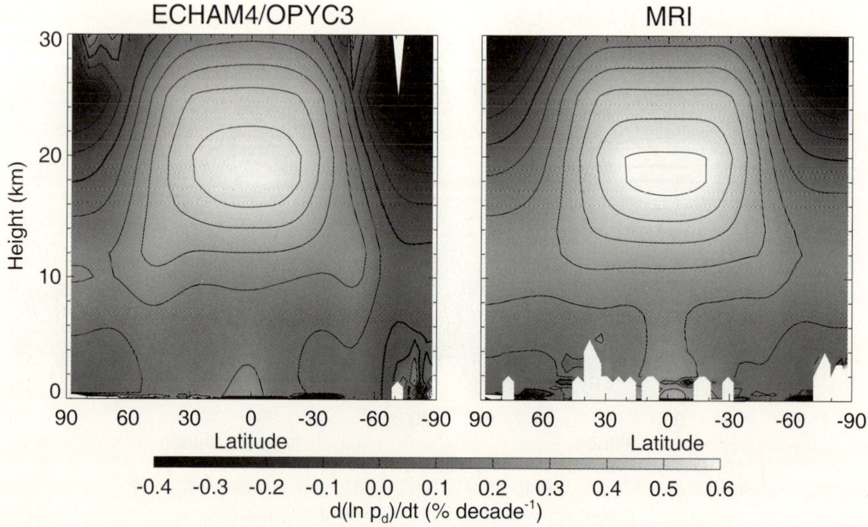

Fig. 5. Trends in microwave integrated refractivity, or dry pressure (p_d), as simulated by the ECHAM4/OPYC3 and MRI climate models. The trends in zonal average integrated microwave refractivity as computed by $1\,\%\ \mathrm{yr}^{-1}$ carbon dioxide forced runs of the ECHAM4/OPYC3 and MRI climate models are contoured as a function of latitude and geopotential height.

mentioned before, integrated refractivity is the same as pressure in the stratosphere and upper troposphere. A fractional change in pressure at a fixed height multiplied by a scaleheight is the same as the change in height of a constant pressure surface because, by the hydrostatic equation, $(\partial h/\partial t)_p = H(\partial \ln p/\partial t)_h$ where h, H, and p are geopotential height, the local scaleheight, and pressure. For example, the maximum trend in integrated refractivity according to the MRI model is $0.6\,\%$ decade^{-1}, which corresponds to an increase of the 100 hPa surface of $0.6\,\%$ of a scaleheight per decade, or ≈ 30 m. The abrupt, discontinuous structures in the lowest 2 km of the MRI model are a reflection of the MRI model having an overly variable surface pressure field. The decrease of the integrated refractivity trend with increasing height above the tropopause is the signature of stratospheric cooling.

Observing trends in integrated microwave refractivity will strongly constrain the responses of the Brewer-Dobson circulation, the tropical hydrological cycle, and the poleward migration of the midlatitude storm track to increasing greenhouse gases (see Fig. 6). The MRI climate model shows integrated refractivity (approximately pressure at constant height/height at constant pressure) increasing slower than ECHAM4/OPYC3 at low latitudes and faster at high latitudes in the stratosphere. This is the signature of the Brewer-Dobson circulation increasing faster according to MRI than according to ECHAM4/OPYC3. Temperature trends and integrated refractivity reveal

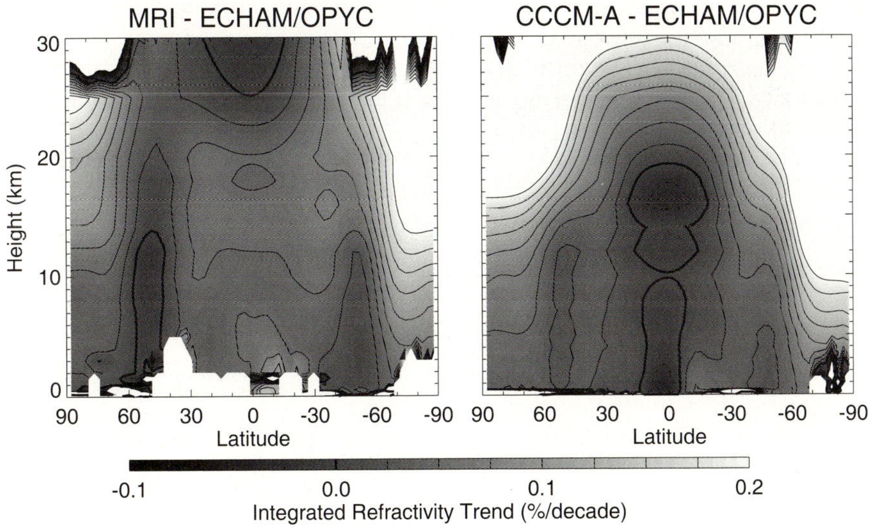

Fig. 6. Difference between model-forecast trends of integrated microwave refractivity. These plots show the difference between the forecast trends of integrated microwave refractivity from different models when each is forced by carbon dioxide increasing at 1 % yr^{-1}. The plot on the left shows the difference between MRI and ECHAM4/OPYC3 model trends, and the plot on the right shows the difference between the CCCM and ECHAM4/OPYC3 model trends. The scale below the plots shows the contours used in plotting.

this because an accelerated Brewer-Dobson circulation implies warming of the stratosphere at high latitudes and cooling at low latitudes, which means heights of constant pressure surfaces increase at high latitudes and decrease at low latitudes. Finally, the ridges in integrated refractivity/pressure trends seen at 50°N and S in the troposphere indicate sensitivity to the midlatitude storm track position. Such a ridge comes about because the storm track migrates northward in response to forcing by increasing carbon dioxide. In this example, the MRI model shows much more rapid poleward migration of the storm track than ECHAM4/OPYC3 as does CCCM, yet CCCM less so.

In order to deduce requirements for climate monitoring, one needs to know how long before a signal, such as that seen in Fig. 6, emerges above the background noise of natural interannual fluctuations of the climate. The calculation which needs to be performed is that described in the previous section wherein one defines the data $\mathbf{d}$ to be the Jacobian $\nabla_{\boldsymbol{\mu}}\mathbf{F}$ and finding the amount of time required before $|\boldsymbol{\Sigma}_{\boldsymbol{\mu},\mathrm{mp}}| \ll |\boldsymbol{\Sigma}_{\boldsymbol{\mu}}|$. This is the requirement for Bayesian learning, i.e., when significant improvements can be made in our estimates of the model's parameters. If we assume that a strong test can be performed in two decades, corresponding to a growth in carbon dioxide concentration of approximately 10 %, we can derive approximate requirements for climate

Table 1. Science requirements for climate monitoring with GNSS occultation. GNSS occultation has clear sensitivity useful for improving forecasting ability of the phenomena listed below. The accuracy is given in units of a percentage of microwave refractivity. The temporal resolution required for each phenomenon is seasonal. The ITCZ is the intertropical convergence zone.

Phenomenon	Accuracy	Spatial resolution
Brewer-Dobson circulation	0.2 % in stratosphere	hemispherical, scaleheight
Hydrological cycle	0.01 % in troposphere	ITCZ, $\sim$ 4 km
Storm track	0.01 % in troposphere	meridionally 5°, scaleheight

monitoring using GNSS occultation. Those numbers correspond to the decadal response of the climate models forced by 1 % yr^{-1} in the CMIP2+ perturbed runs. They are listed in Table 1. The quantities in Table 1 are taken directly from Fig. 6. The temporal resolution is chosen to be seasonal in order to maximize the averaging time without washing out climate signals which tend to be most enhanced during winter. Achieving 0.01 % accuracy in the troposphere promises to be challening in light of recent studies [7].

5 Benchmarking GNSS Occultation

We have arrived at scientific requirements for monitoring climate using GNSS occultation predicated on the assumption that we must learn something about the climate in the process of monitoring it. But what do these requirements mean in practice? What kind of monitoring system is needed? How must the data be processed? The solution must satisfy the needs of a future user who will difference data sets widely separated in time to detect meaningful trends in the climate system over that time. The solution is to make GNSS occultation a *benchmark* measurement. A key component of making a benchmark measurement lies in the methodology of making climate monitoring traceable to the international system of units, otherwise known as "S.I. traceability."

GNSS occultation has claimed absolute accuracy for years, but it is not really absolutely accurate. But how accurate can it be made? The Système International d'Unités (International System of Units, international abbreviation S.I.) provides a recommended system of metrological units with well-determined uncertainties ideally suited to assessing the accuracy of physical measurements. The primary realizations of the S.I. base units provide the most accurate foundation possible for measurements made in the most familiar physical units: the second, the meter, the kilogram, the ampere, the kelvin, the candela and the mole. For example, if one wants to measure the mass of an object, the most accurately it can ever be done is by use of the S.I. prototype kilogram as a counterweight and collecting enough information on the balance system to remove as many potential systematic errors as possible. This technique of calibration, which has a clear relationship between the S.I. foundation and the presented measurement, has demonstrable "S.I. traceability" [18].

In the case of GNSS occultation, much of the work in guaranteeing S.I. traceability has already been done. The Doppler delay of the GNSS signal induced by the atmosphere is the basic measurement in a GNSS occultation (which can be turned into refractivity profiles in retrieval), and, hence, GNSS occultation is a timing measurement. Guaranteeing S.I. traceability means tying the atmospheric Doppler delay to the international standard for the second. This is done through the double-difference methodology of processing GNSS occultation [9, 10], wherein the atmospheric path delays can be calibrated by Cs-133 atomic clocks [1] by measuring phase to a reference GNSS satellite during an occultation event and observing both the occulting and reference GNSS satellites with an atomic clock on the ground. Even if the atomic clock on the ground is not directly traceable to a Cs-133 clock, S.I. traceability is maintained as long as the ground atomic clock has been calibrated against the Cs-133 clocks with accuracy sufficient for radio occultation.

Even with double-differencing, hurdles remain in tracing a GNSS occultation to the international definition of the second. These hurdles include (but are not limited to) unusual ionospheric activity and uncertainties in GNSS and receiver spacecraft orbits. While GNSS ordinarily provides two frequencies for the removal of ionospheric effects, it is possible for the ionosphere to become so active that its effects on occultation cannot be completely removed using the two GNSS frequencies. To maintain S.I. traceability, it becomes necessary to collect information on the ionosphere sufficient to compensate for this unwanted noise or sufficient to know when an occultation cannot be used. In addition, bending angles in GNSS occultation can only be found with precise information on the orbits of the GNSS and receiver satellites. The orbits themselves are determined only after assumptions are made on the impact of the exosphere on the satellites' orbits, making the orbit solutions untraceable to S.I. standards. To maintain S.I. traceability, it is necessary to not only archive the satellites' precise orbit solutions, but also the occultation receiver measurements of phase to overhead GNSS satellites and the measurements of GNSS phase by reference ground networks used in the determination of precise orbits.

Creation of a benchmark measurement requires an archive of the S.I. traceable observations and external information on physical properties of potential error, but not a perfectly accurate retrieval algorithm. As long as an archive of the S.I. traceable components exists for all time, any potential future user will have enough information to implement his own retrieval algorithm. The investigator would use the same retrieval algorithm for every occultation data set, and this guarantees the consistency required for evaluating trends in the climate system over long time baselines. In GNSS occultation, the data required to make it a benchmark measurement are

- atmospheric phase delay as a function of time as calibrated by the double-differencing methodology,
- the Keplerian orbital elements as a function of time used to deduce the bending of the occultation signal by the atmosphere,
- the ground station and receiver satellite phase measurements used in the calculation of orbits, and
- data on the activity of the ionosphere in the vicinity of the occultation, especially activity related to ionospheric turbulence.

The platforms exist to collect and archive the information of the first three items, but not the fourth item. This points to the necessity of tracking global scale ionospheric activity in the form of a scalar as a function of time and possibly deploying an ionospheric monitoring system which can map ionsopheric activity as a function of space and time in the lower ionsophere.

Acknowledgements. We thank the contributors to CMIP2+ for making the outputs of forced runs of their models available to us for analysis.

References

[1] Bureau International des Poids et Mesures (1997) The International System of Units. Paris, 7th edition

[2] Covey C, AchutaRao KM, Cubasch U, Jones P, Lambert SJ, Mann ME, Phillips TJ, Taylor KE (2003) An overview of results from the Coupled Model Intercomparison Project. Global and Planetary Change 37(1–2):103–133

[3] Douville H, Chauvin F, Planton S, Royer JF, Salas-Melia D, Tyteca S (2002) Sensitivity of the hydrological cycle to increasing amounts of greenhouse gases and aerosols. Climate Dyn 20(1):45–68

[4] Folland CK, Karl TR, Coauthors (2001) Observed Climate Variability and Change. In: Houghton JT, Ding Y, Griggs DJ, Noguer M, van der Linden PJ, Dai X, Maskell K, Johnson CA (eds) Climate Chance 2001: The Scientific Basis, Contribution of Working Group I to the Third Assessment Report of the Intergovernmental Panel on Climate Change. Cambridge University Press, New York, pp 99–181

[5] Forest CE, Stone PH, Sokolov AP, Allen MR, Webster MD (2002) Quantifying uncertainties in climate system properties with the use of recent climate observations. Science 295(5552):113–117

[6] Gauch HG (2003) Scientific Method in Practice. Cambridge University Press, New York

[7] Gobiet A, Kirchengast G (2004) Advancements of Global Navigation Satellite System radio occultation retrieval in the upper stratosphere for optimal climate monitoring utility. J Geophys Res 109:D24110

[8] Goody R, Anderson J, North G (1998) Testing climate models: An approach. Bull Amer Meteor Soc 79(11):2541–2549

[9] Hajj GA, Kursinski ER, Romans LJ, Bertiger WI, Leroy SS (2002) A technical description of atmospheric sounding by GPS occultation. J Atmos Solar Terr Phys 64(4):451–469

[10] Hardy KR, Hajj GA, Kursinski ER (1994) Accuracies of atmospheric profiles obtained from GPS occultations. Int J Sat Comm 12(5):463–473

[11] Haynes PH, Marks CJ, McIntyre ME, Shepherd TG, Shine KP (1991) On the downward control of extratropical diabatic circulations by eddy-induced mean zonal forces. J Atmos Sci 48(4):651–679

[12] Houghton JT, Ding Y, Griggs DJ, Noguer M, van der Linden PJ, Dai X, Maskell K, Johnson CA (2001) Climate Chance 2001: The Scientific Basis, Contribution of Working Group I to the Third Assessment Report of the Intergovernmental Panel on Climate Change. Cambridge University Press, New York

[13] Kalnay E, Kanamitsu M, Kistler R, Collins W, Deaven D, Gandin L, Iredell M, Saha S, White G, Woollen J, Zhu Y, Chelliah M, Ebisuzaki W, Higgins W, Janowiak J, Mo KC, Ropelewski C, Wang J, Leetmaa A, Reynolds R, Jenne R, Joseph D (1996) The NCEP-NCAR Reanalysis Project. Bull Amer Meteor Soc 77(3):437–471

[14] Kursinski ER, Hajj GA, Schofield JT, Linfield RP, Hardy KR (1997) Observing Earth's atmosphere with radio occultation measurements using Global Positioning System. J Geophys Res 102(D19):23429–23465

[15] Leroy SS (1998) Detecting climate signals: Some Bayesian aspects. J Climate 11(4):640–651

[16] Leroy SS, North GR (2000) The application of COSMIC data to global change research. Terr Atmos Ocean Sci 11(1):187–210

[17] Melbourne WG, Davis ES, Duncan CB, Hajj GA, Hardy KR, Kursinski ER, Meehan TK, Young LE, Yunck TP (1994) The application of spaceborne GPS to atmospheric limb sounding and global change monitoring. Technical Report 94–18, Jet Propulsion Laboratory, California Institute of Technology

[18] Pollock DB, Murdock TL, Datla RU, Thompson A (2003) Data uncertainty traced to SI units. Results reported in the International System of Units. Int J Rem Sensing 24(2):225–235

[19] Smith EK, Weintraub S (1953) The constants in the equation for atmospheric refractive index at radio frequencies. Proc IRE 41:1035–1037

[20] Stott PA, Tett SFB, Jones GS, Allen MR, Ingram WJ, Mitchell JFB (2001) Attribution of twentieth century temperature change to natural and anthropogenic causes. Climate Dyn 17(1):1–21

[21] Yuan LL, Anthes RA, Ware RH, Rocken C, Bonner WD, Bevis MG, Businger S (1993) Sensing climate-change using the Global Positioning System. J Geophys Res 98(D8):14925–14937

Global Climatologies Based on Radio Occultation Data: The CHAMPCLIM Project

U. Foelsche[1], A. Gobiet[1], A. K. Steiner[1], M. Borsche[1], J. Wickert[2], T. Schmidt[2], and G. Kirchengast[1]

[1] Wegener Center for Climate and Global Change (WegCenter) and Institute for Geophysics, Astrophysics, and Meteorology (IGAM), University of Graz, Austria
ulrich.foelsche@uni-graz.at
[2] GeoForschungsZentrum Potsdam (GFZ), Germany

Abstract. The German/US research satellite CHAMP (CHAllenging Minisatellite Payload for geoscientific research) continuously records about 230 radio occultation (RO) profiles per day since March 2002. The mission is expected to last at least until 2007, thus CHAMP RO data provide the first opportunity to create RO based climatologies on a longer term. CHAMPCLIM is a joint project of the Wegener Center for Climate and Global Change (WegCenter) in Graz and the GeoForschungsZentrum (GFZ) in Potsdam. It aims at exploiting the CHAMP RO data in the best possible manner for climate research. For this purpose, CHAMP excess phase data provided by GFZ are processed at WegCenter with a new retrieval scheme, especially tuned for monitoring climate variability and change. The atmospheric profiles which pass all quality checks (~150 profiles/day) are used to create climatologies on a monthly, seasonal, and annual basis. Here, we focus on dry temperature climatologies from the winter season (DJF) 2002/03 to the summer season (JJA) 2004, obtained by averaging-and-binning. The results show that useful dry temperature climatologies resolving horizontal scales >1000 km can be obtained even with data from a single RO receiver. RO based climatologies have the potential to improve modern operational climatologies, especially in regions where the data coverage and/or the vertical resolution and accuracy of RO data is superior to traditional data sources.

1 Introduction

The provision of accurate, long-term, consistent data to sustain and expand the observational foundation for climate studies is one of the high priority areas for action to improve the ability to detect, attribute and understand climate variability and changes (e.g., IPCC 2001). A promising climate monitoring tool meeting these requirements is the Global Navigation Satellite System (GNSS) Radio Occultation (RO) technique. The self-calibrated nature, high accuracy, all-weather capability, and global coverage of RO data suggest them as well suited for global short- and long-term monitoring of atmospheric change.

The German/US research satellite CHAMP was launched in July 2000 into low Earth orbit (LEO), since March 2002 it continuously records about 230 RO profiles per day (Wickert et al. 2001, 2004). Out of these ~230 daily profiles, about

160 can be successfully calibrated and are of sufficient data quality, ~150 of these pass the quality checks during the WegCenter retrieval. The CHAMP mission is expected to last until 2007, CHAMP RO data thus provide the first opportunity to create RO based climatologies for a multi-year period of >5 years.

In Section 2 we summarize the properties of RO data, with respect to the CHAMP mission and climate monitoring. Section 3 deals with appropriate atmospheric parameters for climate monitoring. In Section 4 the CHAMPCLIM project is presented, with selected results in Section 5, followed by concluding remarks.

2 Properties of Radio Occultation Data

Many satellite-derived data records are degraded by problems such as changes in instrumentation and satellites' orbits, insufficient calibration, instrument drift, and poor vertical resolution (Anthes et al. 2000). Because of these shortcomings, the magnitude of temperature trends in the troposphere is still under debate (e.g., Christy and Spencer 2003; Vinnikov and Grody 2003; Mears and Wentz 2005; Stendel 2006, this issue). GNSS RO data have the potential to solve many of these problems due to their combination of specific properties.

Highest quality of RO observations is achieved in the upper troposphere/lower stratosphere region (UTLS), a domain re-acting very sensitive to climate change (see Section 3). Compared to weather analyses CHAMP RO temperature data show an ensemble mean agreement of <0.4 K between 10 km and 35 km height with a standard deviation of ~1 K at 10 km increasing to ~2 K at 30 km height (Wickert et al. 2004). The active use of L-band signals with wavelengths of 19.0 cm and 24.4 cm (in case of GPS), respectively, allows for measurements during day and night and for the penetration of clouds.

2.1 Long-Term Stability due to Intrinsic Self-Calibration

Regarding climate monitoring, the long-term stability of RO data is of particular importance (see also Leroy et al. 2006, this issue). It can be achieved since atmospheric profiles are not derived from absolute values (phase delays) but from Doppler shift (phase change) profiles. Therefore, RO measurements require no external calibration and only short-term measurement stability over the occultation event duration (1 – 2 min), which is guaranteed by very stable oscillators onboard the GNSS satellites. Given this "self-calibration", data from different sensors and different occultation missions can be combined without need for inter-calibration and overlap, as long as the same data processing scheme is used.

The long-term stability of RO data could not be tested so far due to the lack of long-time observations. An intercomparison study by Hajj et al. (2004) based on data from CHAMP and SAC-C (Satélite de Aplicaciones Científicas-C), however, showed a remarkable consistency of the data obtained from these two different satellites with temperature profiles found consistent to 0.1 K in the mean between

5 and 15 km. While CHAMP and SAC-C are equipped with very similar receivers, leaving the possibility of common systematic errors, future RO missions will help assess whether these results can also be obtained with data from completely different receivers, like the GRAS instrument (GNSS Receiver for Atmospheric Sounding) onboard MetOp (Meteorological Operational satellite, launch expected for April 2006) (Loiselet et al. 2000).

2.2 Spatial and Temporal Coverage

The number of RO events depends primarily on the number of available transmitters and receivers. A single receiver in low Earth orbit (LEO), which is capable of tracking GPS signals during setting occultations (like on CHAMP) can collect ~250 RO profiles per day (for a nominal constellation of 24 GPS satellites). LEO satellites with an additional antenna for rising events can achieve twice that amount. The 6 COSMIC satellites (Constellation Observing System for Meteorology, Ionosphere, and Climate), which are scheduled to be launched in March 2006, can be expected to obtain ~3000 setting and rising occultations per day, providing a valuable database for RO based climatologies (Rocken et al. 2000). With the upcoming European Galileo system (nominal constellation of 30 satellites), the number of transmitters will more than double; the operational status of Galileo is expected to be reached in 2008/09.

The geographic distribution of the RO events depends on the geometry of the satellite orbits. Global coverage can only be obtained with a high-inclination orbit of the LEO satellite. This orbit geometry leads, however, to a high RO event density at high latitudes with comparatively fewer events at low latitudes. Figure 1 shows, as an example for this situation, the typical coverage of CHAMP RO data during one season. LEO satellites with a low inclination orbit, on the other hand, provide a better sampling at low latitudes, but do not reach global coverage.

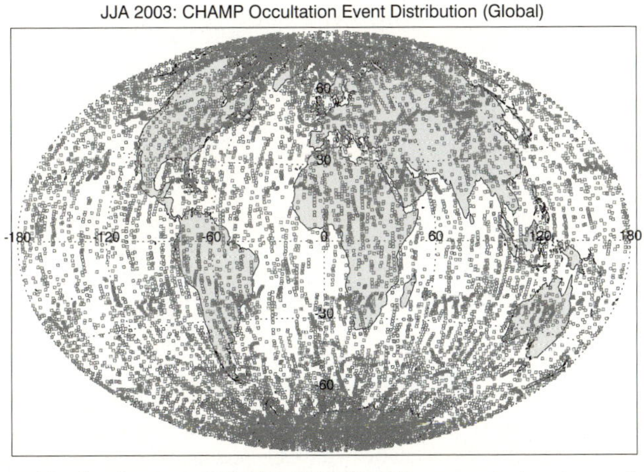

Fig. 1. Geographic distribution of 13553 CHAMP RO events during the northern summer season (June-July-August) 2003 (orbit inclination = 87.3°).

The LEO orbit geometry determines furthermore the local times of the RO events. Satellites in sun-synchronous orbits, like MetOp, always cross the equator at the same local times. As a consequence, MetOp RO events will be clustered around 9:30 a.m. and 9:30 p.m. local time, respectively. All non-sun-synchronous LEO orbits are subject to a drift in equator-crossing time. The resulting local time drift of the CHAMP RO data is about 1 hour in 11 days.

When attempting to build RO climatologies, we have to consider that any uneven spatial and temporal sampling of the "true" evolution of the atmospheric fields can lead to sampling errors (see Section 5.2).

3 Atmospheric Parameters for Climate Monitoring

In contrast to applications of RO data in numerical weather prediction, where the focus is clearly on RO products which are as close as possible to the raw measurements (e.g., Poli 2006, this issue), ideal parameters for climate monitoring are those which change most in a changing climate. Refractivity (Vedel and Stendel 2004) and geopotential height (Leroy 1997) have recently been identified as good indicators for climate change.

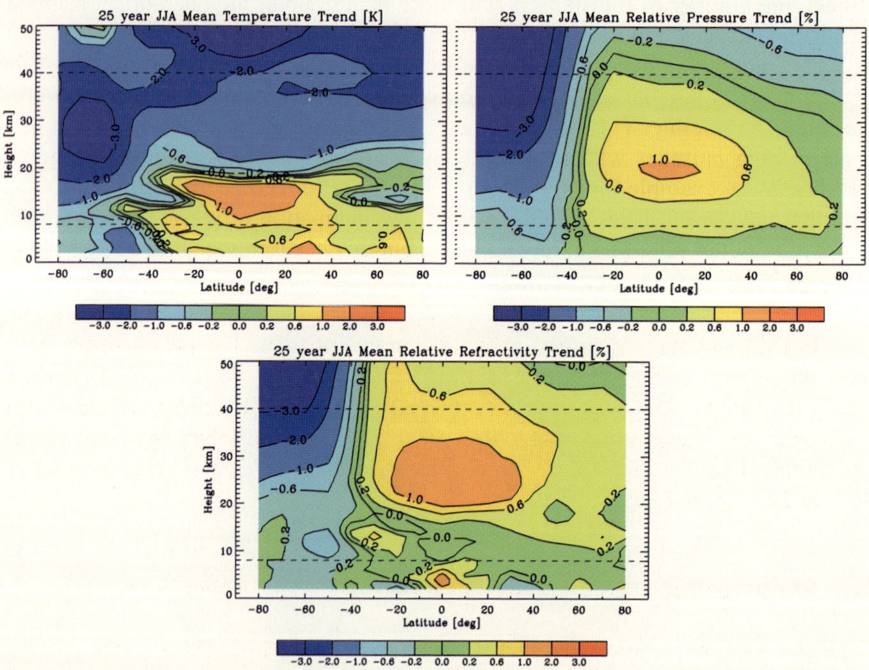

Fig. 2. Trends in atmospheric parameters over the 25-year period 2001–2025 as modeled with ECHAM5 (IS92a emission scenario with CO_2 concentration doubling between 1990 and 2100). Temperature trends (upper left panel), relative pressure trends (upper right panel), and relative trends in microwave refractivity (lower panel).

Results from climate model runs can be used as indicators for expected trends in atmospheric parameters. As an example, Figure 2 shows 25-year-trends (2001–2025) of temperature, pressure, and microwave refractivity as results of runs of the ECHAM5 model (Roeckner et al. 2003) in "middle atmosphere mode" with anthropogenic forcing. For this experiment, the vertical domain of the model was extended to 80 km. A dominant feature, which is only partly visible in "normal" climate model runs with a vertical domain up to 30 km, is the pronounced cooling in the stratosphere. Given the accuracy of RO data in the lower stratosphere it is likely that "global cooling" will be the first consequence of anthropogenic climate change that can be detected with the aid of the RO technique.

An interesting feature of Fig. 2 is the lack of change in refractivity in the upper tropical troposphere. Microwave refractivity N is related to temperature T, total pressure p, and water vapor partial pressure e, via (Smith and Weintraub 1953):

$$N = k_1 \frac{p}{T} + k_2 \frac{e}{T^2} \qquad (1)$$

where k_1 is 77.6 K/hPa and k_2 is $3.73 \cdot 10^5$ K^2/hPa. When atmospheric humidity is small, the second term on the right-hand-side of Eq. 1 can be neglected. We immediately see that in this case, the same relative increase in T and p will result in no change in refractivity. Figure 2 shows that different atmospheric parameters are sensitive in different regions of the atmosphere. Climate monitoring with RO data should therefore, in principle, comprise all parameters that can be retrieved with the RO technique.

4 The CHAMPCLIM Project

CHAMPCLIM is a joint project of the Wegener Center for Climate and Global Change (WegCenter) in Graz and the GeoForschungsZentrum (GFZ) in Potsdam. The overall aim of CHAMPCLIM is to contribute to the best possible exploitation of CHAMP RO data, in particular for climate monitoring. The results of this project provide a starting point for RO based climatologies, which can be continuously expanded with data from other RO missions. The main objectives of CHAMPCLIM and some initial results have been described in Foelsche et al. (2005). Here we just briefly recall the three main objectives (Sections 4.1 to 4.3) and focus on new results.

4.1 RO Data Processing Advancements for Optimizing Climate Utility

The essential outcome of this work was a robust CHAMPCLIM retrieval scheme (WegCenter/CCRv2), building on the EGOPS4 software tool (Kirchengast at al. 2002) and a reasonable error characterization for CHAMP/GPS RO data in meeting the aim to improve the maturity and utility of the data products especially for climatological purposes. Main aspects of WegCenter/CCRv2 are:

- Geometric optics retrieval. The implementation of a wave optics retrieval scheme for the troposphere is planned for CCRv3.
- Ionospheric correction via linear combination of bending angles.
- Transparent input of a priori information via statistical optimization of bending angles. For the results shown here, ECMWF (European Centre for Medium-Range Weather Forecasts) data have been used for background bending angles.
- No further background information for initialization of the hydrostatic integral.
- Dry air retrieval (Syndergaard 1999). A 1DVar retrieval for humidity and temperature in the troposphere is planned for CCRv3.

Further details of the CCRv2 retrieval can be found in Gobiet and Kirchengast (2004), Steiner et al. (2004), and Borsche et al. (2006, this issue). Results of a related error analysis can be found in Steiner and Kirchengast (2005) and of an error analysis using CHAMP refractivity profiles in Steiner et al. (2006, this issue).

4.2 RO Data and Algorithms Validation Based on CHAMP/GPS Data

The WegCenter/CCRv2 retrieval scheme was validated against a modeled atmosphere in an end-to-end simulation study, the GFZ operational RO retrieval scheme, numerical weather prediction analyses from ECMWF, and remote-sensing instruments onboard ENVISAT (MIPAS and GOMOS). Results of these validation studies can be found in Gobiet et al. (2004, 2005a).

4.3 Global RO Based Climatologies for Monitoring Climate Change

In this part we focus on building global climatologies based on the validated datasets obtained by advanced retrievals of atmospheric parameters from CHAMP RO data. In a first approach, we perform direct (model independent) monitoring of the evolution of climatological atmospheric fields through averaging and binning of RO profiles. The setup for these climatologies is described in Borsche et al. (2006, this issue). Seasonal dry temperature climatologies are presented in Section 5. In a second approach we assimilate CHAMP RO-derived refractivities into ECMWF short term forecast fields (via 3D-Var) to obtain global climate analyses with higher horizontal resolution. Results of this approach are the focus of Löscher and Kirchengast (2006, this issue). The current record of RO occultations is still too short to actually monitor trends, but comparison with other climatologies shows the value and the potential of the climatologies based on RO data.

5 Seasonal CHAMP Dry Temperature Climatologies

For the results shown here, we sampled CHAMP profiles in 18 latitude bands (10°latitudinal extent). As examples, Fig. 3 shows the zonal mean dry temperatures for the boreal winter season (Dec-Jan-Feb) 2003/04 (left panel) and the summer season (Jun-Jul-Aug) 2004 (right panel), respectively.

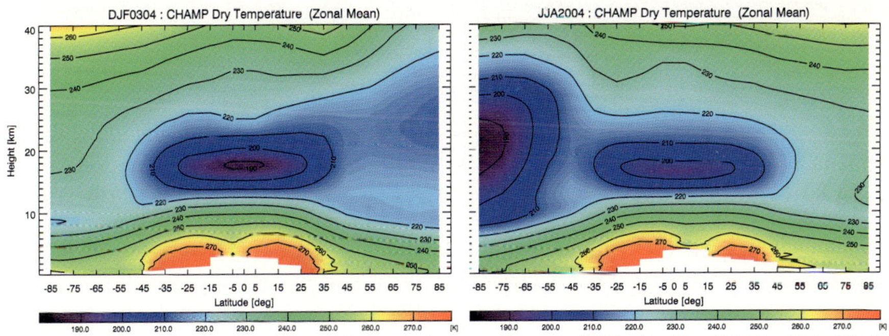

Fig. 3. CHAMP seasonal and zonal mean dry temperature fields for Dec 2003 – Feb 2004 (left panel) and Jun – Aug 2004 (right panel).

5.1 Observational Error

Latitudinally and vertically resolved difference statistics have been computed by comparing each CHAMP RO profile with a co-located ECMWF analysis profile. The systematic differences (sampling errors excluded) for two (northern) winter and two summer seasons (DJF 2002/03, JJA 2003, DJF 2003/04, JJA 2004) are shown in the left panels of Fig. 4 (taking ECMWF as reference). While differences in the lower troposphere can be clearly attributed to RO errors, the differences above 30 km are most probably due to errors in both CHAMP and ECMWF. In the height range, where RO data have the highest quality (~8 km to ~30 km), the agreement between CHAMP and ECMWF is, in general, very good: The absolute bias is <0.5 K, occasionally peaking at 1 K. However, two features are prominent:

- The tropical tropopause region in the CHAMP-derived fields is consistently warmer than the ECMWF analyses. This difference is probably caused by a weak representation of atmospheric wave activity and tropopause height variability in ECMWF fields, but work is ongoing to explain the discrepancies in detail.
- The wave-like bias structure with a magnitude of several degrees in the southern winter polar vortex region (JJA 2003 and JJA 2004) is caused by deficiencies in the representation of the austral polar vortex in the ECMWF analyses. A detailed analysis can be found in Gobiet et al. (2005b). During JJA 2004 this bias structure is less pronounced, due to the addition of new data to the ECMWF analysis scheme in October 2003 (AIRS radiances) and changes in the assimilation scheme like bias adjustments of satellite data (A. Simmons, ECMWF, pers. communication, 2005).

The apparent observational error in the UTLS region is therefore not only caused by errors in CHAMP RO data but contains also considerable errors of the reference dataset (ECMWF).

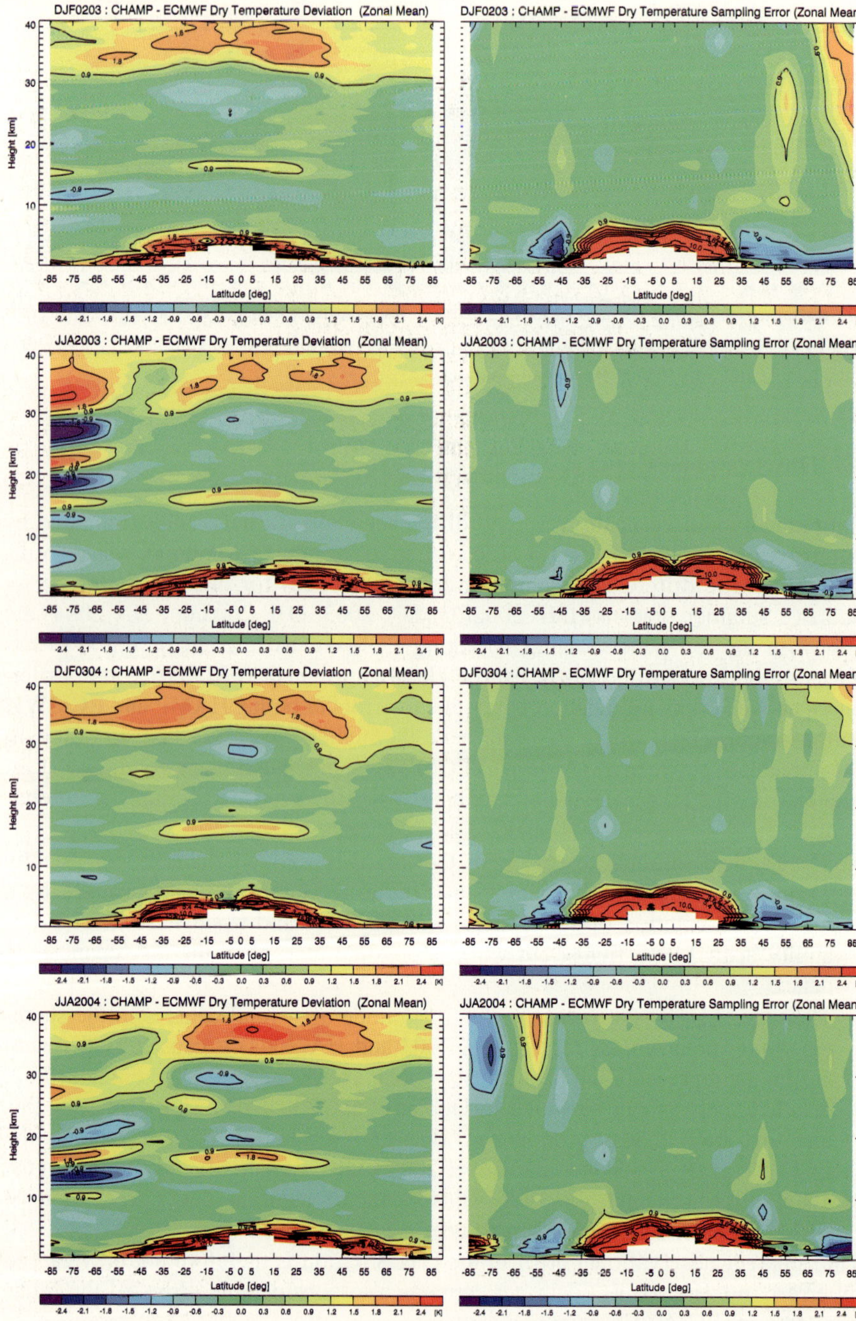

Fig. 4. Seasonal mean and zonal mean dry temperature deviations between CHAMP and ECMWF for DJF 2002/03, JJA 2003, DJF 2003/04, and JJA 2004 (left panels). Estimated CHAMP sampling errors for the same seasons (right panels).

5.2 Sampling Error

The error due to spatial and temporal undersampling of the true evolution of atmospheric fields has been identified as a potential major error source for single-satellite climatologies with the aid of simulation studies (Foelsche et al. 2003). Even with perfect observations at the occultation locations the "measured" climatologies would differ from the "true" ones as the sampling through occultation events is discrete and not dense enough to capture the entire spatio-temporal variability of the atmosphere. Under the assumption that the ECMWF analysis fields (4 time layers per day) represent the true state of the atmosphere, we can estimate the sampling error by comparing climatologies derived from the "true" ECMWF profiles at the RO locations with climatologies derived from the "true" 3D ECMWF fields using the complete field grid. The results are displayed in the right column of Fig. 4. Above ~8 km the sampling error is, in general, <0.5 K.

In the lower troposphere at low and mid-latitudes, however, there is a large "warm" sampling error for dry temperatures. This feature can be interpreted as a selective "dry sampling error". The tracking of CHAMP signal and the geometric optics retrieval tends to stop at higher altitudes in moist compared to dry conditions. The lowest part of the RO ensembles is therefore biased towards dry conditions, resulting in a systematic under-representation of the true mean refractivity (see Eq. 1). When the refractivities are converted to dry temperatures, this systematic error maps into warm-biased mean dry temperatures. This effect is most pronounced at low latitudes, where the event density is particularly low (see Fig. 1) due to the high inclination of the CHAMP satellite (87.3°). The implementation of a wave optics algorithm in the WegCenter/CCR retrieval will reduce this "dry sampling error", but it will remain an important error source for RO based climatologies at low latitudes below ~8 km. Operational CHAMPCLIM dry-retrieval climatologies will therefore be provided down to 8 km at low latitudes and down to 4 km at high latitudes (see Borsche et al. 2006, this issue).

The total climatological error, which can be estimated by computing differences of RO based and reference climatologies, is a combination of sampling and observational error (not shown).

6 Concluding Remarks

Our results show that accurate zonal mean seasonal climatologies between 8 km and 30 km height can be obtained even with data from a single RO receiver. Future RO missions like the Taiwan/US FORMOSAT-3/COSMIC constellation with 6 LEOs will provide thousands of RO profiles per day, but already now RO based climatologies have the potential to improve modern operational climatologies in regions where the data coverage and/or the vertical resolution and accuracy of RO data is superior to traditional data sources.

CHAMPCLIM activities will continue in the future. Climatologies of other atmospheric parameters (like refractivity and geopotential height) are currently being prepared and will be validated. Error models for these parameters (currently available for refractivity) will be developed. The next version of the WegCenter/CCR retrieval will include wave-optics based tropospheric bending angle retrieval, and 1D-Var retrieval of temperature and humidity in the troposphere. A provision of CHAMPCLIM climatologies (including error estimates and tropopause parameters) to the scientific community is planned for early 2006.

Acknowledgments. The authors thank A. Simmons (ECMWF Reading), and S.S. Leroy (Harvard University) for valuable scientific discussions, and E. Manzini (INGV Bologna) and L. Kornblueh (MPI-M Hamburg) for their support and the provision of ECHAM5 data. ECMWF kindly provided validation data. Funding support for the research related to the CHAMPCLIM project is received from the START Programme Y103-N03 of the Austrian Science Fund (FWF) and from Contract ASAP-CO-004/03 of the Austrian Space Agency (ASA).

References

Anthes RA, Rocken C, Kuo Y (2000) Applications of COSMIC to meteorology and climate. Terr Atmos Oceanic Sci 11(1):115–156

Borsche M, Gobiet A, Steiner AK, Foelsche U, Kirchengast G, Schmidt T, Wickert J (2006) Pre-Operational Retrieval of Radio Occultation Based Climatologies. This issue

Christy JR, Spencer RW (2003) Reliability of satellite data sets. Science 301:1046–1047

Foelsche U, Kirchengast G, Steiner AK (2003) Global climate monitoring based on CHAMP/GPS radio occultation data. In: First CHAMP mission results for gravity, magn. and atm. Studies. Reigber C, Lühr H, Schwintzer P (eds), Springer, pp 397–407

Foelsche U, Gobiet A, Loescher A, Kirchengast G, Steiner AK, Wickert J, Schmidt T (2005) The CHAMPCLIM Project: An overview. In: Earth Observation with CHAMP: Results from Three Years in Orbit. Reigber C, Luehr H, Schwintzer P, Wickert J (eds), Springer, Berlin, pp 615–619

Gobiet A, Kirchengast G (2004) Advancements of GNSS Radio Occultation Retrieval in the Upper Stratosphere for Optimal Climate Monitoring Utility. J Geophys Res 109, D24110, doi:10.1029/2004JD005117

Gobiet A, Steiner AK, Retscher C, Foelsche U, Kirchengast G (2004) Radio Occultation Data and Algorithms Validation Based on CHAMP/GPS Data. IGAM/UniGraz Tech Rep for ASA 1/2004, 46 pp

Gobiet A, Kirchengast G, Wickert J, Retscher C, Wang D-Y, Hauchecorne A (2005a) Evaluation of Stratospheric Radio Occultation Retrieval using Data from CHAMP, MIPAS, GOMOS, and ECMWF Analysis Fields. In: Earth observation with CHAMP - Results from three years in orbit, Reigber C, et al. (eds), Springer, Berlin, pp 531–536

Gobiet A, Foelsche U, Steiner AK, Borsche M, Kirchengast G, Wickert J (2005b) Climatological validation of stratospheric temperatures in ECMWF operational analyses with CHAMP radio occultation data. Geophys Res Lett 32, L12806, doi:10.1029/2005GL022617

Hajj GA, Ao CO, Iijima PA, Kuang D, Kursinski ER, Mannucci AJ, Meehan TK, Romans LJ, de la Torre Juarez M, Yunck TP (2004) CHAMP and SAC-C atmospheric occultation results and intercomparisons. J Geophys Res 109, D06109, doi:10.1029/2003JD003909

IPCC (2001) Climate change 2001: The scientific basis. Cambridge Univ Press, 881 pp

Kirchengast G, Fritzer J, Ramsauer J (2002) End-to-end GNSS Occultation Performance Simulator Version 4 (EGOPS4) Software User Manual (Overview and Reference Manual), Tech Rep ESA/ESTEC-5/2001, IGAM, Univ of Graz, Austria, 472 pp

Leroy SS (1997) The measurement of geopotential heights by GPS radio occultation. J Geophys Res 102(D6):6971–6986

Leroy SS, Dykema JA, Anderson JG (2006) Climate Benchmarking Using GNSS Occultation. This issue

Loiselet M, Stricker N, Menard Y, Luntama J-P (2000) GRAS – MetOp's GPS-based atmospheric sounder. ESA Bulletin, 102:38–44

Löscher A, Kirchengast G (2006) Optimal Fusion of Radio Occultation Profiles with GCM Fields within a 3D-Var Framework. This issue

Mears CA, Wentz FJ (2005) The effect of diurnal correction on satellite-derived lower tropospheric temperature. Science 309:1548–1551, doi: 10.1126/science.1114772

Poli P (2006) Assimilation of GNSS radio occultation data into numerical weather prediction. This issue

Rocken C, Kuo Y-H, Schreiner WS, Hunt D, Sokolovskiy S, McCormick C (2000), COSMIC system description. Terr Atmo Oceanic Sci 11(1):21–52

Roeckner E, Baeuml G, Bonaventura L, Brokopf R, Esch M, Giorgetta M, Hagemann S, Kirchner I, Kornblueh L, Manzini E, Rhodin A, Schlese U, Schulzweidea U, Tompkins A (2003) The atmospheric general circulation model ECHAM5 Part 1. MPI Rep, 349, Max-Planck-Inst f Meteorol, Hamburg, Germany

Smith EK, Weintraub S (1953) The constants in the equation for atmospheric refractive index at radio frequencies. Proc IRE 41:1035–1037

Steiner AK, Kirchengast G (2005) Error analysis for GNSS radio occultation data based on ensembles of profiles from end-to-end simulations. J Geophys Res 110, D15307, doi:10.1029/2004JD005251

Steiner AK, Gobiet A, Foelsche U, Kirchengast G (2004) Radio Occultation Data Processing Advancements for Optimizing Climate Utility. Tech Rep ASA 3/2004, Inst for Geophys, Astrophys, and Meteorol, Univ of Graz, Austria, pp 87

Steiner AK, Löscher A, Kirchengast G (2006) Error Characteristics of Refractivity Profiles Retrieved from CHAMP Radio Occultation Data. This issue

Stendel M (2006) Monitoring Climate Variability and Change by Means of GNSS Data. This issue

Syndergaard S (1999) Retrieval analysis and methodologies in atmospheric limb sounding using the GNSS radio occultation technique. DMI Sci Rep 99-6, Danish Met Inst, Copenhagen, Denmark, 131 pp

Vedel H, Stendel M (2004) GPS RO refractivities as a climate change measure. Clim Change, submitted

Vinnikov KY, Grody NC (2003) Global warming trend of mean tropospheric temperature observed by satellites. Science 302:269–272

Wickert J, Reigber C, Beyerle G, König R, Marquardt C, Schmidt T, Grunwaldt L, Galas R, Meehan T, Melbourne W, Hocke K (2001) Atmosphere Sounding by GPS Radio Occultation: First results from CHAMP. Geoph Res Lett 28:3263–3266

Wickert J, Schmidt T, Beyerle G, König R, Reigber C, Jakowski N (2004) The radio occultation experiment aboard CHAMP: Operational data analysis and validation of vertical atmospheric profiles. J Met Soc Japan 82:381–395

Pre-Operational Retrieval of Radio Occultation Based Climatologies

M. Borsche[1], A. Gobiet[1], A. K. Steiner[1], U. Foelsche[1], G. Kirchengast[1], T. Schmidt[2], and J. Wickert[2]

[1] Wegener Center for Climate and Global Change (WegCenter) and Institute for Geophysics, Astrophysics, and Meteorology (IGAM), University of Graz, Austria
michael.borsche@uni-graz.at
[2] GeoForschungsZentrum Potsdam (GFZ), Department 1, Geodesy & Remote Sensing, Potsdam, Germany

Abstract. CHAMPCLIM is a joint project of WegCenter/UniGraz and GFZ Potsdam. The overall aim of the project is to exploit the CHAMP (CHAllenging Minisatellite Payload for geoscientific research) radio occultation (RO) data in the best possible manner for climate monitoring. This paper focuses on describing the pre-operational status and technical aspects of the CHAMPCLIM processing system at WegCenter/UniGraz. For creating RO based climatologies we ingest, on the one hand, the complete CHAMP RO dataset provided by GFZ at excess phase level (GFZ level 2, ~180 profiles/day), which is processed to obtain atmospheric profiles of refractivity, geopotential height, and temperature (in future also humidity). On the other hand, we use operational atmospheric analysis fields from the European Centre for Medium-Range Weather Forecasts (ECMWF), at T42L60 resolution, as reference for quality control and evaluation. For delivering climatologies operationally, which will be prepared at monthly, seasonal, and annual time scales, our aim is to provide them with a delay of at most two weeks after the last measurement (e.g., JJA 2003 seasonal climatology available by September 14, 2003, latest). The climatologies are set up in overlapping equal-area and non-overlapping almost equal-area grids. In order to monitor the error characteristics of the climatologies, various types of error statistics (vs. ECMWF analyses) are performed. The main emphasis of this paper lies on processing the complete 2002–2004 data – starting from March 2002 when the CHAMP data stream became stable and quasi continuous – and on creation of climatologies including error estimates. The spatial set up of the climatologies, exemplary seasonal climatologies (as far as processed) as well as preliminary climatological error estimates are presented.

1 Introduction

The main advantages of probing the Earth's atmosphere with the self-calibrating radio occultation (RO) sounding method using the Global Positioning System (GPS) are long term stability, high vertical resolution and accuracy, all-weather capability, and global coverage. These characterizations make the RO method near-ideal for climate monitoring.

In July 2000 the German research satellite CHAMP (CHAllenging Minisatellite Payload) was launched and, since March 2002, it continuously records up to 280 RO events per day (Wickert et al. 2004, 2006). Since its lifetime is projected to last until at least 2007, this mission provides the first opportunity to create RO based climatologies over a longer time period.

The CHAMPCLIM project, which is undertaken at WegCenter/UniGraz (Foelsche et al. 2005) in close cooperation with the GFZ Potsdam, aims at exploiting the CHAMP RO data in particular for use in climate monitoring and research. The main focus lies on optimizing RO retrieval algorithms for climate utility and creating global RO based climatologies.

In Section 2, a description of the pre-operational CHAMPCLIM processing system currently under development at WegCenter is given. Section 3 explains the two different binning modes of the climatologies, which are the main product of the pre-operational CHAMPCLIM processing system. Furthermore, the summer season (JJA) 2003 is depicted as an example for the climatology. Finally, in Section 4 the conclusions of the paper and an outlook to further work are given.

2 Pre-Operational Retrieval Status

The CHAMPCLIM retrieval chain starts with CHAMP data provision by GFZ Potsdam. GFZ provides operational occultation measurements and corresponding analysis results such as atmospheric excess phases and vertical profiles of refractivity, temperature, and water vapor via the CHAMP data center (http://isdc.gfz-potsdam.de/champ; for details see Schmidt et al. (2005) and Wickert et al. (2006)). As input data for the WegCenter processing chain serve the calibrated atmospheric excess phase data (level 2, on average ~180 files per day). The GFZ-internal calibration process is described by Wickert et al. (2004).

As reference data, 6-hourly analysis fields of the European Centre for Medium-Range Weather Forecasts (ECMWF) are used (four time layers per day; 6 UTC, 12 UTC, 18 UTC, 24 UTC), which are downloaded directly from ECMWF.

The RO retrieval scheme applied to these data is developed at WegCenter (Gobiet and Kirchengast 2004; Steiner et al. 2004) in close connection with developments of the End-to-end GNSS Occultation Performance Simulator (EGOPS) (Kirchengast et al. 2002). The retrieval is especially focused on minimizing the bias of atmospheric parameters. Table 1 below provides an overview on the main aspects of the retrieval scheme. The retrieval implements statistical optimization in two different ways, one of which uses the MSISE-90 climatology (Hedin 1991) (WegCenter/MSIS) and the other operational ECMWF analysis fields (WegCenter/ECMWF) as background information. This background information is integrated into the retrieval only at one point (bending angle level), resulting in well defined error characteristics and allowing to initialize the hydrostatic integral at very high altitudes (120 km), where the upper-boundary initialization has no effect on the retrieved atmospheric parameters in the height interval under consideration below 50 km (Gobiet et al. 2004).

Table 1. Summary of some main aspects of the CHAMPCLIM retrieval (CCR v2) (after Steiner et al. (2004)).

	WegCenter/MSIS	WegCenter/ECMWF
Outlier Rejection and Smoothing	"3σ" outlier rejection on phase delays and smoothing using regularization (third order norm, regularization parameter = $10^{(\text{sampling rate}/10)}$) (Syndergaard 1999).	Like WegCenter/MSIS
Bending Angle Retrieval	Geometric Optics retrieval.	Like WegCenter/MSIS
Ionospheric Correction	Linear combination of bending angles (Vorob'ev and Krasil'nikova 1994). Correction is applied to low-pass filtered bending angles (1 km sliding average), L1 high-pass contribution is added after correction (Hocke et al. 2003). L2 bending angles < 15 km derived via L1-L2 extrapolation.	Like WegCenter/MSIS
Bending Angle Initialization	Statistical optimization of bending angles from 30 km to 120 km. Vertically correlated background (corr. length $L = 6$ km) and observation ($L = 1$ km) errors. Observational error estimated from observed profile > 65 km. Background error: 15 %. Background information: MSISE-90 (Hedin 1991) best fit-profile, bias corrected (Gobiet and Kirchengast 2004).	Like WegCenter/MSIS, but co-located bending angle profile derived from ECMWF operational analysis as backg. information (above ~60 km: MSISE-90). No further pre-processing.
Hydrostat. Integral Initialization	At 120 km: pressure = p(MSISE-90).	Like WegCenter/MSIS
Humidity Retrieval	Optional: 1D-Var using ECMWF short-range forecasts as background.	Like WegCenter/MSIS
Quality Control	Refractivity 5 km to 35 km: $\Delta N/N < 10$ %; Temperature 8 km to 25 km: $\Delta T < 20$ K. Reference: ECMWF operational analysis (T42L60).	Like WegCenter/MSIS

Recent work at WegCenter focused on the establishment of a pre-operational processing system, which includes data transfer from GFZ and ECMWF, retrieval of atmospheric parameters, quality control, creation of climatologies, and storage of data. At this stage, the processing system operates in an automated way up to the RO retrieval and quality control, whilst the automated creation of climatologies is currently integrated. The aim is to establish a data stream, which contains the most recent 7-days of data provided within a time delay of two days. Furthermore, the climatologies are aimed to be provided within a timeliness of 14 days.

So far, the dataset ranging from March 2002 to February 2004 was transferred from GFZ as well as from ECMWF and processed at WegCenter. More than 80 % of the excess phase profiles analyzed pass the CHAMPCLIM quality control yielding ~150 atmospheric profiles per day.

For validation purposes, subsets of the retrieved CHAMP data were compared at refractivity and dry temperature level to various data sources. Figure 1 shows two comparisons, one with ECMWF operational analysis fields (Fig. 1a) and the other (Fig. 1b) with data from the Michelson Interferometer for Passive Atmospheric Sounding (MIPAS) on the ESA satellite ENVISAT (Gobiet et al. 2004). This analysis was carried out by considering 3160 CHAMP occultation events and 184 co-located ENVISAT/MIPAS events, observed during September 2002. Figure 1 depicts the bias (bold) and bias ± standard deviation profiles (light) of dry temperature.

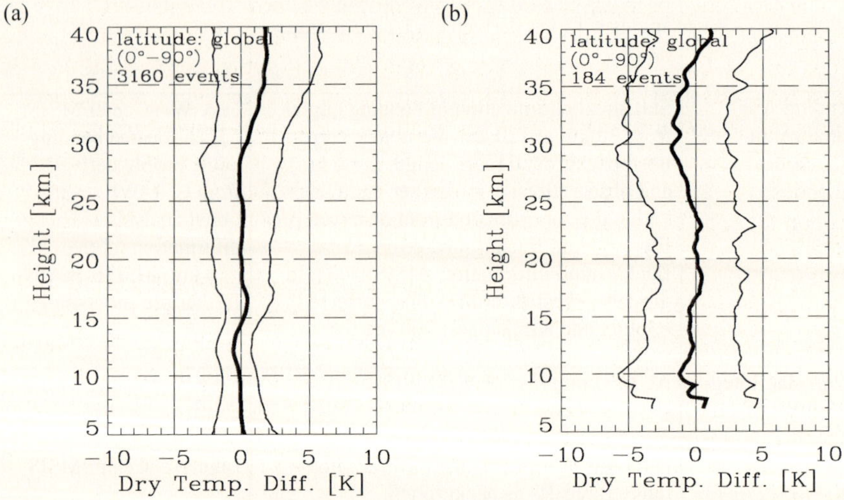

Fig. 1. WegCenter/ECMWF dry temperature profiles retrieval. Comparison with co-located ECMWF analysis profiles (a) and with co-located ENVISAT-MIPAS profiles (b), respectively (after Gobiet et al. 2004).

3 Setup of Climatologies

The implementation of the climatologies as final product is pursued in two different ways. On the one hand, a global 3D-Var analysis by assimilation of CHAMP data into ECMWF short-term forecast fields was developed, which is described by Löscher and Kirchengast (2006). On the other hand, a direct-binning grid strategy is implemented in two modes, which will be the focus of this section.

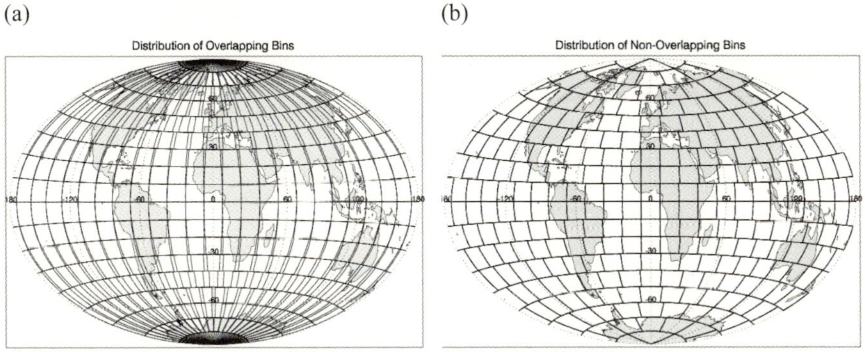

Fig. 2. Two modes of direct binning. Overlapping equal-area bins at regular 18 lat x 24 lon grid (a), and non-overlapping almost equal-area bins at 18 lat x lat-dependent lon grid (b).

The difference between both direct-binning grids is how the equal-area bins are arranged as shown in Figure 2. Along latitude, both modes have the meridian divided into 18 bins of 10° width. Along longitude, the first mode (Fig. 2a) uses 24 fixed bins (baseline) at all latitudes leading to bin overlapping at high latitudes, whilst the second mode (Fig. 2b) uses a latitude-dependent number of bins to obtain non-overlapping almost equal-area bins (within ±0.5° exact; except for polar latitudes where the latitude extension differs up to 7.9° per bin). While climatologies on the regular grid (lat x lon) are most convenient to handle, the second mode allows to avoid horizontal error correlations.

The prime testbed season was arbitrarily chosen to be JJA 2003. Due to the high inclination of the CHAMP satellite the global event distribution varies from sparse sampling in the equatorial region to more dense sampling in the polar region. As shown in Figure 3(a), illustrating the event distribution for a 10° x 15° binning for the testbed season, there are hardly sufficient events in the equatorial region for deriving robust statistics as is required for climatologies. Figure 3(b) depicts the corresponding sampling error climatology field showing sampling errors of up to around ±1.5 K.

There are rare occasions in which an equatorial bin of a 10° x 15° latitude slice does not contain any RO events. In the two year set of CHAMP data there are only a few seasonal latitude slices for which this occurs. In monthly climatologies there are many more latitude slices with no events. Therefore, we have decided to broaden the gridding of the climatologies to 10° x 60° (18 lat x 6 lon grid). Figure 3(c) shows the event distribution of that grid exemplary on the meridional latitude slice in the testbed season. For direct comparison, in Figure 3(d) the corresponding sampling error field is depicted. Here, the sampling error amounts to less than in the previous case ranging up to around ±1 K; additionally, the variability of the sampling error is smaller. If the climatologies are examined as zonal means, the sampling error diminishes even further to below ±0.5 K (compare Foelsche et al. 2006).

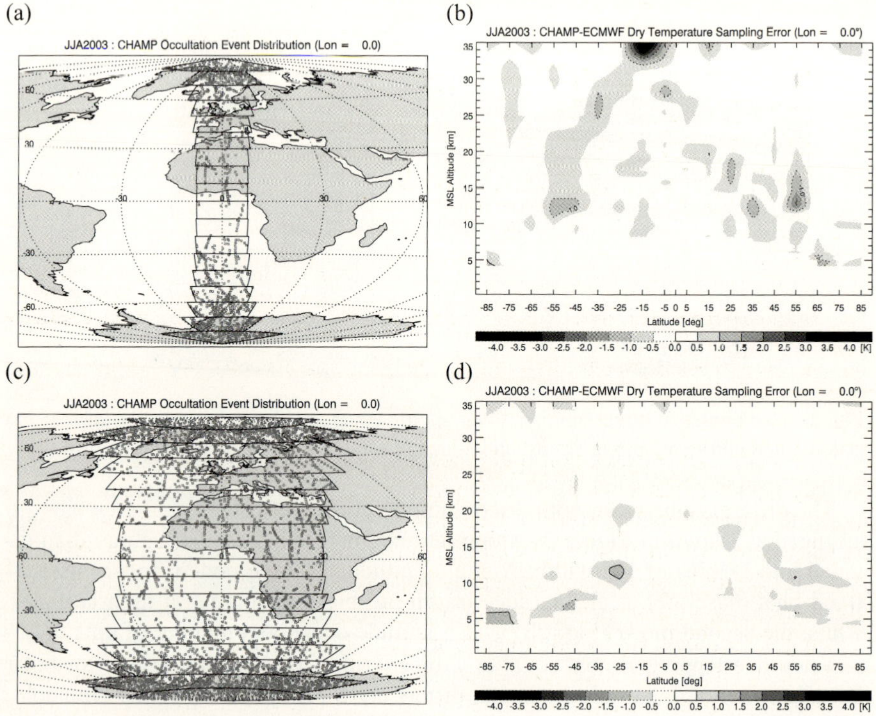

Fig. 3. Event distribution in a latitude slice of 10° x 15° binning (a) and its corresponding sampling error climatology (b) of testbed season JJA 2003; (c) and (d) same as above for 10° x 60° binning.

In Figure 4 an example of a zonal mean climatology is given. Again, the prime testbed season JJA 2003 is depicted for zonal-mean dry temperature (Fig. 4a) and the deviation compared to ECMWF analysis fields (Fig. 4b). The vertical range of the climatologies shown here extends between 4 km at the bottom to 35 km at the top. This range characterizes the area in which the best results for GPS based RO measurements are obtained. Additionally, the climatologies are cut off at the lower end at varying height increasing from the poles towards the equator. From the poles to 60° latitude they reach down to 4 km, the cut-off height then increases over the mid latitude bins to 8 km at low latitudes (equator to 30° north and south). The reason for the cut-off strategy is reasonable due to biased sampling in the region of the lower troposphere and the dry air processing applied. The error in sampling is due to the fact that the tracking of the CHAMP signal tends to stop at higher altitudes in moist compared to dry conditions. Especially in the lower tropical troposphere this can lead to a warm bias in dry temperature in the order of 10 K (compare Foelsche et al. 2006).

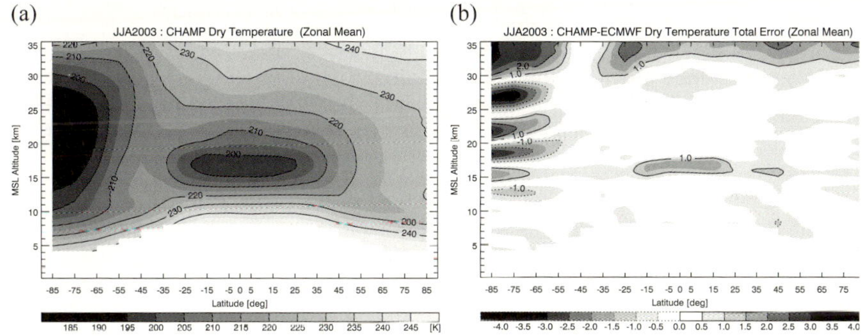

Fig. 4. Zonal mean dry temperatures of JJA 2003 (a), and deviation of CHAMP dry temperatures relative to ECMWF (b).

The temperature deviation field compared to ECMWF analyses (Fig. 4b) mainly equals to zero but there are some distinct features. Above about 30 km a consistent warm bias of about 1 K and more stretches across the whole latitude band. The cause for this bias is ambiguous because at that height the ECMWF analyses are only partly constrained by observational data and the RO measurement errors start to increase. However, validation to independent data (see Section 2) suggest that the CHAMP RO results are still of high quality.

Another feature in the deviation field is a warm bias of up to 2 K in the tropical tropopause region. As has been stated in Gobiet et al. (2005) and is under closer investigation at the moment, this bias may mainly be attributed to the ECMWF analyses and is probably caused by too weak representation of atmospheric wave activity and tropopause height variability.

Probably the most salient feature is located at the southern winter polar vortex. The CHAMP data exhibit a clear alternating deviation in the order of ±3 K compared to the ECMWF data. This deviation can be attributed to be a bias of ECMWF analyses as discussed in and Gobiet et al. (2005), being a good example of the value of RO data as validation.

The processing of the March 2002 to February 2004 time span has been completed in early September 2004 and preliminary results for selected seasons are depicted in Foelsche et al. (2006).

4 Conclusions and Outlook

A description and short report of the pre-operational CHAMPCLIM climatology retrieval scheme was given. It includes data transfer from ECMWF for operational analysis and from GFZ for CHAMP RO data, as well as data provision within a timeliness of at most 14 days.

Two different modes of binning the climatologies were discussed and first results of the summer season 2003 were shown.

In the near future, some already prepared upgrades to the retrieval scheme are planned, which include extraction of tropopause parameters, extraction of ECMWF reference profiles along the 3D tangent point trajectory, geoid reference, wave-optics based tropospheric bending angle retrieval, and 1D-Var based tropospheric processing. Also generation of error-estimate fields and first 3D-Var fields is foreseen.

In 2005 it is planned to set up an operational data stream from GFZ for processing the data within the above noted timeliness targets and to also include processing of the missing time span from March 2004 onwards. Preparations for ingestion of additional data from further missions (e.g., GRACE, Metop/GRAS, COSMIC) will start as well.

Acknowledgments. The authors thank J. Fritzer for continuous support concerning computational issues. Funding support for the research within the CHAMPCLIM Project is received from the START Programme Y103-N03 of the Austrian Science Fund (FWF) and from Contract ASAP-CO-004/03 of the Austrian Space Agency (ASA).

References

Foelsche U, Gobiet A, Loescher A, Kirchengast G, Steiner AK, Wickert J, Schmidt T (2005) The CHAMPCLIM Project: An Overview. In: Reigber C, Luehr H, Schwintzer P, Wickert J (eds) Earth Observation with CHAMP: Results from Three Years in Orbit. Springer, Berlin, pp 615–620

Foelsche U, Gobiet A, Steiner AK, Borsche M, Wickert J, Schmidt T, Kirchengast G (2006) Global Climatologies Based on Radio Occultation Data: The CHAMPCLIM Project. This issue

Gobiet A, Kirchengast G (2004) Advancements of Global Navigation Satellite System Radio Occultation Retrieval in the Upper Stratosphere for Optimal Climate Monitoring Utility. J Geophys Res 109(D24110), doi: 10.1029/2004JD005117

Gobiet A, Steiner AK, Retscher C, Foelsche U, Kirchengast G (2004) Radio Occultation Data and Algorithms Validation Based on CHAMP/GPS Data. IGAM/Uni Graz Tech Rep for ASA No 1/2004

Gobiet A, Foelsche U, Steiner AK, Borsche M, Kirchengast G, Wickert J (2005) Climatological validation of stratospheric temperatures in ECMWF operational analyses with CHAMP radio occultation data. Geophys Res Lett 32(L12806), doi:10.1029/2005GL022617

Hedin AE (1991) Extension of the MSIS thermosphere model into the middle and lower atmosphere. J Geophys Res 96:1159–1172

Hocke K, Igarashi K, Tsuda T (2003) High-resolution profiling of layered structures in the lower stratosphere by GPS occultation. Geophys Res Lett 30(8):1426, doi:10.1029/2002GL016566

Kirchengast G, Fritzer J, Ramsauer J (2002) End-to-end GNSS Occultation Performance Simulator Version 4 (EGOPS4) Software User Manual (Overview and Reference Manual). Tech Rep for ESA/ESTEC, Inst for Geophys, Astrophys, and Meteorol, Univ of Graz, Austria, No 3/2002

Löscher A, Kirchengast G (2006) Assimilation of GNSS Radio Occultation Profiles into GCM Fields for Global Climate Analysis. This issue

Schmidt T, Wickert J, Beyerle G, König R, Galas R, Reigber C (2005) The CHAMP Atmospheric Processing System for Radio Occultation Measurements. In: Reigber C, Luehr H, Schwintzer P, Wickert J (eds) Earth Observation with CHAMP: Results from Three Years in Orbit. Springer, Berlin, pp 597–602

Steiner AK, Gobiet A, Foelsche U, Kirchengast G (2004) Radio Occultation Data Processing Advancements for Optimizing Climate Utility. IGAM/Uni Graz Tech Rep for ASA No 3/2004

Syndergaard S (1999) Retrieval analysis and methodologies in atmospheric limb sounding using the GNSS radio occultation technique. DMI Sci Rep 99-6, Danish Meteorol Inst, Copenhagen, Denmark

Vorob'ev VV, Krasil'nikova TG (1994) Estimation of the accuracy of the atmospheric refractive index recovery from Doppler shift measurements at frequencies used in the NAVSTAR system. Phys Atmos Ocean 29:602–609

Wickert J, Galas R, Schmidt T, Beyerle G, Reigber C, Förste C, Ramatschi M (2004) Atmospheric sounding with CHAMP: GPS ground station data for occultation processing. Phys Chem Earth 29:267–275

Wickert J, Schmidt T, Beyerle G, Michalak G, König R, Heise S, Reigber C (2006) GPS radio occultation with CHAMP and GRACE: Recent results. This issue

Assimilation of GNSS Radio Occultation Profiles into GCM Fields for Global Climate Analysis

A. Löscher and G. Kirchengast

Wegener Center for Climate and Global Change (WegCenter) and Institute for Geophysics, Astrophysics, and Meteorology (IGAM), University of Graz, Austria
armin.loescher@uni-graz.at

Abstract. This paper investigates the application of the 3D-Var methodology within a global climate monitoring framework. It studies the assimilation of GNSS radio occultation derived refractivity profiles into ECMWF analysis or short-term forecast fields as background. The system is tuned for high vertical and moderate horizontal resolution, best suited to the spatial characteristics of these satellite based measurements. The analyses are performed on a GCM-compliant Gaussian grid, comprising 60 model levels up to a height of ~60 km and a horizontal resolution corresponding to a triangular spectral truncation T42 (i.e., T42L60). Within the system two different operational modes are implemented, the first updating a refractivity background, derived from ECMWF analysis fields, the second directly updating the ECMWF temperature, specific humidity and surface pressure fields. First results indicate a significant analysis increment, emphasising the ability of RO data to add independent information to ECMWF analysis fields.

1 Introduction

Relatively new measurement concepts like GNSS RO (Global Navigation Satellite System Radio Occultation) offer the opportunity to develop new processing techniques and strategies to exploit the data in the best possible and most efficient way. The RO experiment on-board CHAMP (e.g., Wickert et al. 2004) is the first system, which delivers continuous observations on a quasi-operational basis, preparing the ground for the first RO-only based global climatologies. With the GRAS (GNSS Receiver for Atmospheric Sounding) sensor on-board the MetOp (Meteorological Operational) satellite (e.g., Loiselet et al. 2000) a fully operational system delivering RO observations will be available from 2006 onwards. On the other hand, the development of NWP systems during the last years improved the forecast skill and the quality of the analyses continuously. Thus it would be interesting to use the same methodology as used by NWP, to introduce data into the model (in our case via 3D-Var), first working with single sets of observations, later with whole climatologies, in order to study the increments of monthly and seasonal mean fields.

2 3D-Var System Implementation

2.1 Coordinate System

A GCM (Global Circulation Model) compliant Gaussian grid corresponding to T42L60, i.e., 64 latitude × 128 longitude grid points, comprising 60 vertical model levels is used. The vertical coordinate system is based on the hybrid pressure coordinate provided by ECMWF analysis fields. From this basic vertical coordinate system, grids of geopotential height and geometric height (over reference ellipsoid) can be derived. The assimilation scheme can be used either with geometric height or geopotential height.

2.2 N and TQP_{surf} Analysis

The whole system can be run in two different modes. The first one performs a refractivity analysis (N analysis scheme), which means that refractivity observations are assimilated into a refractivity background. This background field is derived from ECMWF temperatures, specific humidity, and surface pressure fields using basically the same operators, which are used in the TQP_{surf} version of the assimilation scheme. This operation is performed at the beginning of the procedure, during the assimilation itself only the interpolation operators are used. The TQP_{surf} version of the assimilation scheme directly updates the temperature, specific humidity, and surface pressure background fields, which means that all fields are interpolated separately. At each successful iteration, the new pressure field has to be derived from the updated surface pressure field, and the refractivity has to be calculated at the location of the observation. After comparison of background refractivity (also denoted as model observation) and observation, the gradients of the input fields and observations are calculated and a suitable correction is applied.

2.3 Incremental 3D-Var

The solution of the minimization problem can be performed either in terms of full cost function fields $J(\mathbf{x}_a)$ or in terms of an analysis of increments

$$J(\mathbf{x}_a) = J(\mathbf{x}'_a = \mathbf{x}_a - \mathbf{x}_b), \qquad (1)$$

where $\mathbf{x}_a$ and $\mathbf{x}_b$ are the analysis and the background state, respectively, and the prime superscript denotes the (1st order) increment. We chose the latter solution method, which provides optimal analysis increments, which are added to the unmodified background field. This procedure has a number of advantages like the use of linearized control variable transforms, which allow the straightforward use of adjoints to calculate the gradient of the cost function. Another advantage is that any imbalance introduced through the analysis procedure is limited to the small increments, which are added to the balanced first guess.

2.4 Control Variables

The control variables used in the analysis are temperature, specific humidity, and surface pressure, or refractivity transformed to logarithmic (LOG) space, to get a better conditioned problem, within the refractivity-only assimilation framework. In order to avoid negative specific humidities in the analysis and to get a better conditioned problem, the specific humidity is also transformed to and analyzed in LOG space. The cross correlations between the control variables are assumed to be small enough to be neglected. This assumption serves to effectively block-diagonalize the background error covariance matrix. For each control variable there still remains both, horizontal and vertical correlations. Those are assumed to be separable, which is a reasonable and widely used assumption.

Control Space Transformations

For a model state **x** with n degrees of freedom, minimization of the cost function requires $O(n^2)$ calculations (Bouttier and Courtier 1999), thus becoming prohibitively expensive for usual n's. One practical solution to this problem is to perform the minimization in a control variable space **v** given by

$$\mathbf{x} = \mathbf{U}\,\mathbf{v}. \tag{2}$$

The transform **U** has to be chosen in a way that

$$\mathbf{B} = \mathbf{U}\,\mathbf{U}^{\mathrm{T}} \tag{3}$$

is approximately satisfied. In the control space **v** the number of required minimization calculations is reduced from $O(n^2)$ to $O(n)$. Furthermore, by using the transform Eq. (3), the background error covariance matrix becomes

$$\mathbf{B}_{\mathrm{v}} = \mathbf{I}, \tag{4}$$

hence effectively preconditioning the problem. **I** denotes the identity matrix, $\mathbf{B}_{\mathrm{v}}$ the vertical background error covariance matrix. In terms of increments, the control variable transform can be written as

$$\mathbf{x}' = \mathbf{U}\,\mathbf{v}. \tag{5}$$

The inverse transformation

$$\mathbf{v} = \mathbf{U}^{-1}\,\mathbf{x}' \tag{6}$$

can be specified in different ways. The definition must provide a way to break down the atmospheric state **x** into uncorrelated but physically realistic error modes, which can be penalized in J_{b} according to their estimated error magnitude (Barker 1999). See Löscher (2004) for more details.

Recursive-Filter Representation of Background Error Covariances

The control variable transform uses the identity $\mathbf{BB}^{\mathrm{T}}$ to define a transform $\mathbf{x}' = \mathbf{Uv}$, which relates preconditioned control variables $\mathbf{v}$ to analysis increments in $\mathbf{x}'$ in model space. The horizontal component $\mathbf{U}_{\mathrm{h}}$, defined by

$$\mathbf{B}_{\mathrm{h}} = \mathbf{U}_{\mathrm{h}} \mathbf{U}_{\mathrm{h}}^{\mathrm{T}}, \tag{7}$$

is realized by scaled recursive filters (RF). The RF has to be applied in a non-dimensional space (e.g., Lorenc 1992).

2.5 Minimization

The cost function is minimized by using an iterative descent algorithm, which is in our case the L-BFGS-B routine, a quasi-Newton method. The cost of the analysis is proportional to the number of cost function and its gradient evaluations, denoted as simulations. If a new state $\mathbf{x}$ is found, an iteration is performed, which means that to find a new $\mathbf{x}$, several simulations may be required. See, e.g., Byrd et al. (1994) for details.

2.6 Observation Operators

Horizontal Interpolation

The bi-linear interpolation consists of a weighted average of the four surrounding grid points to determine their interpolated value. Two linear interpolations on opposite sites are performed followed by a consecutive interpolation of these intermediate results. This horizontal interpolation is performed for the atmospheric layer above and below any given observation point.

Vertical Interpolation

Due to the fact of a globally non-uniform vertical grid, the heights of the horizontally interpolated values are also calculated by bi-linear interpolation from the vertical background grid. Given the interpolated background values above and below the spatial location of a given observation, a logarithmic vertical interpolation is finally performed to get the value of the background at the location of the observation. Generally interpolation needs great care to ensure negligible residual interpolation errors.

Vertical Coordinate Operator

The location of the observation is needed to calculate the refractivity. ECMWF provides temperature, specific humidity, and surface pressure fields. The background pressure field is derived by a series of operators. These operators are also used to set up the vertical coordinate system of the assimilation scheme either as a vertical grid of geopotential heights or a vertical grid of geometric heights, respectively. Since the pressure field is also derived, pressure coordinates are possible as well but not implemented at the moment. The values of $T_{i,j,k}$ and specific humidity $q_{i,j,k}$ are given for the Gaussian grid of the latitudes φ_j and the homogenous grid of the longitudes λ_i, and an irregularly spaced height grid $z_{i,j,k}$. For the T42L60 grid, the index ranges are $I = 1...64$, $j = 1...128$, and $k = 1...60$ for full level quantities and $k = 0...60$ for half level quantities. The pressures corresponding to the kth half and full levels are calculated by the means of standard formulae (e.g., Roeckner et al. 2003). The calculation of the geopotential heights is based on the hydrostatic equation and on an interpolation between the half and the full levels (Gorbunov and Kornblueh 2003). The geometrical heights over reference ellipsoid are calculated from the geopotential heights by the approximate formula of the US Standard Atmosphere (Gorbunov and Kornblueh 2003).

Refractivity Operator

To calculate the refractivity N at any point, it is required to know the variables temperature T [K], specific humidity q or water vapor pressure e [hPa], and pressure p [hPa] or dry air pressure p_A [hPa]. There are two standard formulae used, the Thayer and Smith-Weintraub formulae (e.g., cited in Kursinski et al. 1997). The Thayer formula is slightly more accurate, the Smith-Weintraub formula is basically the same but assumes an ideal gas (compressibility factors $Z_{A,W}$ neglected). k_1, k_2, k_3 are the refractivity formula constants (e.g., Kursinski et al. 1997).

Thayer formula:

$$N = k_1 \frac{p_A}{T} \frac{1}{Z_A} + k_2 \frac{e}{T} \frac{1}{Z_W} + k_3 \frac{e}{T^2} \frac{1}{Z_W}. \tag{8}$$

Smith-Weintraub (3 terms) formula:

$$N = k_1 \frac{p_A}{T} + k_2 \frac{e}{T} + k_3 \frac{e}{T^2}. \tag{9}$$

2.7 Calculation of Refractivity Fields from ECMWF Analyses

Given the fields of temperature, specific humidity, and surface pressure (in our case analysis fields of ECMWF), we can calculate the field of refractivity using either the Thayer or Smith-Weintraub formula. A comparison of the two formulae shows no significant differences, which is also true when compared with CHAMP refractivity data, but because of negligible additional computing cost, the more accurate Thayer formula was chosen to calculate the background in the case of pure refractivity assimilation, and as forward operator to calculate refractivity from temperature, humidity, and surface pressure fields. To derive the error characteristics in case of the pure refractivity assimilation scheme, the Smith-Weintraub formula was used.

3 Error Formulation

3.1 Observation Error Covariance

The observation error covariance only takes vertical correlations into account. Due to the separation in space and time between the different RO events (mean distances generally >300 km, time differences >1.5 hrs) this simplification is justified.

Formulation of the Observation Error Covariance Matrix

A simple error covariance matrix formulation was deduced from the empirical estimated matrices following Steiner and Kirchengast (2005). A least-squares method was used to fit analytical functions to the relative standard deviation, which shows a different behavior below and above the tropopause height z_{Tropo}. The empirical relative standard deviation S_z can be approximated with an exponential increase above z_{Tropo}, Eq. (10), and with a decrease proportional to $(1/z)$ below z_{Tropo}, Eq. (11). z_{Tropo} is defined here globally at 15 km; simplified from Steiner and Kirchengast (2005) no constant error range around 15 km is used in this study. To be able to scale the error magnitude, which is receiver dependent, the tropopause standard deviation s_{Tropo} can be tuned, we used 0.4 % based on experience with CHAMP data. Eq. (10) gives the exponential function for the relative standard deviation above z_{Tropo}, with the parameter H_{Strato}, which is the scale height of the error increase over the stratosphere. The value for global H_{Strato} was set to 11.9 km for the error scale height (Steiner and Kirchengast 2005).

$$S_z = s_{\mathrm{Tropo}} \exp[(z - z_{\mathrm{Tropo}}) / H_{\mathrm{Strato}}],\ z \geq z_{\mathrm{Tropo}} \tag{10}$$

Eq. (11) gives the analytical function for the relative standard deviation below z_{Tropo}, where the near-surface (z~1 km) error $s_0 = 4.46$ %.

$$S_z = s_{\mathrm{Tropo}} + s_0\,(z^{-1} - z^{-1}_{\mathrm{Tropo}}),\ z < z_{\mathrm{Tropo}} \tag{11}$$

This formulation of the observation error covariance also accounts for the error of representativeness, thus no additional specification of this is necessary within the assimilation framework. A somewhat improved representation of the observation error is presented by Steiner et al. (2006), based on a CHAMP data error analysis.

3.2 Background Error Covariance

The used background fields of temperature, specific humidity, and surface pressure were provided by ECMWF. Thus the basis of our considerations concerning the background errors and their correlations are based on ECMWF recommendations (M. Fisher, ECMWF Reading, UK, pers. communications, 2003). For the refractivity assimilation scheme, the error characteristics have been derived using error propagation techniques. The relative standard deviation ranges from ~1.5 % at the surface to ~0.8 % at the uppermost levels with a peak of ~2.8 % within the troposphere; see Löscher (2004) for details.

4 First Results

Among a large number of smaller simulation experiments (Löscher 2004), a quasi-operational run was performed for the "testbed month" August 2003. The complete August 2003 was processed day-by-day, dividing each day into four assimilation windows of 6 hours per day, using the corresponding ECMWF analyses as first guess. This translates into 31 independent time slices, around 00 UTC, 06 UTC, 12 UTC, and 18 UTC, delivering 124 analyses over the month. The global quite even distribution of the 4482 CHAMP RO profiles (about 150 profiles per day) of August 2003 (see Fig. 1), comprises 245220 single observational values.

A lower cut-off height of 5 km was chosen, based on other studies suggesting that the CHAMP data quality degrades rapidly below 5 km. This problem has to be solved at retrieval level by advanced wave optics methods (e.g., Jensen et al. 2003) to enable effective use of observations below 5 km. The analysis fields were averaged separately for each time layer of each day and in addition a total monthly mean was derived by averaging the time layer means. These averaged fields were compared to the corresponding monthly mean analysis fields of ECMWF. This procedure was applied for the refractivity and the temperature, specific humidity, and surface pressure assimilation schemes. The minimization process was stopped after 20 cost function and gradient evaluations (a number based on extensive sensitivity and convergence tests), leaving some safety margin, which practically meant that about 5 to 6 successful iterations were found sufficient for completing the analysis within each assimilation time window. The comparison between the two assimilation schemes (N and TQP_{surf}), agrees well, especially above the troposphere.

As an alternative approach concerning the choice of the ECMWF background fields, short-range (24 hr and 30 hr) forecasts and the corresponding error characteristics will be used in future. This will be a better choice of background data for climate monitoring applications and climate studies, since the short-term forecasts will provide physically consistent atmospheric states independent of details of the initial condition analysis, which in future will presumably have assimilated radio occultation data itself routinely via ECMWF's analysis and prediction system

(Healy and Thépaut 2005). Another possible approach might be the derivation of first guess fields from reanalysis projects like ERA40 (ECMWF 40 year Reanalysis), taking a time frame of the last 20 years into account.

As an example of the preliminary assimilation results (see Fig. 2), the increment of the total refractivity assimilation (mean of all time layers) for August 2003 for a model level near 20 km height is shown. It can be seen that any non-persistent deviations are vanishing within the monthly mean delivering a neutral result, leaving only a significant increment over the southern high latitudes. As the RO observations are an independent source of information, these results indicate a systematic deviation within certain regions of the ECMWF analysis fields. A bias within the RO refractivities themselves cannot currently be excluded from the considerations (we applied a 0.4 % correction prior to the assimilation experiments); cf., the error analysis results of Steiner et al. (2006); improved retrieval algorithms are expected to solve this problem. On the other hand, a systematic deviation introduced by the observations would be expected to be globally evenly distributed (see Fig. 1). This not being the case, strengthens evidence of a systematic deviation within certain regions of the ECMWF background data. This evidence is roughly consistent with the significantly more elaborated inter-comparison of seasonal mean RO-only climatologies of dry temperature and refractivity with the corresponding ECMWF fields (Gobiet et al. 2005; Borsche et al. 2006). Next steps of the work are further advancement of the assimilation scheme and larger-scale application to longer radio occultation datasets, using ECMWF short-range forecasts as background.

5 Conclusion

The first results are promising and show that the information content of RO data is not redundant, thus the observations introduce useful information in ECMWF analysis fields. Furthermore, these findings suggest that the operational use of RO data within NWP frameworks yields a promising perspective for the future (see also Healy and Thépaut 2005). In that respect especially the stratospheric temperature information seems to have a significant impact. A variational climate monitoring framework based on RO data will provide the opportunity to study increments over time, and gain insight into the characteristics of the used background fields. First-guess fields from re-analysis data might be another interesting background option in the future, to be independent from operational model updates.

Acknowledgments: The authors are grateful to A.K. Steiner (WegCenter, Univ of Graz, Austria) for valuable inputs regarding the observation error characteristics as well as to M. Fisher (ECMWF, Reading, UK) for supplying background error covariances. The authors also gratefully acknowledge T. Schmidt, J. Wickert and the full team at GFZ Potsdam, Germany, for the provision of CHAMP data and A. Gobiet and M. Borsche (WegCenter, Univ of Graz, Austria) for related refractivity retrieval processing. A.L. was funded from the START research award of G.K. financed by the Austrian Ministry for Education, Science, and Culture and managed under Program Y103-N03 of the Austrian Science Fund.

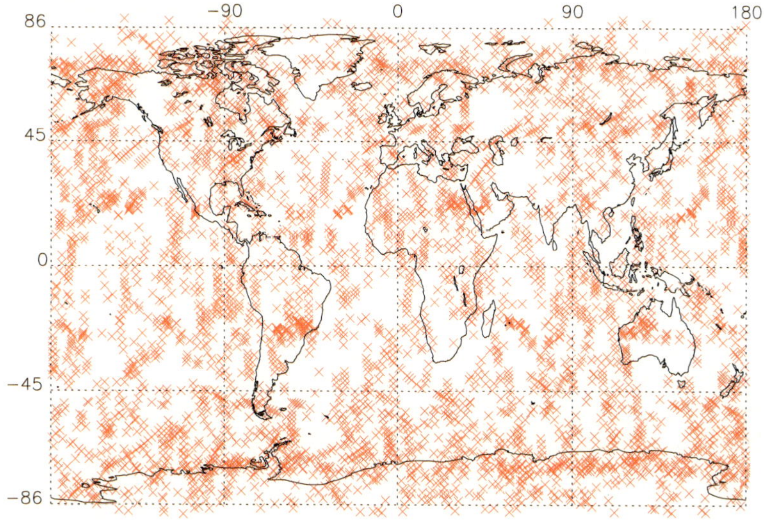

Fig. 1. Geographic distribution (longitude vs. latitude map) of the 4482 occultation profiles used within the assimilation experiment covering August 2003.

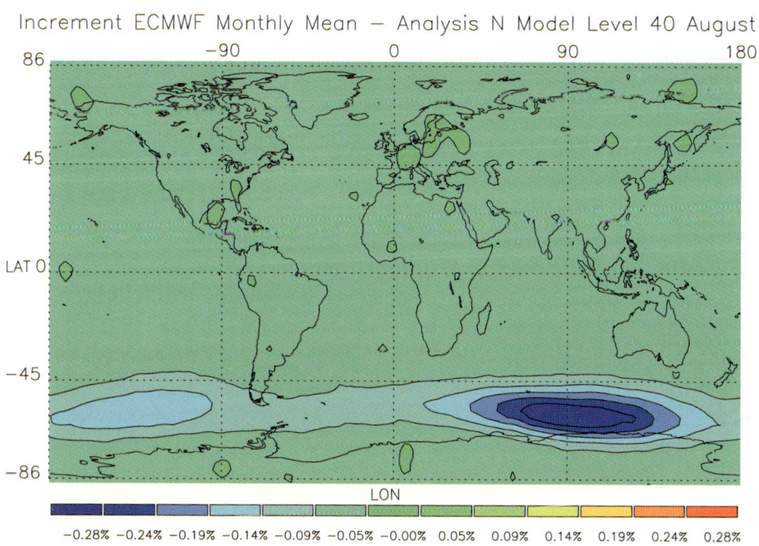

Fig. 2. Increment of the refractivity assimilation of August 2003 between analysis and ECMWF background (monthly mean of all time layers) in a longitude-latitude map for model level 40 (z ~20 km).

References

Barker D (1999) A general formulation for 3D-Var control variables. Part of the WRF package (also available on request from the first author, A.L.)

Borsche M, Gobiet A, Steiner AK, Foelsche U, Kirchengast G, Schmidt T, Wickert J (2006) Pre-Operational retrieval of radio occultation based climatologies. This issue

Bouttier F, Courtier P (1999) Data assimilation concepts and methods. Meteorological Training Course Lecture Series, ECMWF Reading, UK

Byrd RH, Peihuang L, Nocedal J, Zhu C (1994) A limited memory algorithm for bound constrained optimization. Tech Rep NAM-08, Northwestern University, USA

Gobiet A, Foelsche U, Steiner AK, Borsche M, Kirchengast G, Wickert J (2005) Climatological validation of stratospheric temperatures in ECMWF operational analyses with CHAMP radio occultation data. Geophys Res Lett 32(L12806), doi:10.1029/2005GL022617

Gorbunov ME and Kornblueh L (2003) Principles of variational assimilation of GNSS radio occultation data. Tech Rep No 350, MPI for Meteorology, Hamburg

Jensen AS, Lohmann M, Benzon H-H, Nielsen A (2003) Full spectrum inversion of radio occultation signals. Radio Sci 38(3), doi:10.1029/2002RS002763

Healy SB, Thépaut JN (2005) Assimilation experiments with CHAMP GPS radio occultation measurements. Q J R Meteorol Soc, in press

Kursinski, ER, Hajj GA, Schofield JT, Linfield RP, Hardy KR (1997) Observing earth's atmosphere with radio occultation measurements using the Global Positioning System. J Geophys Res 102:23429–23465

Löscher A (2004) Assimilation of GNSS radio occultation data into GCM fields for global climate analysis. Wissenschaftl Ber No 22, IGAM, Univ of Graz, Austria

Loiselet M, Stricker N, Menard Y, Luntama J-P (2000) GRAS –Metop's GPS-based atmospheric sounder. ESA Bulletin 102:38–44

Lorenc AC (1992) Analysis methods for Numerical Weather Prediction. Q J R Meteorol Soc 112:1177–1194

Roeckner E et al. (2003) The atmospheric general circulation model ECHAM5, Part I: Model description. Tech Rep No 349, MPI for Meteorology, Hamburg

Steiner AK, Kirchengast G (2005) Error analysis for GNSS radio occultation data based on ensembles of profiles from end-to-end simulations. J Geophys Res 110(D15307), doi:10.1029/2004JD005251

Steiner AK, Löscher A, Kirchengast G (2006) Error characteristics of refractivity profiles retrieved from CHAMP radio occultation data. This issue

Wickert J, Galas R, Schmidt T, Beyerle G, Reigber C, Foerste C, Ramatschi M (2004) Atmospheric Sounding with CHAMP: GPS ground station data for occultation processing. Phys Chem Earth 29:267–275

Author Index

Number style according to: **First Author** Co-Author

Printing: Krips bv, Meppel
Binding: Stürtz, Würzburg